Brain Art

Anton Nijholt
Editor

Brain Art

Brain-Computer Interfaces for Artistic Expression

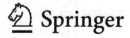 Springer

Editor
Anton Nijholt
Human Media Interaction
Faculty of EEMCS
Universiteit Twente
Enschede, The Netherlands

ISBN 978-3-030-14325-1   ISBN 978-3-030-14323-7   (eBook)
https://doi.org/10.1007/978-3-030-14323-7

This Springer imprint is published by the registered company Springer Nature Switzerland AG
The registered company address is: Gewerbestrasse 11, 6330 Cham, Switzerland

Foreword

Anton Nijholt meets the pivotal need of charting the multiple ways in which artists have strategically challenged existing uses of EEG technology while unveiling its aesthetic and social implications. The term *Brain Art* has come to be associated with the use of Brain-Computer Interfaces (BCI) for artistic expression. It relies on the transmutation of neural signals into realms of sounds and images that render the internal workings of the mind perceptible.

A cornerstone of my career as an artist has been the exploration of what is now known as BCI, brain–computer interface. When I first heard the term, I thought it meant Brain Communication Interaction because that was what I had been working on for so long. In the early 1970s, after experimenting with an alpha wave feedback unit, I sought to create an internal and external video portrait of two people by dissolving images of their brain wave oscillations over their faces as their interaction was being simultaneously videotaped. What intrigued me was showing what often happens beneath the surface as people communicate with each other—at invisible brain level and at gestural level. I began concentrating on expressing visually the synchronous and asynchronous relations established between the brainwaves of people engaged in nonverbal communication. I invited participants in this work to engage in imaginative telepathic experimentation while embracing improvisation and indeterminacy. BCI allowed me to transcend the constraints of linguistic communication.

In 1973, together with collaborator systems engineer, Mike Trivich, we created conceivably the first BCI for two people to communicate nonverbally and tele-pathically via a visual feedback display. Our tools were a Grass Valley EEG connected to a DEC PDP 11 computer and a Heathkit 2 channel oscilloscope, assembled by Mike. We were expected to provide measurable outcomes of brain-to-brain communication upon being granted access to the use of an EEG machine and we fulfilled our promise. I continued to explore the realm of nonverbal communication through the 1980s and into the 1990s when I created the first interactive, multi-participant web-streaming platform in collaboration with Anatole Shaw. Since this time, my work with BCI has continued to expand, encompassing heart rate monitoring coupled with long used audio feedback and the Brain Wave Drawing Game.

Flora Lysen's insightful observation in this collection—"The interface is the work"—is of particular interest to me as it has been fundamental to my own BCI endeavors. It gives us a useful entry point into examining how the medium shapes the social and aesthetic attributes of interpersonal exchanges, and the significance communicated by an artwork that has a technical interface as an essential component.

Artists are the clairvoyants in our society. When they embrace technology as a methodology, they make it visible to a public that has little direct contact with the daily work of scientists. The scientific rigor of artistic BCI has been questioned, but the true value of artists' use of this technology lies in the exploration of the epistemological, emotive, and educational impact of such interfaces. For BCI artists, the goal is always to move beyond the quantitative assessment of the impact of a technological apparatus.

BCI art is grounded in a shared experience of joy and creativity. When artists embrace technology, no matter how rigorous their process, their objective is the creation of an affective space meant to surpass self- or socially imposed boundaries between individuals. Artists are impresarios of technology!

Per Laura Jade in this anthology, we are "hybrid artists." Our work can only be understood through the interaction of artists, technology, and audience. Our passion fills that space and invites participation. Our dependence on scientists for our tools is a love affair that brings both parties closer together and inspires new and more complex collaborations.

Artists deploy BCI technology because it has the ability to render perceptible the inner workings of our deepest emotions. BCI always implies a process of externalization. It is deeply democratic in its ambitions: both demystifying technology and facilitating shared experiences. BCI artists rarely create in isolation. For them, the technological apparatus is foremost a tool of communication—whether between individuals or a multitude.

In editing this collection, Anton Nijholt has performed an invaluable service to scientists, the public and the worldwide creative community whose art finds its expression through a Brain–Computer Interface. This first overview of a field with half a century of groundbreaking collaborations and creations will be a precious guide as we adapt and expand new technologies like mobile–brain interfaces (MoBI), virtual reality (VR), and augmented reality (AR). This summing up is just the beginning!

New York   Nina Sobell
February 2019

Preface

In recent years brain–computer interface (BCI) technology has entered mainstream human–computer interaction (HCI) research for nonclinical applications. BCI has become part of multimodal interaction research as an additional interaction modality for a user of a technological system. BCI has also become part of research in which physiological data provide a system with information about a user's affective and mental state, making it possible to adapt the system, task, and interaction to a particular user.

Artists have been using BCIs for artistic expression since the 1960s. Artistic BCI applications date further back than assistive and clinical BCIs. Many years before Kamiya's and Vidal's influential papers on monitoring and controlling alpha activity and using brain activity for control and communication there were experiments by artists on musical composition, fine art, and other creative applications that required brain activity patterns as input. Early BCI music performances were performed by Alvin Lucier in 1965 and Richard Teitelbaum in 1967. A BCI drawing game was introduced by Nina Sobell in the early 1970s. In the same period, David Rosenboom started his investigations into the use of BCI in musical compositions and real-time performances. He published a wonderful collection on artistic and playful BCI experiments in 1976: "Biofeedback and the Arts: results of early experiments."

At the time artists started using BCI for artistic expression, there was hardly any research into signal processing, pattern recognition, machine learning, and (graphic) display possibilities. Indeed, and this was also mentioned by Vidal in 1973, BCI research was only possible using batch processing of data, rather than real-time processing of brain activity data.

Designers of artistic BCIs are often ahead of more traditional and patient-oriented BCI researchers in their ideas on using BCIs in multimodal and multiparty contexts. Multiple users can be involved in an artistic BCI application. They can have an active role in making a BCI event possible or they can have a role as a 'passive' audience. Communication and control intentions or decisions can be detected from brain activity. This brain activity can be evoked by voluntarily paying attention to specifically designed external stimuli to which a user is exposed.

However, BCI also makes it possible to detect changes in brain activity that have been evoked by conscious actions, for example, performing a mental calculation, imagining a movement, making the decision to relax, or reacting as if angry.

In games, entertainment, and artistic applications there is not always the need for robustness and efficiency that can be asked from other BCI applications. Shortcomings from BCI technology can perhaps be translated into interesting challenges to play and master a game or to create, modify, or experience a piece of BCI art or performance. HCI researchers' interest is increasing, because HCI and computer science have now entered application areas where efficiency is not the main goal or concern. Domestic or public space use of information and communication technology addresses issues that relate more to affect, comfort, family, community, or playfulness, rather than efficiency. In many everyday life situations, efficiency is an 'add-on'.

Interest in nonclinical applications of BCI research is now increasing. Currently, artists can make use of affordable BCI devices and software that does not require them to invest extensive time in getting the BCI to work or tuning it to their application. They may sometimes provoke traditional BCI researchers with their use of BCI hardware and software and their interpretation of brain signals. However, they certainly provide original thoughts about the use of BCI in applications and therefore pave the road for future BCI and HCI applications.

Users of artistic BCI technology can be artists who compose art in real time using BCI signals (usually in a multimodal and multimedia context), performers, audience members, or a full audience. Artists sometimes use reasonably cheap commercial BCI devices to work with or to perform. Or they design installations that require active or passive input from the brain of an individual user or the brains of multiple users or participants in artistic events. Those participants receive feedback from the artistic application that helps them to control their brain activity to create or modify pieces of interactive art. In addition to artistic BCI environments that allow users to play with and modify audio–visual landscapes, animations, and musifications, there are examples of BCI control of instruments and tools for artistic expression and exploration. For example, a BCI-controlled computer painting tool. Adaptive musical interfaces based on a user's brain state have also been introduced.

In this book, we look at current (research) activities in BCIs for artistic expression and to identify research areas that are of interest for both BCI and HCI researchers as well as for artists/designers of BCI applications. More generally, the book is intended for HCI and BCI researchers who are interested in nonclinical BCI applications, in particular, those BCI applications that invite users to interact, play, and to be creative, using BCI. The book addresses an audience that is interested in research that is focused on nontraditional and challenging interactions using BCI as a channel that allows artistic expression of moods, emotions, and other outlets of expressed creativity.

This book grew from the contributions of many colleagues to research activities I co-organized. Together with Hayrettin Gürkök, there was a 2013 paper on affective brain–computer interfaces for artistic expression presented at an affective computing conference. Before that, we used to work on BCI and games. A special issue of the Taylor & Francis BCI journal on artistic brain–computer interfaces, edited by Chang S. Nam and me, appeared in 2015. It was great to meet David Rosenboom when he performed, assisted by Tim Mullen, at the Whitney Museum of American Art in New York in May 2015. A workshop on artistic brain–computer interfaces was organized at the 2016 Asilomar meeting on brain–computer interfaces. Additionally, I was happy to meet Nina Sobell (see picture) at the 2017 Brain on Art conference in Valencia organized by Jose L. Contreras-Vidal. A Handbook on BCI, also from Taylor & Francis and edited by Chang S. Nam, Fabien Lotte and me, with many chapters devoted to nonclinical BCI applications appeared in 2018. In the same year, I organized a workshop at the CHI 2018 conference in Montreal, Canada on BCI for artistic expression. The co-organizers were Rob Jacob, Marvin Andujar, Grace Lesley, and Beste Yuksel. This book would not have been possible without the help of all the people mentioned here.

Enschede, The Netherlands   Anton Nijholt
February 2019

Contents

Chapter 1
Introduction: Brain-Computer Interfaces for Artistic Expression

Anton Nijholt ⓘ

Abstract Capturing brain activity and translating it into multisensorial artistic expressions has been done by artists since the late sixties and early seventies of the previous century. At that time there were only very limited ways to acquire, process, manipulate and transform brain activity. No computing power, no pattern recognition, no machine learning, no graphics, no friendly user interfaces. The results of the transformations were usually presented visually, for example, on an oscilloscope, or auditorily using loudspeakers. In subsequent decades, brain-computer interfacing became a well-established research area that focused on applications in the clinical domain; in particular, applications that aimed at restoring and enhancing communication for the motor-impaired and for rehabilitation purposes. In more recent decades, due to progress in neuroscience, signal processing, and machine learning and progress in sensor technology, we see a growing interest in research and development that aims at clinical and nonclinical users that can use brain-computer interfaces for communication and control in real-life domestic, entertaining, and artistic brain-computer interfaces. This introductory chapter provides some general background on brain-computer interfaces. It mentions some standard paradigms, it provides some historical context and it presents some observations on brain-computer interfacing for artistic expression since the early seventies of the previous century. Currently, there is a market for inexpensive electroencephalographic (EEG) devices and software kits that capture voluntarily and involuntarily evoked brain activity and have this activity translated into control and communication commands for environments and devices. We also see a renewed interest of artists to make use of such devices to design interactive artistic installations that have knowledge of the brain activity of an individual user or the collective brain activity of a group of users, for example, an audience. This chapter provides some background on brain-computer interface technology that can be helpful for understanding the chapters that appear in this book and this chapter provides some context to the developments that are reported and foreseen in this book's chapters.

A. Nijholt (✉)
Human Media Interaction, University of Twente, Enschede, The Netherlands
e-mail: a.nijholt@utwente.nl

© Springer Nature Switzerland AG 2019
A. Nijholt (ed.), *Brain Art*,
https://doi.org/10.1007/978-3-030-14323-7_1

Keywords Artistic expression · Brain-computer interfaces · Multimodal interaction · Multi-agent brain-computer interfaces · Entertainment computing · EEG

1.1 Introduction

Neural activity in our brain causes electric and magnetic fields and blood flow changes in our brain. Changes in this brain activity can be measured and can provide information about which brain regions are involved. Since different brain regions have different functions, it also provides us with information about which cognitive and motor functions are implicated. Neural activity can be activated voluntarily or involuntarily, that is, endogenous or exogenous activation. Making a mental calculation, following a line of reasoning, imagining what to do next, deciding about a movement are examples of voluntarily evoked brain activity. We also experience multisensorial stimuli, sometimes beyond our control. We have less or no direct control over our involuntary brain activity. Different brain regions will show neural activity related to seeing, hearing, feeling, or memorizing. This neural activity is involuntarily involved, although we can have some control by deciding what will have our attention. Fatigue, frustration, excitement, and other affects also involuntarily evoke changes in brain activity, but some regulation of such activity can also take place.

Various methods to measure brain activity changes are available. A functional magnetic resonance imagining (fMRI) scanner detects changes in blood flow associated with neural activity. It has excellent spatial resolution, not very good time resolution, but, more importantly, lying in an fMRI scanner does not allow you to perform your daily activities and therefore cannot support you in your daily activities. However, information about brain activity (not as precise as in an fMRI scan session), can be obtained from sensors that are attached to your scalp. These sensors can measure electrical activity (electroencephalography; EEG) or changes in blood flow (functional near-infrared spectroscopy; fNIRS) and reveal which brain regions are active and where changes in activity are taking place. fMRI scanners can also be used to measure what is happening inside a subject's brain when being exposed to multisensorial stimuli, and such knowledge can be used to improve the detection of the characteristics of particular brain waves using EEG or fNIRS.

fMRI, EEG and fNIRS methods to measure brain activity are 'noninvasive'. There are 'invasive' methods as well. The so-called electrocorticography (ECOG) measurement requires electrodes to be placed on the cortex; hence, this requires surgery. Obviously, better spatial and temporal characteristics of the measurements are then obtained. These measurements can become better when the electrode sensors are inserted into the brain. Presently, such intracortical measurement is performed only in medical laboratory contexts with strict regulations that concern medical and ethical issues.

For domestic applications, one should be able to walk around, do usual things, sit behind a computer screen or a car's steering wheel, enter a home, turn on the lights, or use a smartphone. Nevertheless, there can be situations and environments, for example, being at work, driving a car, playing videogames, or enjoying entertainment, where it is quite acceptable to have a wearable that connects you to a computing device and does not allow you to move around freely. However, present-day wearable brain-computer interface (BCI) technology allows us to be wirelessly connected to computing devices that analyze and interpret our brain activity and translate it to implicit or explicit commands that change our environment and allow us to communicate with our environment, including others that are present in our physical or virtual environment. In this manner, brain-computer interfaces support the notion of people becoming part of an Internet of People, or an Internet of Things, where human beings have become things or nodes in the Internet of Things.

1.2 Measuring and Translating Brain Waves

Research on capturing a person's brain activity and using it to control an environment or an application has become known as brain-computer interfacing (BCI). Control also includes the control of a communication device. Brain activity can be translated into control commands for an application. Control can be explicit, where the user wants to issue a particular command to a brain-controlled device. However, it is also possible that a user's brain activity is monitored and this information is used to determine aspects of a user's mental state, which is then translated into changes in a user's environment that better suits this particular mental state or that aims at changing this mental state.

1.2.1 Extracting Brain Signals

There are various ways to measure brain activity and we should know which brain region is involved in a activity. Is it a region that has to do with pain, a region that has to do with vision, or a region that knows about motion? From neuroscience we have learned about the particulars of the brain. That is, we know where to look for brain signals when we are interested in affect, all kinds of perception, memory, or muscle control. Brain activity can be measured and measurement techniques make it possible to determine which brain regions show activation. It allows us, mediated by a computer's processing power, to know about what is perceived and experienced by a person, but also about someone's intentions to become active in a particular environment.

Recording of electrical activity in the human brain using electrodes attached to the human scalp started with experiments by Hans Berger in the early 1920s (Berger 1929). Before that, such recordings had already been done in animals' brains. Berger needed many more years and publications to have his results accepted in the scientific community. He recorded brain wave patterns, that is, rhythmic repetitive electrical brain activity with a certain amplitude. This repetition is measured in Hertz (Hz), where 1 Hz is one cycle per second. Berger discovered alpha waves (approximately 8–12 Hz) and distinguished them from other activity, not only by looking at the frequency of waves, but also at amplitudes. He experimented with subjects during relaxed wakefulness and subjects who were more consciously aware of what was happening in their environment, that is, having an engaged mind. In this way, he distinguished alpha waves from waves with a higher frequency and smaller amplitude, the beta waves. Usually, alpha activity will increase when a subject closes his or her eyes, and beta activity will then increase when a subject opens his or her eyes. The strongest alpha waves can be observed in the occipital lobe. Beta activity, when our attention is directed towards the outside world, is between 13 and 25 Hz and is most evident in the frontal lobes.

Currently, more is known about brain wave frequencies and where they can be best measured and interpreted in the brain. How can we relate frequencies and amplitudes and disturbances in brain waves to external events that we perceive? Moreover, rather than just sensing, our brain is active in interpreting, making decisions, memorizing, and remembering. Apart from alpha and beta activity, we now know about lower activities such as delta and theta (less than 8 Hz), which are related to, for example, daydreaming, or can be observed in children in a sleeping state, and various forms of gamma activity (25–60 Hz) that are related to higher cognitive functions such as learning and memorizing.

Apart from frequencies and amplitudes it is important to know (and to measure) where and when particular brain waves operate and dominate. How is brain activity measured in the various regions of our brain? The most common noninvasive technique that is used in research laboratories uses electrodes that are embedded in something that looks like a swim cap. The number of electrodes embedded in such a cap may differ, but each of them provides brain activity information detected from different regions of the brain. Different brain regions are involved in different functions. Hence, electrode positions are important. Are we interested in measuring activity related to movement, memory, different types of perception, or emotions? We may have one, two, eight or more electrodes. These are commercial headsets with a few electrodes that do not require such a swim cap, and each electrode can be positioned and adjusted individually. In a research context, it is not unusual to have 32, 64 or even more electrodes in a cap that covers your head. This EEG method to capture brain activity is noninvasive, that is, the electrodes are placed on the scalp, and although your hair may be a disruptive factor, conductive gel or water helps to make the electrodes work and pick up the signals from the brain. Developments in EEG technology have made it possible to use 'dry' electrodes that do not need gel and are much faster to apply. There are systems for defining and naming electrode positions across the scalp. (e.g., the 10–20 system).

With EEG we measure the electrical activity that is related to neuronal firing in our brain. The other noninvasive technique mentioned in the introduction is the fNIRS method. This technique requires a headset that uses infrared light to detect changes in blood (oxygen) flow related to activation in particular brain regions. EEG has a nice temporal resolution, but the spatial resolution is not strong, particularly for deep brain activity. For fNIRS, it is the other way around. fNIRS has slow detection, but it is more precise about which regions are involved.

Whether EEG or fNIRS are being used, there is the problem of artifacts. In a laboratory setting, we can eliminate, reduce or filter brain activity that is not important for what we want to measure, but rather disrupts the measurement of brain activity that we are interested in. We breathe, move, perceive. Muscles are used for eye gaze and changes in facial expressions. We are in a particular mood or experience a particular emotion. If we want to measure brain activity that is related to a specific task that we aim to perform or that the environment nudges us to perform then we should get rid of such non-task related activity or it should be clear how the activity we are interested can be filtered from such non-task related brain activity. Goal-driven further analysis of the measured and filtered brain activity (the brain signals) can ultimately tell us or the application about an underlying aim. Was it meant for muscle control, to utter a thought, a perception, an attempt to remember? Or did we measure mental stress or a particular emotion? Detecting brain signals, analyzing them and extracting the relevant information is one of the main issues of BCI research. The development of BCI technology and designing BCI applications is another main issue. Detecting, analyzing and extracting requires advanced methods of signal analysis, machine learning and pattern recognition (Nam et al. 2018; Lotte 2014).

The noninvasive techniques mentioned above do not require surgery, that is, the placement of electrodes on the surface of the brain or in the brain. Invasive techniques require surgery and the placement of electrodes on the surface of the brain or in the brain. The measurement method where electrodes are placed on the surface of the brain is called electrocorticography (ECoG). When electrodes are positioned inside the brain, we speak of the intracortical measurement of brain activity. At this moment, we can assume that brain-computer interfaces for a general audience will make use of noninvasive technology. This may change in the future. Moreover, such a change may also include the use of devices that stimulate brain activity in addition to detecting and measuring brain activity. There are wireless connections between EEG caps and electrodes that are used for measuring brain activity and computing devices that process and interpret EEG data. This interpretation makes it possible that changes are made to a user's environment as a result of knowing about his or her brain activity or that a user of a brain-computer device explicitly controls his or her environment by issuing commands by manipulating his or her brain activity.

1.2.2 Translating Brain Signals

In BCI research a distinction is made between active, reactive and passive BCI. This distinction is useful, but there is overlap between these three viewpoints.

- In *active BCI*, we assume that a subject is able to manipulate his or her brain activity to issue commands to a brain-controlled device. Can we manipulate our brain activity? Certainly. We can make the decision to relax. When successful, relaxation can be observed in our brain activity. We can act as if we are angry (emotion imagery). Again, when we are able to do this in a convincing way, it will show in our brain activity. We can also imagine that we want to move our body or body parts in a particular direction, for example, move our left hand to the right. Again, what we imagine can be detected by electrodes that pick up signals, in this case from our motor cortex and these signals can be used to, for example, change directions of our wheelchair in a physical environment, or direct our avatar in a videogame to a particular position. We can imagine the rotation of an object that is shown to us. We can perform a mental calculation to evoke certain brain activity. We can be aware and know how this internal and conscious manipulation of our brain activity is used to issue commands to our environment and its devices for communication.

- In *reactive BCI*, it is usually the application that generates stimuli that we are supposed to focus on, which can then give rise to changes in a subject's brain activity. Hence, the subject is asked to pay attention to and choose among artificially evocative stimuli. Paying attention can be seen as a voluntary act. We are engaged in an act that requires us to pay attention to perceptual stimuli. We are asked to pay attention to these stimuli and while doing so our brain emits information about what we perceive. Do we perceive something odd, an incongruity, or is there something we hope to see or do not expect to see, is there an 'Aha' experience? Usually, stimuli are presented visually on a computer screen. However, stimuli can also be presented auditorily or by touch, and in principle, by taste and smell. This reactive viewpoint where the user is explicitly asked to pay attention can be complemented with a reactive viewpoint in which external stimuli are present, their effect is measured, feedback to the BCI application is provided, but the user does not have the explicit task to control the application.

- In *passive BCI* the subject has no intention to control or communicate using BCI. Brain activity is measured and used to make changes to the environment or the task the subject is supposed to perform. A subject's brain activity is measured without him or her being asked to voluntarily evoke a particular kind of brain activity or paying attention to external stimuli that will have an effect on brain activity. The user is simply monitored while performing a task, there are no artificial stimuli that he or she is subjected to, and the user can behave in a natural way, not differently from not having his or her brain activity being measured. Brain activity changes are measured just as we can measure changes in skin conductivity or heart rate changes. Obviously, when a subject is told to wear a BCI cap he or she can become aware and learn how changes are related to a mental state and can turn passive BCI into active BCI by producing different mental states. A subject's active and reactive BCI performance can be dependent on his or her mental state. Therefore, knowledge about a subject's mental state can help in interpreting active and reactive BCI control and communication.

In active and reactive BCI the user is usually expected to know about his or her role to control a system. That is, voluntarily evoke certain brain activity or voluntarily pay attention to external stimuli that are meant to cause changes in brain activity. In active BCI we ask a user to control his or her brain activity. In reactive BCI that is meant to provide a user with control of a device we ask a user to pay attention to stimuli that lead to measurable changes in his or her brain activity. However, a user can also involuntarily experience exogenous stimuli and the corresponding brain activity can be used to adapt an application. In passive BCI the 'ideal' situation is that the user is not aware of being measured at all.

Although there are more ways by which a user's brain waves can be translated into intended commands for control and communication, in clinical BCI research the main paradigms (or markers) for doing this are motor imagery (active BCI) and event related and evoked potentials (reactive BCI). In contrast, see also the next section, passive BCI and turning passive BCI into active BCI has been explored in many artistic interactive BCI applications, long before clinical BCI research took flight.

- Motor imagery (MI) is about movements. Intending to move or beginning to imagine a movement and ending the imagining of a movement leads to changes in the alpha (8–12 Hz) and beta frequency bands that can be measured in the motor cortex; these changes have been called event-related desynchronization (ERD) and event-related synchronization (ERS). This motor imagery BCI, as well as other cognitive imagery tasks, can be used to steer a wheelchair, an avatar in a videogame or a cursor on a screen without making limb movements. MI is an active BCI that requires spatial and spectral information.
- Another paradigm is the steady-state evoked potentials. For example, a steady state visual evoked potential (SSVEP) can be designed such that the user has to pay attention to a screen on which various patterns of repetitive flickering stimuli are displayed. By focusing on one particular pattern, its frequency of flickering can be observed in the occipital region of the brain, and the BCI system can interpret this as a preference or a decision of the user, which can then allow an application to perform a certain task. This is a reactive BCI that requires spatial and spectral information.
- BCI based on event related potentials (ERP) is another reactive BCI paradigm. An example is the P300, a potential that can be elicited using the oddball approach, that is, the user perceives a sequence of stimuli, but only one of them, the target, is relevant for the user. When that stimulus is presented, there is a positive deflection in voltage of the EEG signal occurring with a latency of approximately 300 ms. It can best be measured by electrodes over the parietal lobe. Hence, it requires spatial and temporal information. P300 can be used to choose among stimuli. For example, when presented with a stimulus that represents letters, the target letter, that is, the letter that a subject wants to 'type', elicits a P300. Stimuli can also be auditory or tactile in nature.

We can add these paradigms to the above-mentioned possibilities of controlling alpha activity or consciously moving from a decrease in alpha to an increase of beta and have a BCI system provide an interpretation of such changes that controls an application.

1.2.3   Designing with BCI

An artist or designer can make use of the possibilities these BCI paradigms offer to design an (interactive) artistic system or application. This certainly needs creativity. However, rather than having a design based on only one of these paradigms, we can think of systems that use multiple paradigms or systems where brain wave information is combined with information from other sources. Brain wave input can be integrated into multimodal interactions (Gürkök and Nijholt 2012) that include other physiological information or eye gaze. Artificial intelligence (AI) techniques such as modeling common-sense knowledge, reasoning and pattern recognition can assist in interpreting BCI commands and embedding them in an environment where detailed low-level BCI commands are not necessary.

As a simple example to show how useful it can be to integrate BCI and AI, consider a disabled user who controls a wheelchair using BCI. The AI can have knowledge about the environment in which the user has to navigate. A user can make errors and the BCI device can wrongly interpret the intentions of a user. Thus, if the result of an interpretation is to turn left with the wheelchair and the AI knows that instead of a doorway there are descending steps, it can prevent the wheelchair from taking that turn. As another related example, if a higher-level aim of a wheelchair user can be detected, for example, to go to the kitchen, then no detailed BCI instructions are necessary since the control can be taken over by the AI that knows about the route to go there and the wheelchair can have sensors that use computer vision (another AI technique) to avoid obstacles. As another example, when a BCI user starts a particular action, the AI can predict the most likely next actions and then lets the user make the choice. This can be considered an autocomplete function for actions, rather than for words or letters in a word processor.

An additional issue that should be mentioned is whether an environment can be designed in such a way that issuing particular BCI commands, self-control of brain waves or being receptive to external stimuli that alter brain waves are issues that can be dealt with in a way that is experienced by a user as fitting the environment. If an application requires you to lie down, you will probably expect a peaceful experience, or the environment may expect you to become alert when you see an opponent in a videogame and may notice that you are afraid and want to flee or, that you become angry and aggressive, while preparing to fight this opponent. An environment can be designed in such a way that voluntary and involuntary brain wave information can be made use of in a natural way. Shortcomings of BCI technology can become challenges in designing useful or artistic applications (Nijholt et al. 2009). Or, as mentioned in Novello (2012), "Contrary to science, art can better accept instability and turn it into an interesting parameter."

1.3   Brain Waves and Art

Armed with the knowledge of the previous section we now can investigate how artists have been investigating the use of brain waves for artistic expression. Using brain waves for artistic expression started in the 1960s, many years before Jacques Vidal in 1973 wrote a paper that is currently considered to be the start of BCI research (Vidal 1973). However, Vidal's paper had not yet attracted much attention during the 1970–1990 period. Clinical BCI research still needed to get its start. On the other hand, artists, maybe we should call them brain wave artists, were inspired by neuro-scientific research results, looked for collaboration, and gave brain waves a role in their (interactive) installations. Many observations on this early use of neuroscience research results can be found in the chapters of this book. We will add a few that focus more on the active/reactive/passive viewpoint mentioned in the previous section and on the use of the various paradigms mentioned there. Moreover, we will add observations on the use of brain wave control mechanisms in the artistic domain as presented in the other chapters of this book.

1.3.1   Alpha Waves for Artistic Expression

Edmond M. Dewan of the Air Force Cambridge Research Laboratories experimented with the control of alpha waves. His experiments were reported on in *The US Science News Letter* of October 1964 (MacLaurin 1964). The reporter mentioned: "As with many great scientific advances, the discovery was made unintentionally. During an experiment in which a scientist was measuring his own brain waves, he was suddenly impressed with the fact that he could control their activity." Moreover, "The interesting thing about these waves is that they can be controlled without muscle movement. All a person has to do to turn them on is relax as if going to sleep. To turn them off, all one has to do is concentrate on a scene or object." Hence, "For this reason, Dr. Dewan believes these waves can conceivably be used as a communication device for persons who have lost their ability to move. Once such a person has learned how to manipulate his alpha wave rhythm pattern, it becomes possible for him to communicate through Morse Code, or some other simple response system. Dr. Dewan said one possible drawback to his idea is that all people do not have the ability to turn their alpha waves on and off. It is believed by some experts, however, that the majority of people can."

Dewan and composer Alvin Lucier met in 1965, and Lucier was invited to use Dewan's laboratory equipment to experiment with alpha waves in musical compositions and performances. As a result, Alvin Lucier made his *Music for Solo Performer* in 1965. He performed this composition sitting on a chair, electrodes attached to his head, not being physically active at all, except that his amplified brain waves were routed to loudspeakers (Fig. 1.1). In a documentary on Lucier's work (Rusche and Harder 2012) he mentions: "I could see the cones of the speakers moving in and out,

Fig. 1.1 Alvin Lucier performing *Music for Solo Performer* (1965). Left: Performance in 1976. Still from the film *Music with Roots in the Aether* by Robert Ashley. Right: May 2010 in The Hague, The Netherlands. Still from the 2012 film *No Ideas But in Things* by Viola Rusche and Hauke Harder (2012). Permissions granted

it was very dramatic, the cones of the loudspeakers actually like percussion players, like performers."

In an interview (Lucier 1995; Novello 2012) Lucier mentions that the piece is not truly for solo performers. In fact, Lucier produced the driving energy, an 'assistant' was needed to determine the structure during the performance, including the choice and combination of percussion instruments and the use of tapes with prerecorded alpha waves.

Dewan's ideas were published in scientific papers a few years later (Dewan 1966, 1967). He also introduced a learning aspect. From Dewan (1967): "People can be taught to control voluntarily their own alpha rhythms. This can be used to send messages in Morse code when an electroencephalogram pattern is used as part of a computer programme. Such procedures may help to explain the mechanisms by which the alpha rhythm is 'blocked' or 'unblocked'." Dewan's observations were based on previous work from others, as should be clear from the references in his 1967 paper. Additionally, Hart (1967) reported on the control of EEG alpha, and Joe Kamiya reported, for example (Kamiya 1968), about his earlier experiments (approximately 1962) on the control of alpha waves. Manfred L. Eaton, also in the late 1960s, called this the Voluntary-Involuntary control mode in bio-feedback systems. He also considered three other control modes: Voluntary-Voluntary, Involuntary-Voluntary, and Involuntary-Involuntary (Eaton 1973) and mentions the use of biofeedback for therapeutic and educational purposes. Moreover, listening to music allows feedback to the composer and modifying a composition, "either for himself alone, or as a factor in modifying the signals for everyone listening."

1.3.2 BCIs in the Sixties and Seventies

Artists in the nineteen sixties and early seventies embraced the idea of using brain signals to design and create artistic, playful, and interactive installations. They discovered how these signals can be used in the real-time generation and manipulation of images and sound. They were intrigued by the results of neuroscience research and the possibility to use brain waves in the design and implementation of playful and artistic installations and performances. They cooperated with scientists and introduced ideas that were far away from the research of those days but could nevertheless be illustrated as 'artistic hypotheses' in their playful and artistic applications. Using brain waves in performances and in interactive installations was challenging. It provided new ways of creating and experiencing art and thinking about brain processes and controlling and provoking brain processes in art installations and performances. Interestingly, very often these explorations by artists assumed that two persons had to interact with each other, using their brain waves and getting feedback from the application, to get a certain artistic or playful task done. With two or more subjects, there can be interaction, and the subjects become aware, because of the (audio)visual feedback they receive, of each other's brain activity. Most interests went to the 'discovery' of alpha activity. Other neuroscience discoveries should be mentioned, such as the discovery of different frequencies, stimuli, brain wave distortions, and how brain activity relates to the different functions of the regions in the brain, but with a few exceptions, this research hardly left the (neuro)scientific domain.

In Rosenboom (1976) many projects and experiments on brain waves, biofeedback and the arts from the late sixties and early seventies are collected. Most projects focused on control of alpha waves as may be clear from the names of the projects: *Alpha Bean Lima Brain*, *Alpha Etch-a-Sketch*, *Tai Chi Alpha Tala*, and *Alpha Garden*. However, there is also mention of beta and theta, and evoked responses, for example, flickering lights that stimulate the occipital region of the brain. Suggested experiments address the synchronization of brain waves of two subjects that have to perform a particular task (make a drawing, playful control of a device, etc.). Among these experiments is *Alpha Garden* by Jacqueline Humbert in 1973, who sketched a situation where two persons control the flow of water in a garden hose and sprinkler system by synchronizing their alpha activity. As another example, she designed *Brainwave Etch-a-Sketch* where two participants control horizontal and vertical movements of a dot on a screen to make a drawing. Rosenboom introduced his *Portable Gold and Philosophers' Stones* composition (see Fig. 1.2), in which the brain activity of four musicians was integrated with information about body temperature and galvanic skin response to provide input to a performance. We can even find a composition for brain music that was performed by Rosenboom, John Lennon and Yoko Ono at the Mike Douglas TV show in 1972. "I hope my alpha's all right.", Lennon remarks at the beginning of the performance. In a question-answering session at that time, as reported in Rosenboom (1972), he has to answer the question: "While it seems to be quite a reasonable way to train oneself to attain certain mental states, suppose a girl in a miniskirt sits in the front row during a performance, what happens to the control?" More importantly, in the 1976 book Rosenboom suggests research directions and emphasizes the need for low-cost biofeedback instrumentation.

Artists embraced the idea of using alpha brainwaves to design and create artistic, playful, and interactive installations. *Vancouver Piece* (1972–1973), created for the

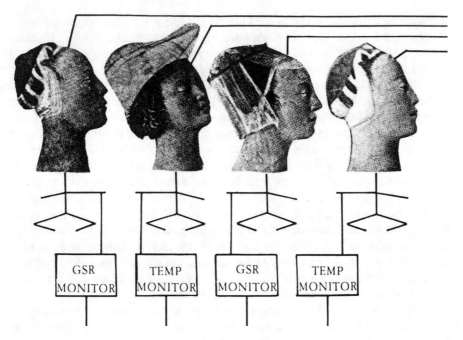

Fig. 1.2 Fragment of the system diagram for Rosenboom's composition *Portable Gold and Philosophers' Stones*: EEG, GSR (galvanic skin response) and TEMP (body temperature) monitoring (1972)

Vancouver Art Gallery's 1972 show (Rosenboom 1975a), had two participants facing a two-way mirror system (see Fig. 1.3).

In *Vancouver Piece*, when participants were able to produce alpha waves that were in phase with each other, musical and lighting effects were produced, and the images of their faces became superimposed on each other's shoulders (more details can be found in Chap. 4 of this book, written by Rosenboom and Mullen).

A similar effect was pursued by Nina Sobell, also in the early 1970s. In her *Brainwave Drawing Game*, she wanted to explore the idea of drawing with one's brainwaves directly on a CRT (cathode ray tube). Her ideas about using brain waves for artistic expression became possible through a cooperation with Dr. Barry Sturman of the VA Hospital neurophysiology laboratory who was willing to allow Nina and her collaborator Michael Trivich access to his laboratory and his equipment if they were able to provide quantitative proof of the existence of nonverbal influences between two people in their brain activity. With the help of the staff and the processing of EEG data using a DEC PDP-10, they were able to show that these influences did indeed exist (Sobell and Trivich 1989; Sobell 2002). Having done that, they implemented a brain game that could be played by museum visitors in the Contemporary Arts Museum in Houston, Texas, 1975. Similar to what was done in the Sturman lab with an oscilloscope, now the participants' faces appeared on a TV screen, superimposed

Fig. 1.3 Two museum attendees participating in Rosenboom's *Vancouver Piece* at the Vancouver Art Gallery in 1972. Photo courtesy of David Rosenboom

with their brain wave drawings (a *Lissajous* pattern that shows the two incoming frequencies and the pattern becomes an ellipse when the frequencies are identical).

In these early years, artists explored new ways of creating playful interactive art and usually involved more than one active participant to create art and, in addition, often assumed an audience that came to experience how the performers' brain waves had an impact on the musical or audio-visual media performance. Some of their activities should be considered to have emerged in the 'beatnik' and Zen movement of the late fifties and early sixties. There was interest in experimenting with the brain by meditation, by using drugs, or by being exposed to devices that were supposed to evoke visual hallucinations (Geiger 2003; Haill 2014; Meulen et al. 2009) using stroboscopic light and having an effect on the EEG-measured alpha waves. The *Dream Machine*, later called the *Dreamachine*, was a flicker machine to provoke alpha waves developed by Brion Gysin in 1960 and later years. According to Gysin: "It's the first art object to be seen with the eyes closed. Just get up close, then close your eyes, and wait for a few minutes. You see kaleidoscopic visions and gorgeous patterns as the light flickers over your eyelids." (Geiger 2003, p. 3). The story of the *Dreamachine* as a cult object can be found in Geiger (2003). Haill (2014) mentions some recent art versions of the *Dreamachine*, including a smartphone application.

1.3.3   More than Waves

Other ways of stimulating and measuring brain activity were discovered: event related potentials (ERPs) and evoked potentials. In his 1973 research paper (Vidal 1973), by many considered to be the start of the BCI research area, Jacques Vidal drew attention to 'evoked responses' of the brain, embedded in ongoing electrical activity, because of exposure to sensory stimuli (visual, auditory, somesthetic). He had suggestions about man-machine communication using BCI: "… such as: recognition of a clue (or matching), its acceptance and rejection, choice between (visual) alternatives, arbitrary positioning of a pointer on a screen, etc.". In his paper, some planned experiments are discussed to distinguish between voluntary and subconsciously evoked responses. For the latter experiments, a space war game was proposed where gamers (one of them is the computer) can fire missiles at opponent's space ships. What difference will be measured in evoked potentials, given a subject's different mental state, by the explosion of either the subject's space ship or that of its opponent?

Influential papers that discussed explicit external stimuli and their impact on brain waves appeared during that time. Their reported research results were picked up by artists, and BCI-based clinical research interest had to wait two decades before it developed, hardly or not at all giving credit to what happened in the previous decades. Changes in potentials because of exposure to external stimuli were reported in Picton and Hillyard (1974) and Nunn (1977), with credits to researchers before them. Various event related potentials were investigated (often called evoked or cerebral-evoked potentials, not to be confused with the steady-state evoked potentials), including the P300 and some subcomponents for pitch perception. In Rosenboom (1975b), the possible role of event related potentials in music perception, composition and performance was discussed and included, in particular, aspects of expectancy and shifts in attention. In the same paper, Rosenboom had observations on brain waves evoked by imagining an event, an expressive action, or an emotion. More theoretical observations on auditory ERPs can be found in Rosenboom (1990, 1997). Rosenboom's ideas were included in the development of an 'attention dependent sonic environment' and in the ongoing development of his already existing composition *On Being Invisible*.

1.4   BCI and Art: Design and Control of Artistic BCI Applications

Employing computers for artistic BCI applications requires cooperation between artists and researchers in computer science, human-computer interaction, neuroscience, and brain-computer interfacing. We saw this cooperation in the early years of BCI for artistic applications. In the period 1975–2000, there are not many interesting artistic applications of BCI. This changed after 2000 when progress in BCI technology made it possible to tinker with electrodes and headsets and make use of

BCI software that was made available by laboratories. Moreover, companies emerged that developed cheap headsets and toolkits that can be used without having detailed knowledge of BCI and brain processes.

There are now many examples of artists who design BCI experiments and use BCI for artistic expression. Often we can recognize similar ideas, as they were presented in the 1970s by Rosenboom, Sobell and others, and usually the focus is on alpha activity. Hence, we see audio-visual representations of brain activity and the possibility to control and modify these representations. When more 'players' are involved, they are made aware of each other's brain activity and have to collaborate or compete to have a satisfying experience or to get a task done. See Nijholt (2015) for a survey of BCI applications in which users are expected to compete or to collaborate.

There is quite a contrast between early brain art that is made visible on an oscillo-scope and present-day brain art that makes use of immersive virtual reality, immersive 360° theatre environments, huge screen display facilities, audio-visual display and other sensorial experiences of extracted brain waves from participants (users, cre-ators, performers, visitors, and audience members), that can be involved in a BCI for artistic expression activity or event. This contrast is not always present in the use of BCI. Currently, commercial grade BCI hardware and software has made it possible that, even with very limited knowledge about BCI and the shortcomings of BCI in general and commercial BCI software and hardware in particular, artistic BCI applications are designed and realized. This certainly does not necessarily affect the artistic quality of the work that is delivered. Artists explore ways of how BCI tech-nology can add to their ideas about artistic expression and how it can influence their way of making art. This does not always require the use of advanced BCI technology.

1.4.1   Designing BCIs for Artistic Expression

In this subsection we present some examples of BCIs for artistic expression that address the general audience, ask for their participation and make them aware of brain waves, not only their own brain waves but also those of others with whom they experience visual and auditory stimuli. Many more examples can be given, but they will appear in many of the other chapters in this book. For example, in Chap. 6 by Suzanne Dikker and colleagues we can read about the *Compatibility Racer*, an installation controlled by brain waves and already displayed at various museum exhibits. In this installation users have to explore and achieve synchronicity in their brain waves to move around in a physical space. Another example can be found in Chap. 7, in which Karen Lancel and colleagues present an environment that visually displays the brain signals of two kissing visitors. Also in this case synchronicity of brain signals can be explored. In Mariko Mori's *Wave UFO* (Mori et al. 2003), exhibited at the 51th Biennale in Venice, June 2008, three players could enter her 'UFO' (Fig. 1.4), have electrodes attached to their scalps, were asked to lie down and relax, and they could see their brain waves projected on walls and ceiling.

Fig. 1.4   Mariko Mori's *Wave UFO* (*Photo credit* Roberto Soncin Gerometta/AFP/Getty Images)

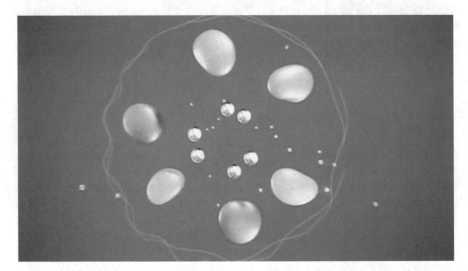

Fig. 1.5   Orbs showing the brain activity of three viewers in Mariko Mori's *Wave UFO*

Their brain waves (left and right hemisphere) were projected in the form of six orbs in different colors onto a screen: yellow (delta/theta) for sleep and dream, alertness (beta) shows in pink and wakeful relaxation (alpha) in blue. Some smaller orbs lighted up when the three viewers were giving off the same brain waves (Fig. 1.5).

Fig. 1.6 *Noor—A Brain Opera.* Photo by Vincent Mak. Photo courtesy of Ellen Pearlman©

More examples of such projects can be given. For example, in *Noor—A Brain Opera* (Pearlman 2017), a performer's affective state, measured with a simple EEG headset, triggers videos, a sonic environment and prerecorded spoken word libretto correlated to four affective states: interest, excitement, meditation, and frustration. The performance takes place in a 360° immersive theatre (Fig. 1.6). Additionally, the performer's affective states are visible as colored bubbles of excitement (yellow), interest (pink), meditation (turquoise), and frustration (red). The performer walks among the audience and can nonverbally interact with audience members. This interaction changes the performer's brain waves.

Again, a wonderful and expensive design, but as mentioned, such projects usually do not introduce the new use of BCI knowledge or new ways of using BCI in interactive art installations. Rather, they have frequency measurements (usually alpha activity) as input to an audio-visual display system that can process input from various users (performers-audience members). Users can be made aware of how the system uses their brain waves and that awareness invites users (performers-audience members) to manage their brain activity, that is, voluntarily vary their alpha waves, to interact with the installation. In addition, as in the case of Mariko Mori's *Wave UFO*, or the *Noor—A Brain Opera* performance, it is the artistic design and narrative that makes it attractive for audience members to be involved in a brain art environment. Later, in this chapter, we will see examples of BCIs for artistic expression where the

Fig. 1.7 Wearable EEG in the artistic responsive environment *Staalhemel*

emphasis is on new and challenging ways of using brain waves for interacting in an artistic environment.

Well-known is Christoph De Boeck's responsive environment *Staalhemel* (De Boeck 2009). Interestingly, the visitor wears a portable EEG set. Using this headset, she walks through a space where large steel plates are hung from the ceiling. Activated by the visitor's brain waves (alpha and beta waves), hammers tap rhythmic patterns on the plates (Fig. 1.7).

Multiple collaborating users associated with Mariko Mori's use of BCI technology and the BCI wearables in Christoph De Boeck's installation. What else could we have expected from artists during that period (2000–2010)? The use of the Internet and the worldwide web (WWW), of course. From Nina Sobell's website (www. brainstreaming.com): "On May 18, 2002, our first transmission operated over the Internet for the first time, letting people in Manhattan and Brooklyn create a collaborative brain-wave drawing on the Web. By making Brainwave Drawings a Web event driven by custom client-server software, a new kind of genre of telepresence will begin to emerge in a multiple-node NetArt performance with accessible physical spaces."

The use of the WWW to distribute and integrate brain waves of remotely present participants in an artistic event was also the topic of a proposal by Sobell to the Rockefeller Foundation. In this proposal, booths were to be installed in three different cities. Each booth was equipped with webcams, monitors, and headgear with EEG and headphones. Participants in the booths see their face on the screen and the color of their faces change in accordance with their brain waves. With the headphones, they hear the output of their brain waves. They could enter a collaborative site where they can hear and see their brain wave output mixed with other participants and can

Fig. 1.8 Integration of water, sound, and brain waves (reprinted with permission from Mann et al. 2008)

create a joint painting and music composition by synchronizing their brain waves. We have included the main part of this proposal in an appendix section to this chapter.

Steven Mann used EEG headsets in various *DECONcert* performances and experiments. One of his experiments was a communal bathing experience in which water waves, sound waves and brain waves were integrated (Fig. 1.8). As mentioned in Mann et al. (2008): "For example, in one *DECONcert*, we had groups of six bathers, at a time, in one rooftop tub, each outfitted with EEG electrodes, connected to bathers, three-at-a-time, in another distant tub that was located on the sidewalk of a busy downtown street. Situating the bath on a busy sidewalk established a juxtaposition of public and private while inviting passers-by to stop, "doff their duds", put on the EEG electrodes, and join in. The different group baths were connected audiovisually, as well as electroencephalically (using EEG sensors), across the World Wide Web, also by way of web cameras, microphones, and various physiological signals such as EEG and ECG (Electrocardiogram). Participants immersed in water and connected to EEG equipment."

There are many more examples of artistic and playful installations that monitor and evoke oscillatory changes in theta, alpha, beta, and low gamma activity. The results have been shown in audio-visual changes in the environment. Visitors of these environments enter a feedback loop, can become aware of the feedback loop and can try to vary this oscillatory activity to experience the installation in a more active and playful way. Some of the chapters in this book provide examples. However, there are also examples of BCIs for artistic expression that explore approaches other than 'playing' with global alpha and other activities and experiencing art and these require a more active role of the BCI user in his or her interaction with an artistic environment or tool. In BCI game development, we see explorations of all kinds and combinations of active and reactive BCI paradigms together with more traditional human-computer interaction modalities. Although there are examples of using BCI to measure engagement, anxiety, boredom, and dominance (feeling of being in control), BCI use in videogames is mainly about control. BCI tools have been designed to create paintings and compositions. In performance art and staged works in which the audience can participate it is usually the EEG-measured experience that is given a role in the performance rather than the use of explicit active or reactive BCI control mechanisms.

In the paragraphs above, we already provided examples of brain art where involuntarily evoked oscillatory brain activity can be turned into voluntarily evoked oscillatory brain activity. In the next section (Sect. 1.4.2), we will look at tools that are controlled by voluntarily generated brain activity (active control, direct manipulation) and then (Sect. 1.4.3) look at the use of voluntary and involuntary control possibilities in media art, interactive installations and performances (staged works).

1.4.2 BCI Tools for Artistic Expression

Tools to create art need to be controlled by the artist. Hence, the active and reactive control of a toolset is the first requirement. Creating or making changes to a digital painting is interactive art. The tool is experienced as well and affects the result, not only in a technical way but also in an emotional way. Obviously, this is not only true for (digital) painting; other ways of creating art can be looked upon in a similar way. Whether it is the artist or the viewer, EEG detected voluntarily and involuntarily evoked brain waves can play a role in creating, experiencing, and interacting with a piece of art.

Artistic performances can be 'brain wave' interactive, where both passive, reactive and active BCI paradigms are involved. BCI can be used to control a tool that creates a piece of art, BCI can measure how a piece of art is experienced, BCI can be used to interact with a piece of art, and through these interactions, BCI makes the piece of art alive. Changes can be virtual, they can be real-time while we are experiencing art, they can be off-line, that is, BCI experiences can be collected and used to decide about changes later, whether done by the artist or by the (digital) art itself.

Brain painting tools have been developed with and for ALS patients (see Chap. 15 of this book). Figure 1.9 shows an exposition of paintings created by ALS patients using tools that can be controlled by the P300 paradigm. Brain painting tools can also be developed for artists and others to provide them with a nontraditional tool and therefore can be used to explore new ways of artistic expression (van de Laar et al. 2013).

Similarly, composers can be offered tools for music composition that can be controlled by their brain waves (see Eaton and Miranda 2014 and Chap. 5 by Duncan Williams in this book). Tools require decision making, the selection of functions and how functions should be used and tuned to the artist's preferences and aims. Not

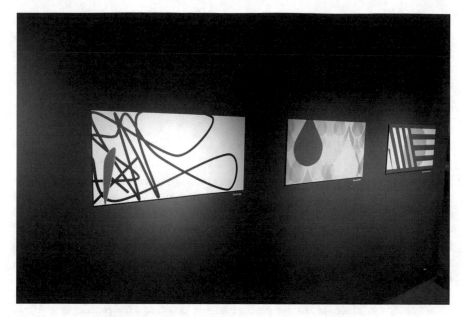

Fig. 1.9 Exposition *Brain Paintings* 2018–2019, LWL-Museum für Naturkunde, Münster, Germany (Photo by Anton Nijholt)

surprisingly, we see the use of ERP, SSVEP and sometimes MI to issue commands to the tools. In Grierson and Kiefer (2014), we find examples of the use of ERPs in a composer tool. Brain painting and brain composing tools need direct BCI control of their functions by the artist. Nevertheless, we can design such tools incorporating EEG-measured knowledge about the mood or emotions of the artist while creating. The tool can make suggestions and know about previous preferences of the artist. However, emotions (and stress and fatigue) also affect the quality of the active and reactive EEG signals that can be detected (Garcia-Molina et al. 2013).

Brain painting and brain-composition tools are usually meant to be used by an individual artist. Obviously, joint use and joint-decision making with other artists or cocreators from an audience can also be considered. Interactive installations and staged works can also be equipped with BCI tools that can be used to voluntarily create changes to an artistic interaction, environment, and experience. Some examples can be given where players actively contribute to a live music performance through their brain control of tools. *Multimodal Brain Orchestra* (MBO) presented in Le Groux et al. (2010) has four performers and a conductor. P300 and SSVEP are used by the performers to trigger sound events and earlier recorded MIDI sequences. The conductor decides when and which events have to be triggered and decides about tempo modulations.

Examples of performances where brain waves of performers and audience members are combined include *The Space Between us* (Eaton et al. 2015) in which the affective states (valence and arousal) of a performer and an audience member are measured and then used to select musical phrases for real-time feedback. An aim of the selection can be to achieve affect matching. In *Enheduanna—A Manifesto of Falling* (Zioga et al. 2017), an actress and two audience members are involved. In part of the live act, the brain activity of actress and an audience member are merged to control the color of the live visuals.

1.4.3 Controlling Artistic BCIs

BCI can provide us with knowledge of how someone experiences visuals and music and responses to such events. What oscillatory changes are evoked? Can we detect the attention shifts? What are the reactions to auditory stimuli, such as a change in rhythm and other repetitive sounds, or musical cues during improvisation? Evoked ERPs during a live performance can be used to make decisions about how to continue. Performers or audiences can take control over performance and composition. Earlier, we mentioned the use of ERPs in the development of Rosenboom's composition *On Being Invisible*. Unfortunately, ERPs usually require averaging over a number of trials, which makes live-performance decision making problematic. However, averaging can also be done over a group of performers or audience members, and it can be accepted that precision in brain decisions, as it should be in more traditional human decision making, is not always 'perfect', as it can, for example, challenge the improvisation by musicians and the listening experience of the audience. Nevertheless, as has been shown in *On Being Invisible* and in the more recent composition/performance *Ringing Minds* by David Rosenboom, Tim Mullen, and Alexander Khali (Mullen et al. 2015), evoked ERPs by musicians and EEG-detected ERPs of audience members and oscillatory shifts make it possible to include audience or co-performer expectancies, anticipations, and incongruities into decision-making during a live performance. See also Chap. 4 by Rosenboom and Mullen in this book.

In Wadeson et al. (2015) we distinguished four main control types of artistic BCIs:

- Passive: Heavily reliant on preprogrammed artistic material. These systems are built to respond to certain brainwave signals, which do not require interaction or intention from the user to create the desired signals.
- Selective: Allow for interaction of the user by way of controlling emotion, levels of relaxation or excitement, etc. to affect the end artistic result. However, the user is not directly responsible for the output, due to the application relying on preprogrammed artistic material.
- Direct: Allows users to choose specific output from the toolbox-style application (e.g., similar to MS Paint). Users can directly choose musical notes or shapes or brush styles, etc. Still limited by the number of options the application displays and the current technology available for use.
- Collaborative: Allows multiple users to interact with each other or individually but collectively to create unique artistic experiences. Collaborative control generally also falls under the categories of passive or selective control, but with multiple users.

This distinction was based on a survey of artistic BCIs in 2015. In this book, more recent examples of artistic BCIs can be found. Again, we can find examples of each of these control or agency (Mirjana Prpa and Philippe Pasquier, Chap. 3 of this book) paradigms in this book. So, we see control by voluntarily and involuntarily generated alpha waves, active control of painting and composition tools by using SSVEP and P300, and the use of motor imagery in rehabilitation. BCI Hackathons are often the cradle of playful and artistic BCI applications and experiments with control paradigms. Hence, in Chap. 17 of this book by Christoph Guger et al., on recent BCI hackathons, we can find lots of examples that show students experimenting with P300, motor imagery, changes in frequency bands, alpha/beta ratio, SSVEP, and mental arithmetic. Interesting observations on voluntary and involuntary control and the role of awareness and attention can be found in Chap. 14 of this book by Richard Ramchurn and colleagues. Attempts to reach synchronization of particular brain activity for two or more participants can also be considered as a control paradigm (Sobell and Trivich 1989) and Suzanne Dikker and co-authors in Chap. 6 of this book. Lancel and coauthors (Chap. 7) report on measurements in their EEG KISS project, including measurements from the motor cortex, that sometimes show synchronization. Again, achieving synchronization can trigger (i.e., control) an application.

Although the distinction in main control types mentioned above does not truly change by including the creative tools, interactive systems, and staged works that are discussed in the chapters of this book, some additional observations are useful. First of all, due to progress in computational power, sensor and actuator technology, multimedia tools and toolkits, and augmented and virtual reality, we can expect more complex brain art. That is, artistic installations that allow multiple users, and that will also allow the use of more than one control strategy by one or more users.

Some refinements of the four control types of artistic BCIs should also be considered. Above, we looked at the ERP P300 signal as a signal that denotes an 'oddball'. Traditionally, this view requires that we consciously pay attention to what is presented to us by a BCI system and identify, among many other things presented to us, the one particular item we are interested in. Other than P300 ERP brain signals can be evoked by unexpected and not anticipated events. We can have repetitions of events displayed to us visually, auditorily (as mentioned in the discussion on David Rosenboom's *On Being Invisible* and *Ringing Minds*), or in a mixed multimedia display. These repetitions will generate expectations that can be disrupted and can lead to ERPs. Hence, there is a reactive ERP BCI that occurs because of an event that we did not and could not anticipate, let alone that before it happened it belonged to our interests. We are not in a process of paying attention to a particular 'oddball', but we nevertheless notice it, and therefore it can also be detected in our brain. Once detected, it can be used to bring about changes in performance or art installation. Voluntary change of attention by viewers or listeners during a performance can also be considered as a control paradigm. Shifting attention by listeners to voluntarily direct features of a sound texture, as discussed in Chap. 4 by Rosenboom and Mullen in this book, is an example of this control paradigm. *Listening as performance* is the name they have reserved for this paradigm.

In Chap. 9 of this book, by Zakaria Djebbara and colleagues, detected ERPs are not used for control of an environment, but for obtaining knowledge about how people experience physical spaces while moving around.

Changes that are detected in the 8–12 Hz alpha band in the motor cortex (ERDs and ERSs, as mentioned in Sect. 1.2.2) caused by movement-related events can also be used to trigger actions that change the feedback to performers and participants of an artistic event. We can have voluntarily (motor imagery) and involuntarily evoked ERDs, they can be responses to visual stimuli, shifts in attention, and movement. In Chap. 11 of this book, Eric Todd and colleagues discuss their design of an art installation in which movement related ERDs influence the sonification of the environment and the movement of LED panels attached to a ceiling.

1.5   More About This Book

In this introductory chapter we touched upon the many aspects that are discussed in greater detail in the next chapters. In this book we have the following sections.

After this introductory chapter, we start with Section 1 (History, State of the Art, and Developments of Brain Art) which contains chapters on the history, the state of the art, and developments in the area of brain art research. Chapter 2 is by Flora Lysen on the rise of real-time brainmedia in the 1964–1977 period. Chapter 3 by Mirjana Prpa and Philippe Pasquier provides us with a survey of brain-computer interfaces in contemporary art. It is followed by two chapters on BCI for musical expression, one by David Rosenboom and Tim Mullen (Chap. 4), focusing on the work of composer and performer David Rosenboom, and one on the evaluation of musical expression by Duncan Williams (Chap. 5).

Section 2 of this book explores our emotions and shares our emotions with the help of BCI. Suzanne Dikker and coauthors (Chap. 6) write about exploring synchronous brain activity between interacting subjects. A research and art project on human-human (kissing) interactions using BCI is discussed in Chap. 7 by Karen Lancel and her co-authors. In Chap. 8 Laura Jade reports her project on exploring emotions using BCI.

Section 3 of this book is on 'Your Brain on Art'. It includes chapters on BCI and experiencing architecture (Chap. 9), on artistic creativity (writing, dancing, music making) in Chap. 10, and on an immersive environment that tracks our behavior as it can be measured using EEG (Chap. 11).

Section 4 discusses BCI and therapy. The chapters in this section (Chaps. 12 and 13) explore how BCI-based expressive arts can help in rehabilitation, and therapy.

The chapters in Section 5 of this book are about tools and BCI control of tools in brain-computer interfaces. Chapter 14 discusses the BCI control of cinema, Chap. 15 discusses the control of a tool that allows ALS patients to paint, and Chap. 16 provides us with a survey on how BCI and virtual and augmented reality can be combined to design new artistic BCI applications and environments. Finally, Chap. 17 presents many examples of playful BCI applications that have been designed in

BCI hackathons. These hackathons have made many young people familiar with BCI and have generated much positive publicity about BCI.

Artists and BCI researchers have contributed to this book. All of them have been involved in producing art works and art installations or other applications that involve the use of BCI for artistic expression. The applications are meant for users, viewers, and participants. They can accept the invitation of an interactive BCI installation and its designers to become active and use that installation. All that is necessary is a brain wave detecting EEG device.

Having an easel and pallet does not make us artists. Neither does BCI. As mentioned in Gürkök and Nijholt (2013), BCIs cannot create art on our behalf. Art is a means for the expression of emotions. BCIs can recognize our emotional states and with the help of designers and artists that use BCI research and technology our emotional states can be expressed in a multimedia display that gives us pleasure or otherwise helps us to address our emotions. As mentioned in Gürkök and Nijholt (2013), "if art is a means to express emotion, then through art we might understand what emotion we are experiencing." Moreover, BCI-generated displays of emotions can also allow us to explore our emotions. We can extend this view to artistic BCIs that provide us with opportunities to explore not only our artistic, hedonic, and emotional needs and preferences, but also the use of BCIs for artistic expression to explore other affective and mental capabilities such as empathizing, remembering, fantasizing, imagining, and anticipating.

Appendix

Nina Sobell, The Brain Streaming Project (2003): Proposal to the Rockefeller Foundation

In 2003 Nina Sobell submitted a proposal (N. Sobell. The Brain Streaming Project. Project proposal for the Rockefeller Foundation New Media Fellowships.) to the Rockefeller Foundation with a view on how Internet and the World Wide Web (WWW) could be used to communicate, share brain activity and engage with remote others to perform a collaborative painting task. Underlying the proposal was an earlier experiment. As mentioned on the website of the Brain Streaming Project (www.brainstreaming.org): "On May 18, 2002, our first transmission operated over the Internet for the first time, letting people in Manhattan and Brooklyn create a collaborative brain-wave drawing on the Web. By making Brainwave Drawings a Web event driven by custom client-server software, a new kind of genre of telepresence will begin to emerge in a multiple-node NetArt performance with accessible physical spaces." The costs of this proposed 'Brain Streaming Project' were estimated to be $35,000. Unfortunately, it was not approved by the Rockefeller Foundation. The text of the main part of the proposal, slightly adapted, follows below.

Proposal

The Brain Streaming Project presents the means for people to connect and collaborate with one another by using only their brain waves. This non-verbal communication will be represented as a continuously evolving aural and visual expression, accessible to anyone logging on. The Brain Streaming Project will premiere with a 1-h international performance at physical and virtual locations including pocket computers and cellphones at www.brainstreaming.org. The transformed photo booth installations for the premiere performance will remain at each location for participation for the duration of the Fellowship term. For the duration of the performance, participants will be connected to electroencephalographs that amplify and identify their brain waves. The individual logon and brainwave data will be sent to the project server over the Internet, and entered into the server's database. The server then streams this information to the project's Web page, along with sounds and images that change dynamically as new input is received and viewed on touch screen monitors inside the booths, and on the Web. Brain Streaming is a metaphor for universal human consciousness. It reflects our similarities through the transformation of our converging thought patterns into the creation of a collaborative virtual collage.

My collaborators, M. E. Trivich, a systems engineer, Dr. John Dubberstein, a neurosurgeon and a musician, Sun Qing, a programmer and a theoretical physicist are committed to building our new form of universal language. Now that our custom client/server software has been successfully developed, we can devote our attention to discovering new ways of representing the EEG data into dynamically changing sounds and images, and designing a meaningful graphical user interface. Together we seek to create the meeting point at the intersection of art and technology.

Installation Plan: The Brain Streaming project will take place on pocket computers, cellphones, home computers and installed at three art/academic spaces located in three different cities. Three typical passport photo booths will be installed at the physical spaces. The exterior visual appearance of the booths will remain the same, the interior of the photo booths will appear much like their initial form but be equipped with web cams, monitors, and headgear with EEG (wireless electrodes) and headphones.

- A visitor approaches the photo booth with typical instructions posted outside that include extra features about the piece. They enter by pulling back the curtain, and adjusting their seat, so the web cam can see them. They put on the headgear equipped with EEG (wireless electrodes) and headphones.
- They enter the name of the city and the country they are from on the touch screen monitor facing them; longitude and latitude are automatically registered.
- They see their web cam image placed on their registered geographic location.
- They see the color of their faces change in accordance with the output of their own brain waves, and hear the output of their brain waves through the headphones.
- On the same login screen, they touch a button, which brings them to a collaborative painting and composing site.

- Then they hear and see the transformation of their brain wave output mixed with other participants. Volume as dynamics; Beta as rhythm; Alpha as tempo; Theta as pitch, and other parameters expressing the complexity of harmonics.
- The installation will remain open during gallery hours, and 24/7 for those with the headgear on the web at home, cellphones or pocket computers with built-in webcams, and those who logon to observe.
- Those who logon to observe will be identified by a color they choose from a color wheel.

In this interface, three typical passport booths in three cities are represented by the yellow circles. The yellow circles are placed at locations where photo booths will be hooked to EEGs. Surrounding them are expanding and contracting circles which illustrate amplitude through size and brain wave types and color. Red squares indicate the locations of visitors. Dimming and brightening of the map is a rough illustration of one way to arrogate the data from the EEG participants. In this illustration, when the circle is brighter, it is intended to illustrate when their brain waves are in sync. This diagrammatic page leads both physical and virtual participants to a full screen painting and music composition they are creating together, as illustrated in the model below.

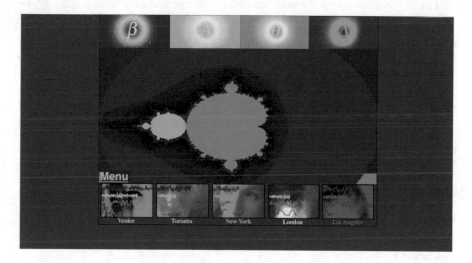

Brain Streaming Project 2003

References

Berger H (1929) Über das elektrenkephalogramm des menschen. Arch f Psychiat 87:527–570
De Boeck C (2009). https://vimeo.com/7746875
Dewan EM (1966) Communication by the voluntarily control of the electroencephalogram. In: Proceedings of the symposium on biomedical engineering, Marquette University, No 1

Dewan EM (1967) Occipital alpha rhythm eye position and lens accommodation. Nature 214:975–977

Eaton ML (1973) Bio-music. Something Else Press, Barton, VT, USA

Eaton J, Miranda ER (2014) On mapping EEG information into music. In: Miranda ER, Castet J (eds) Guide to brain-computer music interfacing. Springer, London

Eaton J, Williams D, Miranda E (2015) The space between us: evaluating a multi-user affective brain-computer music interface. Brain Comput Interfaces 2(2–3):103–116. https://doi.org/10. 1080/2326263X.2015.1101922

Garcia-Molina G, Tsoneva T, Nijholt A (2013) Emotional brain-computer interfaces. In: Nijholt A, Heylen D (eds) Int J Auton Adapt Commun Syst (IJAACS) Spec Iss Affect Brain Comput Interfaces 6(1):9–25

Geiger JG (2003) Chapel of extreme experience. A short history of stroboscopic light and the dream machine. Soft Skull Press, Brooklyn, NY, USA

Grierson M, Kiefer C (2014) Contemporary approaches to music BCI using P300 event related potentials. In: Miranda E, Castet J (eds) Guide to brain-computer music interfacing. Springer, London

Gürkök H, Nijholt A (2012) Brain-computer interfaces for multimodal interaction: a survey and principles. Int J Hum Comput Interact 28(5):292–307

Gürkök H, Nijholt A (2013) Affective brain-computer interfaces for arts. In: Nijholt A, D'Mello S, Pantic M (eds) Proceedings 5th biannual Humaine Association conference on affective computing and intelligent interaction (ACII 2013), Geneva, Switzerland. IEEE Computer Society, pp 827–831

Haill L (2014) ICT & art connect: revelations by flicker, dreamachines and electroencephalographic signals in art. In: Proceedings of the 50th anniversary convention of the AISB. Symposium on "The future of art and computing: a post-turing centennial perspective"

Hart J (1967) Autocontrol of EEG alpha. Presented at the seventh annual meeting of the society for psychophysiological research, San Diego, Calif. Also: Hart JT (1968) Autocontrol of EEG alpha. Psychophysiology 4(4):506

Kamiya J (1968) Conscious control of brain waves. Psychol Today 1(11):56–60

Le Groux S, Manzolli J, Verschure PFMJ, Sanchez M, Luvizotto A, Mura A, Valjamae A, Guger C, Prueckl R, Bernardet U (2010) Disembodied and collaborative musical interaction in the multimodal brain orchestra. In: Conference on new interfaces for musical expression (NIME 2010), pp 309–314

Lotte F (2014) A tutorial on EEG signal processing techniques for mental state recognition in brain-computer interfaces. In: Miranda ER, Castet J (eds) Guide to brain-computer music interfacing. Springer, pp 133–161

Lucier A (1995) Reflections, interviews, scores, writings, 1965–1994. Musik-Texte

MacLaurin W (1964) Talk via brain waves. Sci News Lett

Mann S, Fung J, Garten A (2008) DECONcert: making waves with water, EEG, and music. In: Kronland-Martinet R, Ystad S, Jensen K (eds) Computer music modeling and retrieval. Sense of sounds. CMMR 2007. Lecture notes in computer science, vol 4969. Springer, Berlin, Heidelberg, pp 487–505

Mori M, Bregenz K, Schneider E (2003) Mariko Mori: wave UFO. Verlag der Buchhandlung Walther König, Köln

Mullen T et al (2015) MindMusic: playful and social installations at the interface between music and the brain. In: Nijholt A (ed) More playful user interfaces. Gaming media and social effects. Springer, Singapore, pp 197–229

Nam CS, Nijholt A, Lotte F (eds) (2018) Brain-computer interfaces handbook: technological and theoretical advances. CRC Press, Taylor & Francis Group, Oxford, UK

Nijholt A (2015) Competing and collaborating brains: multi-brain computer interfacing. In: Hassanieu AE, Azar AT (eds) Chapter 12: Brain-computer interfaces: current trends and applications. Intelligent systems reference library series, vol 74. Springer International Publishing Switzerland, pp 313–335

Nijholt A, Oude Bos D, Reuderink B (2009) Turning shortcomings into challenges: brain-computer interfaces for games. Entertain Comput 1(2):85–94. Elsevier, Amsterdam, The Netherlands

Novello A (2012) From invisible to visible: the EEG as a tool for music creation and control. Based on my Master's Thesis for the Institute of Sonology, Den Haag

Nunn CM (1977) Biofeedback with cerebral evoked potentials and perceptual fine-tuning in humans. Master's thesis, York University (1976); and in J Exp Aesthet 1: 11–100. Aesthetic Research Centre of Canada Publications, Vancouver

Pearlman E (2017) Brain opera. Exploring surveillance in 360-degree immersive theatre. PAJ: J Perform Art 39:2(T116):79–85

Picton TW, Hillyard SA (1974) Human auditory evoked potentials, I: Evaluation of components, and II: Effects of attention. Electroencephalogr Clin Neurophysiol 36:179–199

Rosenboom D (1972) Method of producing sounds or light flashes with alpha brain waves for artistic purposes. Leonardo 5(2):141–145

Rosenboom D (1975a) Vancouver piece. In: Grayson J (ed) Sound sculpture: a collection of essays by artists surveying the techniques, applications and future directions of sound sculpture. Aesthetic Research Centre of Canada Publications, Vancouver, pp 127–131

Rosenboom D (1975b) A model for detection and analysis of information processing modalities of the nervous system through an adaptive, interactive, computerized, electronic music instrument. In: Proceedings of the second annual music computation conference, Part 4, information processing systems. Office of Continuing Education in Music, University of Illinois, Urbana, IL, pp 54–78. Archived by International Computer Music Association—ICMA. Also included in Rosenboom (1976), pp 69–81

Rosenboom D (ed) (1976) Biofeedback and the arts: results of early experiments. A.R.C. Publications, Vancouver

Rosenboom D (1990) The performing brain. Comput Music J 14(1):48–66

Rosenboom D (1997) Extended musical interface with human nervous system. Assessment and prospectus. Leonardo monograph series. International Society for the Arts, Sciences and Technology (ISAST), San Francisco, California, U.S.A. Original 1990, Revisions to Parts 5 and 6 and additional Appendixes in 1997

Rusche V, Harder H (2012) No ideas but in things. The composer Alvin Lucier. Documentary, Germany, 2012, see: http://alvin-lucier-film.com, published as DVD: Wergo. Scott Music & Media, Mainz, Germany, 2013

Sobell N (2002) Streaming the brain. IEEE Multimed 9(3):4–8

Sobell N, Trivich M (1989) Brainwave drawing game. In: Delicate balance: technics, culture and consequences, 20–21 Oct 1989, Los Angeles, CA, USA. IEEE Xplore, pp 360–362. https://doi.org/10.1109/tcac.1989.697094

ter Meulen BC, Tavy D, Jacobs BC (2009) From stroboscope to dream machine: a history of flicker-induced hallucinations. Eur Neurol 2009(62):316–320. https://doi.org/10.1159/000235945

van de Laar B, Brugman I, Nijboer F, Poel M, Nijholt A (2013) BrainBrush, a multimodal application for creative expressivity. In: Miller L (ed) Proceedings sixth international conference on advances in computer-human interactions (ACHI 2013). IARIA XPS Press, Nice, France, pp 62–67

Vidal J (1973) Toward direct brain-computer communication. In: Mullins LJ (ed) Annual review of biophysics and bioengineering, vol 2. Annual Reviews, Inc., Palo Alto, pp 157–180

Wadeson A, Nijholt A, Nam CS (2015) Artistic brain-computer interfaces: current state-of-art of control mechanisms. Brain Comput Interfaces 2(2–3):70–75. Taylor & Francis

Zioga P, Chapman P, Ma M, Pollick F (2017) Enheduanna—a manifesto of falling: first demonstration of a live brain-computer cinema performance with multi-brain BCI interaction for one performer and two audience members. Digit Creat 28(2):103–122. https://doi.org/10.1080/14626268.2016.1260593

Part I
History, State of the Art, and Developments of Brain Art

Chapter 2
The Interface Is the (Art)Work: EEG-Feedback, Circuited Selves and the Rise of Real-Time Brainmedia (1964–1977)

Flora Lysen

Abstract This chapter examines the rise of EEG-feedback research in the period between 1964 and 1977, the time between the first EEG-feedback setup that gained public attention and the subsequent waning of the explosive enthusiasm for EEG-feedback in the late 1970s. Studying both artistic and scientific experiments of EEG-feedback during this period, the chapter traces the emergence of a new direction within this subdomain of EEG-research—beyond an interest in the meaning of measured brain wave states, towards the significance of the design of brain-feedback situations that perform and emphasize the relationality and mutability of brain activity. By examining research cultures and practices of EEG-feedback, the chapter traces conditions of possibility for a shifting epistemological commitment, revolving around the idea that 'the interface is the work.' Research cultures of EEG feedback were impacted by both artistic and scientific experiments with media environments and the idea of a 'circuited self'. In turn, artists and researchers were actively engaged with the public manifestation of EEG-feedback in popular news reports and television broadcasts, which created a particular sphere of resonance for the emphasis on playful and spectacular demonstrations of circuits. When computing was introduced in EEG-feedback after 1970, it brought notions of 'on-line' and 'real-time' into the circuit. These developments were not only understood as technological advancement through faster feedback, but they also brought an emphasis on the social potential of computing: self-insight, augmenting the self and connecting with others. The chapter ends with a reflection on the resonance of histories of performance and design-oriented approaches in neuroscientific research today.

Keywords Interface · EEG-feedback · Real-time · Art-science interaction · Brainmedia

F. Lysen (✉)
Mediastudies Department, University of Amsterdam, Amsterdam, The Netherlands
e-mail: f.c.lysen@uva.nl

© Springer Nature Switzerland AG 2019
A. Nijholt (ed.), *Brain Art*,
https://doi.org/10.1007/978-3-030-14323-7_2

2.1 Introduction

Are artworks that employ Brain Computer Interfaces (BCI's) a mystification of tech-nological measurements or even a sham-version of neuroscientific research? For researchers and artists working with EEG in art, the 'scientificity' of the measured active brain is of concern, because at the basis of many audio-visual installations are recordings of electrical brain activity by consumer-grade headsets for EEG (electro-encephalography), which can be bought relatively cheap but that are often not very precise (Chu 2015; Maskeliunas et al. 2016). Hence, while artistic EEG-set ups may awe the public with technological ingenuity and the promise of revealing the invisi-ble workings of the human brain, these mediations may be based on muddled data. Despite such reservations however, the measured EEG-data, spectacularly mediated in visual and auditory form, lend significant scientific authority to many artistic BCI-pieces and performances. EEG-art is not only considered mind-blowing and gorgeous, but also (cool!) science.

How do artistic BCI creators themselves reflect on this conundrum? In a 2015 arti-cle on musical brain-computer interfaces, a team of thirteen BCI creators, neuroscien-tists and artists (Mullen et al. 2015) mention the contested status of consumer-grade BCI for the arts. Even though interactive artworks using consumer-grade EEG-data often portend to show valid representations of mental states, many of these audio-visual mediations are not very precise; and thus "while these may be entertaining playthings, they hold no particular scientific value." (Mullen et al. 2015, p. 216). Yet in turn, when the authors discuss one example of a BCI-artwork to which this critique may pertain, they stress instead that if subjects can "consciously modulate" a set-up "in real-time" through their behavior, this designed situation is nevertheless valuable because it enables new attitudes by participants and new forms of artistic expression (new methods of music composition, for example). Hence, for a brief moment in the article, the argument moves from data measurement to the value of interaction. Arguably, this move signals a momentary bracketing of the scientific meaning of the recorded data that is at the basis of these artistic brain-computer pieces, shifting attention instead to the relational effects of the installation. In this line of reasoning, the most significant contribution of such pieces is the design of reciprocity: the real-time interaction between brain-subject-machine then becomes the true inventiveness of artistic BCI's, an observation that is underlined by the authors' statement that "the interface is the work" (Mullen et al. 2015, p. 217).[1]

In this chapter, I trace genealogies of artistic and scientific set ups to give feed-back on brain activity, arguing that the aforementioned dictum—"the interface is the work"—has emerged as a characteristic and guiding research principle in the field of (artistic) BCI. I examine the rise of EEG-feedback research in the period between 1964 and 1977, the time between the first EEG-feedback setup that gained public attention (by the U.S. physicist Edmond Dewan) and the subsequent waning of the

---

[1]In the article by Mullen et al. 2015, the statement 'the interface is the work' particularly relates to the views of composer Richard Warp and his work *Spukhafte Fernwirkung* ('spooky action at a distance,' 2013).

explosive enthusiasm for EEG-feedback in the late 1970s. Studying this period, I observe the rise of a new direction within this subdomain of EEG-research: beyond an interest in the meaning of measured brain wave states, towards an interest in the design of brain-feedback situations that perform and emphasize the relationality and mutability of brain activity. Examining research cultures and practices of EEG-feedback, I trace the conditions of possibility for this shifting epistemological commitment. I show that throughout this period, artistic and scientific approaches to EEG-feedback were closely entangled. Artists and other researchers shared a particular aesthetics of research practice, which affected a move from an interest in precise correlations between electro-physiological measurements and psychological states, towards an emphasis on the designed interfaces and particular situations in which subjects can engage with (their own) EEG. This shift was constituted by a number of emerging art- and research practices and changing perspectives that I outline in this chapter.

On the basis of examples, I show that particular forms of performing and demonstrating intercross between different spheres of producing knowledge. In the 1960s and throughout the 1970s, cybernetic theories and countercultural ideas influenced both scientific and artistic research cultures. New approaches to human augmentation, to media environments and alternative forms of communication, but also the explorations of inner worlds and inner selves all contributed to EEG-feedback research. The exchanges between artists and scientists in this field then, are most indicative of an ambiguous and complex cyber-countercultural entanglement in the increasing importance attributed to the interface.

Historical research of early EEG- and BCI-set-ups is important because it can unearth core assumptions and specific forms and norms in research that are still important in the hybrid and broad field of the neurosciences today. At present, electro-encephalography (EEG) is (still) an important measuring technology in scientific and clinical research into the activity of the brain and is applied in a variety of fields, ranging, for example, from fundamental research into visual- and motor functions to the understanding of epilepsy and brain injuries. My historical research contributes to a better understanding of the development of today's field of consumer-grade BCI and neurofeedback devices, which is a hybrid domain of clinical- and user-experimentation with non-invasive EEG-set-ups, sometimes dubbed "lifestyle BCI systems" or "neuro toys" (Brenninkmeijer and Zwart 2016). The performative identity of EEG-feedback and BCI that I uncover in this chapter gives insight into (spectacular) demonstration-oriented research cultures of current research that employs EEG, a point to which I will return in my conclusion.

Following my introduction (1), this chapter is structured into separate sections. Section 2.2 gives a general introduction to the rise of EEG-feedback in art and science, feeding into Sect. 2.3 which covers the ambiguous import of cybernetics in EEG-feedback artworks in the 1960s and 1970s, pointing particularly to the interpretative flexibility of these feedback situations. Section 2.4 describes how research into EEG-feedback persisted in the 1970s, despite mounting critique, because of a shift in emphasis from researching the mind-brain, towards researching the 'self'. Section 2.5 points to the rise of what I call the 'circuited self', which is particularly evident in artists working with EEG-feedback as part of a broader media ecological discourse. Section 2.6 shows how this aesthetics of research practice—a design- and performance-oriented approach to the creation of circuited systems—the idea that the 'interface is the work'—was also part of scientific research into EEG-feedback and I note how these researchers were closely connected to popular news media. Section 2.7 recounts the first (hypothesized and realized) experiments with brain-computer-interfaces for EEG-feedback in the 1970s and the technological promises of computing (real-time speed, but also a social potential) that were part of a new focus on interface design. Section 2.8 draws a brief bridge from the late 1970s to the re-emergence of neurofeedback in the early 1990s, culminating in a reflection, in Sect. 2.8, on the resonance of histories of performance and design-oriented approaches in neuroscientific research today.

2.2   The Rise of EEG-Feedback: From Brain-Controlled Machines to Exploring the Inner Self

When did EEG-researchers start to think of giving feedback to their experimental subjects, to device set-ups in which wired-up participants would be able to perceive their own fluctuating brain activity? Anton Nijholt, in his historiographical notes on BCI, remarks that artists and composers such as Alvin Lucier, Pierre Henry, Richard Teitelbaum and David Rosenboom had already been experimenting with EEG-feedback circuits before the explosion of laboratory-based EEG-feedback research in the late 1960s (Nijholt 2015). In fact, elements of feedback in EEG-set ups had already been part of EEG-experimentation since its inception around 1930, and EEG-feedback had been envisioned as a practical warning system for pilots and motor cyclists as early as the 1940s (Adrian and Matthews 1934; Kornmuller 1945 cited in Borck 2018, p. 243; Walter 1953). Yet, the exponential growth of EEG-feedback research took place roughly in the period between Lucier's pioneering 1965 EEG-music piece *Music for Solo performer* and Rosenboom's famous 1970 EEG-installation *Ecology of the Skin*. In this timespan, the number of EEG-feedback experiments grew rapidly and feedback research became a particularly mediagenic and much-reported topic in the popular press. The borders between artistic and scientific experimentation with EEG were blurry in this period. Lucier and Rosenboom for example, developed their works in close collaboration with academic researchers: Lucier worked with physicist Edmond Dewan (who was much influenced by Norbert Wiener's cybernetics) and Rosenboom's early scientific informers were Neil Miller, Lesther Fehmi and Edgar Coons, neurophysiologists working at various academic institutions around New York.[2]

---

[2]E. E. Coons at New York University, Neil Miller at Rockefeller University, Lester Fehmi at Stony Brook. In 1966, Richard Teitelbaum had started to work on alpha-wave feedback with Lloyd Gilden at the Pyschology Department of Queens College.

The rise of both scientific and artistic experimentation on EEG-feedback must be viewed within a cultural- and scientific environment that was dominated by (cybernetic and biological) systems theory, media-ecological thinking and prominent countercultural and scientific attention to altered states of consciousness. EEG-feedback's popularity, in the late 1960s, was partly caused by the increasing association between recorded alpha brain waves and meditative and spiritual states. Various researchers were interested in EEG-feedback for its perceived potential to train the brain for states of 'Zen' relaxation and envisioned feedback as a way to give insight into the inner self (Luce and Pepe 1971). Artists built on these interests, in 1968, for example, composer Richard Teitelbaum created the piece *In Tune* (1968) combining the sounds of sonified alpha wave feedback with Tibetan Buddhist chanting (Teitelbaum 1976). These varying research directions drew a motley crowd to EEG-research. Neurophysiologist Joe Kamiya characterizes the hybrid public at the the the first official biofeedback conference in Santa Monica, California in 1969, "it was a mixture of uptight scientific types of all types, and people barefooted, wearing white robes, with long hair." (Kamiya cited in Robbins 2008, p. 65).

Between the "uptight scientific types" and the "barefooted white robes," the meaning of 'feedback' and its operationalization in EEG-research seemingly underwent a shift in the last years of the 1960s. In a 1964 set-up designed by Edmond Dewan, alpha wave activity was employed as part of a technological set-up that was (rather hyperbolically) called a "thought-controlled device," a simple lamp shade could now be switched on or off according to a subject's learned concentration (Dewan 1964). Here 'feedback' simply meant the visible result of light-control as part of a technical set-up. In contrast, in 1968, Joe Kamiya provided auditory feedback to train his subjects to prolong certain brain wave states and also asked subjects to describe what they felt during alpha- and non-alpha, he was interested in feedback as serving the "awareness of an internal state" (Kamiya 1968, p. 58).

Juxtaposing the approaches of Dewan and Kamiya intuitively leads to the conclusion that there was a change of emphasis in EEG-feedback research in the last years of the 1960s, from the prominence of control and the utilization of brain waves in man-machine interaction, towards attention to the inner mind and self-exploration. Indeed, while Dewan's cybernetic-inspired research was funded by the research lab of the United States Air Force, Kamiya would cite research into the brain wave states of practiced Zen meditators. Yet, even though the shift from 'control' to 'insight' might describe a general change in vocabulary, notions of regulation (control) and sensitivity (insight) were in fact always intermingled. Kamiya claimed that EEG-research would reveal how yogi's actually *controlled* brain activity, which would turn meditative practices into "something measurably real," and "might strip it of much of its mystical quality." (Kamiya 1968, p. 59). At the same time he contended EEG opened up "immense possibilities" for "intensive exploration of the consciousness of man."

Kamiya's utterances are exemplary for the fundamental ambiguities and contradictions at the heart of EEG-feedback research up to the present. In the next two sections, I sketch some of these disputations that cross-cut different elements of neuro-feedback research. First, I outline the ambiguous relation between (artistic) neuro-feedback research and cybernetics, arguing that different interpretations of cybernetics are at the basis of disparate readings of historical EEG-artworks. Secondly, I turn to the objections raised to EEG-feedback since the late 1960s, which raises the question how this practice sustained considerable scientific and popular attention.

2.3 The Cybernetic Ambiguities of (Artistic) EEG-Feedback

The history of EEG-feedback experiments is characterized by an oscillating and sometimes counterintuitive intertwining of notions of control and play, management and insight, mystery and the "measurably real." Such fluctuating notions cause an interpretative flexibility of EEG-feedback artworks—contemporary accounts of historical pieces have offered remarkably disparate evaluations. Consider, for example, art historian Branden Joseph's analysis of the musical biofeedback experiments of composer Manford Eaton around 1970 (Joseph 2011). Eaton used EEG (among other physiological parameters such as galvanic skin response, heartbeats and respiration) to envision what he called *Biomusic*: sounds based on measured physiological signals, which would reciprocally alter the sensory activities of participants, thus creating a feedback medium, in Eaton's terms, for "real-time," "multi-directional communication on a physiological level," a "spontaneous" form of communication (Eaton 1971). Branden Joseph, reviewing Eaton's work, has pointed to *Biomusic*'s indebtedness to (defense department-funded) research into cybernetic human servomechanisms, to research in subliminal messaging and Eaton's invocation of possible applications of his invention in altering ("programming" or "controlling") the behavior of individuals. Ultimately, in Joseph's view, Eaton's *Biomusic* was a (mind-)control-oriented project, serving "as but one more means by which a false notion of the efficacy and controllability of individuals by such techniques is propagated throughout culture at large" (Joseph 2011, p. 143).

Yet, consider, on the other hand a wholly different interpretation of EEG-feedback experimentation: Andrew Pickering's evaluation of Alvin Lucier's famous EEG-feedback work *Music for Solo Performer* (1965), who interprets the piece as a "reciprocal and open-ended interplay between the performer and the performance, with each both stimulating and interfering with the other—a kind of reciprocal steersmanship" (Pickering 2010, p. 85). For Pickering, Lucier's work must be viewed as part of a cybernetic research culture that was interested in designing open-ended machines and circuits with a particular 'black box' quality—systems of which researchers did not exactly know how they would behave. Such experiments were precisely valued for their "performative materiality"—a cybernetics, in Pickering's words of "anti-control" (Pickering 2010, pp. 26, 31). Hence, while some scholars underline that EEG-feedback art works served visions of the brain as an electrical circuit that could be modelled, controlled and applied to serve (military) command, others argue that (artistic) EEG-feedback set-ups may engender a non-hierarchical, ever-changing circuit of energetic flows between machines, subjects and environments.

Ultimately, these different interpretations of EEG-feedback artworks can be traced back to divergent strands (and readings) of cybernetic experimentation (from first-order cybernetics starting in the 1940s to second-order (neo-)cybernetics of the 1970s), which come with varying notions of 'control,' 'feedback' and 'black boxes' (Hayles 1994; Clarke and Hansen 2009). It is beyond the scope of the current chapter to elaborate on the entanglement of cybernetics and EEG-(feedback) research, yet it is important to stress that when cybernetic researchers such as Norbert Wiener (Wiener 1961; see Borck 2018) or William Grey Walter (Walter 1953) employed EEG to research the behavior of the brain, clear-cut dichotomies between control and autonomy, decoding and black-boxing become untenable. Using EEG-measurements to research the behavior of the brain inevitably meant that conceptually, the cerebral 'black-box' was pried open to extract electrical measurements of brain activity, i.e. to mine the brain for data that could potentially 'decode' mental behavior. Yet, in the act of opening the black-boxed brain, new black boxes could arise.

A characteristic illustration is William Grey Walter's attempt, in the end of the 1940s, to design an EEG-visualization apparatus called the 'toposcope'. Augmenting conventional EEG-records, the device was created to show researchers a real-time, spatio-temporal record of electric activity in the brain. Affectionately called "topsy," this oval, brain-shaped visualization display did not yet work satisfactorily, but was, according to the researchers, "most pleasing" (…) "for demonstration purposes" (Walter and Shipton 1951, p. 282) (Fig. 2.1). In turn however, this model of the active brain functioned itself as a black-box with indeterminate behavior; Topsy generated visualizations with a complexity that was mesmerizing, but that could be hardly understood.[3] Control fed back into anti-control.

These ambiguities of cybernetic EEG-research become even more clear in several artworks of the 1960s. On the one hand, the EEG-feedback works of artists such as Alvin Lucier and Richard Teitelbaum can be valued particularly for the indeterminate and non-hierarchical nature of their artistic circuits. At the same time, Lucier's brain wave-based work, for example, was made in close collaboration with the military-oriented and Wiener-inspired EEG-research of physicist Edmond Dewan (Dewan 1964; see Kahn 2013). While Dewan's brain-controlled lamp switch did little to actually decode the precise 'messages' of EEG, he had nevertheless plugged alpha waves into a (hypothesized and speculative) military application for Morse coding through the brain.

Artists were aware of such control-oriented affiliations. In 1974, Richard Teitelbaum warned that EEG-research allowed brains to be opened up for commanding applications: "with some of the most technically 'advanced' psychology work currently being carried out in our prisons [under] the guise of aversion therapy and the

---

[3]Topsy fitted with a general model-making heuristic of cybernetics, whereby a designed model ideally superseded established theory to reveal something new about the behavior of a system (Schlimm 2008). Historian Cornelius Borck has aptly described Topsy as a kind of "occult alpha-wave radar system" (Borck 2018, p. 260).

Fig. 2.1 Toposcope by William Grey Walter and Harold Shipton, 1951, printed in Walter, W. Grey, and H. W. Shipton. "A New Toposcopic Display System." *Electroencephalography and Clinical Neurophysiology* 3, no. 3 (August 1951): 281–92, 282

like, there is clearly great cause for concern" (Teitelbaum 1976). What these examples show us is that EEG-feedback art works of the 1960s and 1970s are characterized by what Andrew Pickering has called a particular "double valuedness," based on ambiguous cybernetic positions that interlaced both control and anti-control, reductionism and performativity (Pickering 2010). Hence, it is important to approach these artwork as particularly hybrid entities, as Caroline Dunbar puts it, "neither fully determined nor fully autonomous" (Dunbar-Hester 2010). This double valuedness results in the interpretative flexibility of feedback artworks: though artistic experiments with brain waves may easily fall prey to visions of (military) mind management, such works could also serve ideas of indeterminacy and playful, open behaviour.

2.4 'Disproportionate Excitement' Over a 'Fresh Approach'

To understand the rapid rise of EEG-feedback in the 1960s, it is important to understand this enthusiasm as resonating with an increasing emphasis on the 'self'—a self that was based on the premise of new self-insight and a self that could be modulated by different techniques and technologies (McGee 2005; Crowley 2011). Andrew Pickering and Jonna Brenninkmeijer have employed Foucault's concept of "technologies of the self" to position the rise of EEG-feedback (later called neurofeedback) as part of a 1960s movement that was particularly interested in new structures of self-exploration, served by new technologies of the self, such as flotation tanks, stroboscopic devices, LSD, mutable architectures, flicker, video therapy and breathing techniques (Foucault 1988; see Pickering 2010; Brenninkmeijer 2016). Such technologies were shaped by specific ideas of the self and were in turn also shaping particular ways of being a self. This cultural embeddedness of 'self-oriented' EEG-feedback partly explains how a considerate number of scholars and practitioners continued until the mid-1970s to experiment with EEG-feedback (particularly alpha-wave feedback) despite intensifying objections to this type of research. Popular attention also remained steady, biofeedback methods including EEG, enjoyed an "unusually long period of publicity," as researcher Barbara Brown noted in 1974 (Brown 1974, p. 43).

Objections to EEG-feedback scholarship, starting in the late 1960s, veered in two directions: a concern about the EEG-research' reductionist or dualist approach to the mind, but also methodical doubts about the scientific validity of the experimental designs used in research. A report by psychologist John Grossberg of the University of San Diego, published in 1972, is exemplary in this respect (Grossberg 1972). Grossberg sketched the contemporary enthusiasm for new EEG-feedback research, such as Kamiya's claim for a new strategy to the study of consciousness. Constituting this perceived success was Kamiya's perception of a "fresh approach" to introspection combined with the manipulation of mental life by control over brain wave activity (Stoyva and Kamiya 1968, p. 203). Kamiya argued that a newly possible triangulation between introspection, mental life and brain waves strengthened the proposed intimate and more direct correlation between EEG and the mind (or could perhaps do away with this distinction all-together). In a similar vein, the famous psychologist Abraham Maslow concluded in 1969 that with EEG-feedback, "the mind-body problem, until now considered insoluble, does appear to be workable after all" (Maslow 1969, cited in Lynch and Paskewitch 1979, p. 326). For Grossberg, such utterances were evidence that EEG-feedback was based on a "sophisticated modern version" of a fraught mind-brain dualism, an enthusiasm that was based on the imagined closing of the gap between psychology and biology, a promise, Grossberg emphasized, that did not have any basis in experimental evidence.

Writing in 1972, Grossberg's major objection to EEG-feedback was however, more practical. By that time, he could list a considerate number of reviews that had criticized the experimental designs of EEG-feedback research and the weak scientific proof for their proposed results. Various studies questioned the reliability of correlations between alpha-feedback and reported feelings (and observations) of behavioral change, suggesting that many non-specific factors could also be involved (Lynch and Paskewitz 1971; Cleeland et al. 1971). Considering the mounting contestation of EEG-feedback, Grossberg dryly concluded that "(t)he disproportionate excitement generated by such modest evidence is intriguing" (Grossberg 1972, p. 247).

Grossberg's observation on the "disproportionate excitement" over EEG-feedback, despite emerging critique, can be better understood in relation to the rising importance of the self, a notion that afforded, as the science studies scholar Jonna Brenninkmeijer has argued, an ambiguous brokering between minds, brains, bodies, individuals and communities (Brenninkmeijer 2016). EEG-feedback offered practitioners a way of "working on the self by working on the brain," which resulted not in a vision of a self (as) reduced to the brain, but rather an extended or multiplied self that could be simultaneously a brain-based self and a kind of spiritual self (Brenninkmeijer 2016, pp. 46, 70). The next section recounts the characteristic kind of self—what I call a 'circuited self'—that was called into being through EEG-feedback in the 1960s and 1970s, and which became particularly evident in artistic set-ups with feedback circuits.

2.5   Circuited Selves, EEG-Environments and *Radical Software*

Rising, in experimental communities and heterogenous discourses around 1970, was a notion of the self that could not only be captured through introspection, but a self that was envisioned as connected with—or plugged into—broader circuits or systems. EEG-feedback set-ups generated opportunities that resonated with this new interest: recording and amplification technologies were viewed as offering new self-insight but also new abilities for the self to connect to wider circuiting systems, to become part of what was envisioned as a unitary whole. A closer look at artistic feedback works around 1970 particularly reveals this emergence of what I call a 'circuited self': a self that was dispersed in media installations with various assembled, circuited elements that together modelled a vision of the world as a total environment. These installations—understood as media 'environments' or media 'ecologies'—incorporated EEG as one of a variety of ways to melt media and 'selves' to create circuited selves. Particularly indicative of such instances of the 'circuited self,' were EEG-feedback works such as David Rosenboom's 1970 *Ecology of the Skin* and Nina Sobell's *Interactive Brainwave Drawing: EEG Telemetry Environment* (1975), which conjured an imaginary of electric flows through which selves (including brains, minds and bodies) became part of a broader media circuitry which opened up new horizons of (group) communication and even 'synchronization'.

In 1970, Rosenboom's feedback work *Ecology of the Skin* opened in the New York based venue Automation House, combining, among other elements, Alpha-feedback headbands, synthesizers, a closed-circuit television system, 'phosphene stations', and oscilloscope displays for brain wave activity (Fig. 2.2). Together, these varying parts created a "group encounter brain bio-feedback performance system" for ten participants, which Rosenboom regarded a "systems procedure" in progress (Rosenboom 1972, p. 143). In tandem with other artists, engineers and organizers involved in Automation House exhibits in the 1970s, Rosenboom was interested in exploring concepts of systems and environments: his installation showed couplings of humans and technologies that were envisioned as part of a broader, all-encompassing environment. The title of Rosenboom's work was emblematic, in the 1960s and 1970s, for a new interest by artists in 'ecological systems.' Here, 'ecology' did not primarily refer to nature, but any form of natural, technological, social or other interrela-

Fig. 2.2 Rosenboom, *Ecology of the Skin*, 1970, printed in Rosenboom, David. "Method for Producing Sounds or Light Flashes with Alpha Brain Waves for Artistic Purposes." *Leonardo* 5, no. 2 (1972): 141–45, 143. (*Photo* P. Moore, New York)

tion between elements (Kaizen 2008; see Benson 2014). Ultimately, Rosenboom's brain-media-self-circuit aimed to create, in the artist's words, "information-energy exchange rituals," a "mediational language, a coherent energy," he wanted to "stimulate non-centralized expression and more profound interactions so needed in our mechanistically functional world" (Rosenboom 1970, p. 141, 1972).

Rosenboom's EEG-feedback work must be understood as part of a new (media) ecological thinking, a framework that became influential in artistic discourses in the 1970s specifically through Gregory Bateson's writings on a 'cybernetics of the self' (Bateson 1971). Bateson's ideas were prominently circulated in the pages of the arts and culture journal *Radical Software* (published between 1970 and 1974, closely connected to artists affiliated with the Automation House) (see Collopy 2015). A Bateson-influenced media ecology dissolved the brain and the mind into the idea of a self that was unbounded by the body, a circuited self that stood in a continuous relation to the world and arose through changing flows of communication.[4] In *Rad-*

---

[4]As William Kaizen explains, Gregory Bateson proposed a radical understanding of the self and the mind, "the self as an expanded mental field in which the subject and object are no longer separable," and with a mind that was "no longer bounded by the individual body, becoming a conjunction of self and world produced through communicative ecologies" (Kaizen 2008, p. 87).

ical Software, these media ecological visions of a circuited, cybernetic self were the underlying philosophy for artistic experiments with portable video equipment, computers and closed-circuit monitoring networks, technologies that where envisioned as DIY forms for creating systems—"expanded media" (see Joselit 2007)—which ultimately connected multiple selves into a kind of a "world brain" (Benson 1970) and that stood in opposition to mass forms of communication such as broadcast television.

Feedback, within this (video) art universe, denoted both literal techniques of real-time experimentation—live and delayed playback, for example—offered by new video equipment, but also resonated with more abstract notions of feedback as the basis for an environment of direct communicative flows. EEG-feedback perfectly fitted with these notions of feedback and introduced a novel (mediated) dimension to this larger circuited self. Hence, in the pages of *Radical Software*, EEG feedback devices and feedback training were discussed as "providing people with a chance to explore the internal, and in a socially constructive way" (Ezios 1971). Artists such as Woody Vasulka and Richard Lowenberg, working together with psychophysiologist Peter Crown, conceptualized new "Techno-Sensory Interface Projects" that included 'video-moog' and audio-video systems triggered by brain wave-alpha rhythm-readings' to study, for example, "human control of purely contemplative creative processes" (Vasulka and Lowenberg 1970; Lowenberg and Crown 1971). Drawings in *Radical Software* showed heads sprouting with meandering nerves connected to television screens (Fig. 2.3). The magazine also listed examples of consumer-oriented alpha wave detectors and trainers, such as the *Toomin Alpha Pacer*, between $150 and $275 ("just a fraction of the costs of a portapak") ("Getting Wired," 1970). A *New York Times* article remarked in 1971 that these alpha devices were "hot sellers among an introspective generation" (Luce and Peper 1971).

In 1974, when artist Nina Sobell saw a friend carry such a consumer audio-alpha wave device, she started to include EEG as part of her video installations with elements of time-delay and closed-circuit systems (Stermitz and Sobell 2007). Sobell experimented with feedback video images in the EEG-laboratory of Barry Sterman and subsequently created *Interactive Brainwave Drawing: EEG Telemetry Environment* (1975) (Harzell and Sobell 2001) (Fig. 2.4). In the latter, two subjects placed in a living room-setting could look at direct video feedback of themselves on a monitor while their EEG-activity was recorded and translated into a zigzagging line figure, a visualization of the measured combined brain activity of both participants. Outside the room, five television monitors displayed their EEG recordings as well as activity from previous participants, which was superimposed on the live video.

Both in Sobell's work, as in Rosenboom's work, notions of 'synchrony' of brainwaves become intuitively aligned with cooperation in person-to-person communication, the circuited self was envisioned as enabling new versions of communicative harmony, circuits allowed for the playing, connecting, tuning and merging of different selves. Moreover, the circuited selves proposed by artists in the early 1970s are indicative of the importance that was placed on participation, experiencing and performing by users, more so than the elucidation of particular EEG-measurements. In these installations, different disembodied 'energies' were extracted from partic-

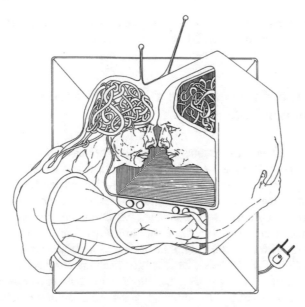

Fig. 2.3 Richard Lowenberg, *Environetic Synthesis*, drawing in *Radical Software* 1972, volume 2, nr. 1, page 47, courtesy of Richard Lowenberg

Fig. 2.4 Nina Sobell & Michael Trivich, stills from *A Video Brainwave Installation*, taped in the laboratory of Barry Sterman, Veterans Hospital Sepulveda, California, 1974, part of "Electro-encephalographic video drawings," video produced by Nina Sobel and LBMA Video, 1992, https://archive.org/details/XFR_2013-08-23_1B_16, courtesy of Nina Sobell

ipants and re-routed, displaced, manipulated, and merged into a bigger circuited whole. Writing one decade after his first EEG-experiments, Rosenboom emphasized the way that artists had thus become instrumental in designing the "interface" of these "transformations of information," to "create observable, symbolic representations of activity" through communications media (Rosenboom 1983, p. 32). Hence, artists' performative approaches to circuited selves in media ecological installations around 1970 must be understood as giving shape to notions of 'interfacing' and 'interfaces', and were part of an increasing interest in questions of designing the experimental environments that supported EEG-feedback research.

2.6 "The Interface Is the Work" and New Modes of Communication

Artistic visions of feedback systems and circuited selves employed the term 'interface' (such as the "Techno-Sensory Interface Projects") as a notion that intuitively fitted with (the vocabulary of) new visions of reciprocity and new forms of 'communication' between selves, brains and machines. At present, the notion of the 'interface' may have become more narrowly associated with 'graphic user interfaces' or even simply with the screen, i.e. with the visual appearance of a structure that allows users to interact with a digital environment. Yet as several scholars have recently argued, the notion of the interface has changed meaning at several moments in history, including a shift, poignant since the 1960s, with 'human factors research' and the rapid development of more complex, black-boxed computing apparatuses that necessitated comprehensible mediators between men and machine (Cramer 2011; Hookway 2014; Hadler and Irrgang 2015). The interfaces discussed in the emerging computing era had a particular speculative and futurist dimension, they often took the form of "inspiring visions and prototypes," as Jonathan Grudin argues: discourses on human-computer interaction with titles such as "Man–Computer Symbiosis," "Augmenting Human Intellect," and "A Conceptual Framework for Man–Machine Everything" described a world that did not exist" (Grudin 2005, p. 48). For artists as well as EEG-technology researchers, 'interfaces' were envisioned and framed as the crucial node for new forms of 'communication,' or 'dialogues' between humans, bodies, brains and machines.

This emphasis on dialogue and communication also effected the framing of EEG-feedback practices as a new experimental field: as one newspaper explained, communication was at the basis of this type of research, not only between "man and his inner being", but also between scientists and lay people (Kirsch 1974). The popular best-seller *New Mind, New Body* (1974) by biofeedback researcher Barbara Brown characteristically demonstrates the anti-authoritarian atmosphere that surrounded biofeedback research. Brown framed bio-feedback as a science of self-taught practitioners and experimenters (distinct from an older generation of scientists), who actively sought to demonstrate their science to interested laymen and did not shy away from popular media. Interaction with system designs and interfaces was at the heart of public attention to EEG-feedback research. "the laboratories were invaded by television film crews and reporters, newspaper and magazine writers. The settings were ideal for the visual media: brain-wave recordings, sound and colored-light displays and even tiny trains or racing cars which could be hooked up and energized at will by the subject's alpha waves" (Kirsch 1974).

From the mid-1960s to the mid-1970s, new EEG-feedback research experimented with custom-engineered experimental systems built with amplifiers, lights, oscilloscopes, timers, speakers and other elements, through which subjects would receive visual or audio-feedback on brain wave activity. These intricate feedback set-ups—dark rooms with brain-controlled switches, differently colored lights or changing, ringing tones—were a principal element to EEG-feedback experimentation. While researchers did not always explicitly use the word 'interface' to denote the structures that allowed for a feedback between human beings and technology, it was the carefully designed circuit of feedback, and the performative interactions of the human with elements of this circuit, that became the center of the work of EEG-feedback research.[5]

In 1970, Brown devised an EEG-feedback interface that used three EEG frequency ranges of theta, alpha, and beta to operate red, yellow and green lights (Brown 1970). Building on this research, she subsequently experimented with ways to move away from the "abstractness" of such visualized EEG-signals to create cues that had "some symbolic meaning or interest" to subjects (Brown 1974, p. 42). To do so, she designed two playful devices to make alpha waves more vivid to subjects: the 'Alpha train' and the 'Alpha wave racetrack,' which received much press attention. The Alpha train would start or stop according to the alpha waves produced by a subject, while the Alpha wave racetrack could be played by two players competing for alpha-wave control.[6] Edmond Dewan's earlier brain-controlled electric lamp had been a similar favorite for popular attention. Already in the lab, the set-up was demonstrated in front of audiences who could give orders to the performing subject (Dewan 1964).[7] The performative potential of this experiment became even more potent when *CBS News* featured a lab demonstration of Dewan's set-up and concluded their report with a scene in which the subject's EEG-signals supposedly switched on a television set and returned the viewer to the CBS studio headquarters (Fig. 2.5).

These popular imaginaries of brains plugged into everyday (media-) technological circuits were common, as Barbara Brown recounts, "scientists and tv crews alike began plotting brain wave electric companies" (…) Almost everyone decided that in the future it would be possible to have the brain start the coffee pot" (Brown 1974, p. 42). As a humorous response to the media 'hype' of alpha-feedback, artist Richard Teitelbaum and his collaborators created *Alpha Bean Lima Brain*, in which alpha waves in California were transmitted to New York to activate water sprinklers in pots containing dry white beans. "Naturally," Teitelbaum recounts, a "film of this performance was featured on the NBC evening news in Los Angeles" (Teitelbaum

---

[5] See for example Joe Kamiya's notes on his improvement of the feedback sounds and presentation of the subject's performance in Schwartz et al. (2017).

[6] A similar line of thought was pursued by artist David Rosenboom in 1971, who designed the game 'Alpha Checkers,' that showed a checkers board on a computer screen only when two players exhibited simultaneous alpha. The game, played on an IBM 360 Model 40 with computer graphics, was rather impossible to play, since perception of the board often immediately blocked alpha (Rosenboom 1976).

[7] Dewan agreed that EEG was imprecise. "The use of EEG for communication purposes therefore seems to have very little likelihood of practical application at this time" (Dewan 1964).

Fig. 2.5 Edmond Dewan, brain-controlled lamp set-up, still of CBS news item, 1964, https://www.youtube.com/watch?v=nCGcY6sQjcM

1976).[8] Whether ironic or poetic, both in scientific and artistic EEG-feedback installations there was a definite emphasis to the performative, playful, interesting or even humorous aesthetic of the designed interfacing between EEG-measurements and the measured subjects. Both in art and in EEG-science, the interface became the (art) work.

2.7 Computer Augmentation and the *Spacewar* Brain (1970–1977)

If interfaces became the (art)work, this also meant that the aesthetics, interaction potential and speed of technological interfacing between brains-minds-selves and machines became vitally important. In the early 1970s, novel computing systems started to be used in the circuits of EEG-feedback experimentation. One of the arguments to use computers, in early accounts of these brain-computer feedback-designs, was the speed of calculation and transmission offered by computing machines. As David Rosenboom put it in 1972, "through the use of computers as appendages of man's brain and methods of learning with bio-feedback, rates of information processing will be achieved that approach the speed of light" (Rosenboom 1972, p. 141). Two years before, researchers Robert Kantor and Dean Brown projected that existing EEG-feedback experiments, which involved the "speed of thinking" indeed could only be rightfully accomplished by "employing the speed of the computer" (Kantor

---

[8] Another example is Jacqueline Humbert's 1973 proposal for *Alpha Garden*, an installation which a "synchrony detector" (for brain wave synchrony between two subjects) would control a sprinkler system installed on a plot of land (Rosenboom 1976).

and Brown 1970, p. 270).[9] In 1970, they conceptualized the first brain-computer-interface ("on-line computer augmentation of bio-feedback processes") for EEG-feedback to assist "man's communication with himself."

Yet, closer examination of early brain-computer feedback designs shows that often it was not primarily the actual (realized) speed of 'on-line' and 'real-time' processes in feedback that was most important, but instead it was the imagined potential of computers as a social tool and a source of augmentation of human processes. Kantor and Brown's work is exemplary in this respect, as it arose from the educational and information science institutes of the Stanford Research Institute (SRI) at Menlo Park. Their research was closely connected to the now infamous Augmentation Research Center (ARC, led by Douglas Engelbart), with a broader research vision for 'augmentation' of human processes. Computers and computing processes were envisioned at SRI as enabling a move from the individual towards the collective, they had, as historian Fred Turner expounds, an important "social potential," i.e. the computer not as a calculating machine, but as a tool for collaboration and for sharing knowledge by means of its capacity to facilitate collective feedback (Turner 2006).[10]

Kantor and Brown's EEG-BCI fitted with SRI's research on new computer-assisted education to students, with a strong emphasis on students' insight and inner awareness. If the newly SRI-developed 'mouse' pointing device had offered a new experiential 'feel' of graphics shown on displays, direct EEG-feedback was a perfect next step, in Kantor and Brown's words: "machine augmentation of self-education in the inner states will help us to know and develop new dimensions of the self" and allowed one to "penetrate to levels much closer to the source of creativity insight and artistic expression – to the source of consciousness itself (…)" (Kantor and Brown 1970, p. 272). These new interactions with computer interfaces were at the heart of a general project to augment the intellect of human individuals by improving information processing and self-reflexive learning (Bardini 2000). Importantly, the researchers emphasized, when BCI's were used for the field of education, "content is almost secondary:" interaction with the computer was about "awareness, openness to experience, curiosity, mental agility and insight" (Kantor and Brown 1970, p. 272). It was the interaction with the BCI, not the content of the feedback data, that was at the core of this research: *the interface is the work*.

The first actual design and coinage of the term 'brain-computer interface', three years after Kantor and Brown, has been attributed to computer engineer Jacques

---

[9]Robert E. Kantor was a researcher at SRI's educational policy center and had previously published on schizophrenia and educational psychology. Dean Brown was a researcher at the information Science Laboratory of SRI. Kantor and Brown listed the areas in which they projected computers could improve existing EEG-experimentation: operant conditioning, the correlation of physiological and subjective qualities, the monitoring of patients in treatment, the control of machinery by signals from the brain and the feedback of psychological states to subjects (Kantor and Brown 1970).

[10]As Fred Turner recounts, research at SRI was of a typical hybrid cybernetic affiliation: it was both closely connected to the control visions of U.S. defense research (who funded various project at the institute) but researchers were also partaking in LSD-experiments, commune life, EST-training and generally interested in notions of community and creativity (Tuner 2006) and see (Markoff 2006; Rosenzweig 1998).

Vidal, developed at UCLA's Brain Research Institute in Los Angeles and published in his 1973 paper—"Toward direct brain-computer communication" (Vidal 1973). Vidal described his (partially defense department-funded) BCI as a "real-time system," referencing James Martin's 1967 book *Design of Real-Time Computer Systems*.[11] Ultimately, this direction of work aimed for a true "man-computer dialogue," (a term Vidal might have drawn from Martin's book *Design of Man-Computer Dialogues* published in 1973) that would turn the computer into a "genuine prosthetic extension of the brain," enabling humans to control other external apparatuses such as, Vidal imagined, a prosthetic device or a spaceship. Vidal envisioned BCI as part of a project to "decode" the "fuzzy signals" in "neuroelectric language," to provide new "channels of communication between man and machine," an endeavor he would later call "neurocybernetics" (Vidal 1975).

In contrast to Kantor and Brown's online augmentation of EEG-feedback, Vidal's brain-computer interface (which consisted of over a dozen circuited displays, control terminals, A-D-converters, buffer controllers and other components), seemingly shifted attention from the exploring the 'self' back to decoding the brain. Yet, here again, we see the ambiguities of cyber-countercultural entanglement. When Vidal wanted to test not just attention or perception but also subjects' emotional states, he designed a BCI-circuit that integrated the computer game *Spacewar*. Within the laboratory research culture of the early 1970s, *Spacewar* had become omnipresent and was much-loved; the game was so addictive that some laboratory managers had tried to limit playtime by the lab workers (Markoff 2002). *Spacewar* helped Vidal's BCI-research by offering stimuli that were identical (space ships), but that generated very different emotional responses in subjects: witnessing the destruction of your own spaceship versus that of an opponent would make varying "cognitive influences" visible in EEG-patterns (Fig. 2.6). *Rolling Stone* magazine reported about the game in 1972 in an article by Stewart Brand, who characterized it as originating as much from the "disestablishmentarianism of the freaks who design computer science" as well as research from the "very top of the Defense Department" (Brand 1972). The article described the hybrid hacker-culture from which *Spacewar* originated, with an ethos that was ultimately geared towards community formation and a global revolution (Turner 2006; Bryant 2006). Within this frame, *Spacewar* could be understood as a "communication device between humans," and the game "served *human interest*, not machine," Brand reported.[12]

Brand's *Rolling Stone* article shows how *Spacewar* had particular symbolic purchase in the early 1970s and hence, through Vidal's appropriation of the game into a BCI-assisted EEG-feedback set-up, it becomes clear how here computing too, was envisioned as turning experimental subjects into circuited selves as part of (play-

---

[11] In the paper, Vidal described an EEG-feedback set-up that contained various computers that allowed "on-line, real-time feature extraction" of the incoming EEG-signals, feeding the results back to subjects on visual output display. The analysis that could be performed in Vidal's 'real time' experiment was less detailed than non-real time analysis of EEG, which used batch processing and averaging. Vidal remarks that one would need a round trip of less than 0.5 s to 'directly communicate brain messages'. Because averaging made the round trip too slow, Vidal's BCI project made use of a reference set of typical EEG waveforms for various responses (Vidal 1973). Vidal's research was

Fig. 2.6 The *Spacewar* Olympics at Stanford Research Institute, 1972, printed in Brand, Stewart, "Spacewar: Fanatic Life and Symbolic Death among the Computer Bums." *Rolling Stone,* December 7, 1972. (*Photo* Annie Leibovitz)

ful yet serious) media ecologies. This framing of the brain computer interfaces as 'augmentation for self-education' or a 'man-computer dialogue' also demonstrates how much computing was part of a promissory narrative for the future of EEG-research. The computer fitted with the existing importance in EEG-feedback circuits to design a set-up for circuited selves, augmentation and playfulness—yet, more than anything else, computers served as a next step in interface construction. Vidal's brain-computer interface was a prospective project, it was (using Grudin's aforementioned terms), foremost an "inspiring vision" or "prototype" of a future interface, one that was particularly predicated on (and a driving force for) the eventual success of real-time feedback. 'Real' 'real-time' feedback could only be partially obtained in computer-aided set ups. Vidal would spend the next years trying to achieve better real-time results that would, he projected, ultimately reveal how brain processes were correlated to mental processes (Fig. 2.7).

Writing in 1977, Vidal claimed that "to bring about real-time discrimination would provide a quantum increase in the power of the psychophysiological method" (Vidal 1977). 'Real' real-time feedback, in Vidal's view, would ultimately bridge the mind-brain problem: a specific visualized brain activity could actually be understood as a "behavior" that could then be consciously modulated. Yet, during the 1970s, his experiments did not yield these results, real-time feedback could not provide this 'quantum leap.' Ultimately, it was in the field of "adaptive user interfaces" that his

partly funded by a broader Defense research (DARPA) research program on Biocybernetics, see Daly (1981).

[12]"It was part of no one's grand scheme. It served no grand theory. It was the enthusiasm of irresponsible youngsters." In Brand's view, ultimately, *"Spacewar serves Earthpeace.* So does any funky playing with computers or any computer-pursuit of your own peculiar goals" (Brand 1972).

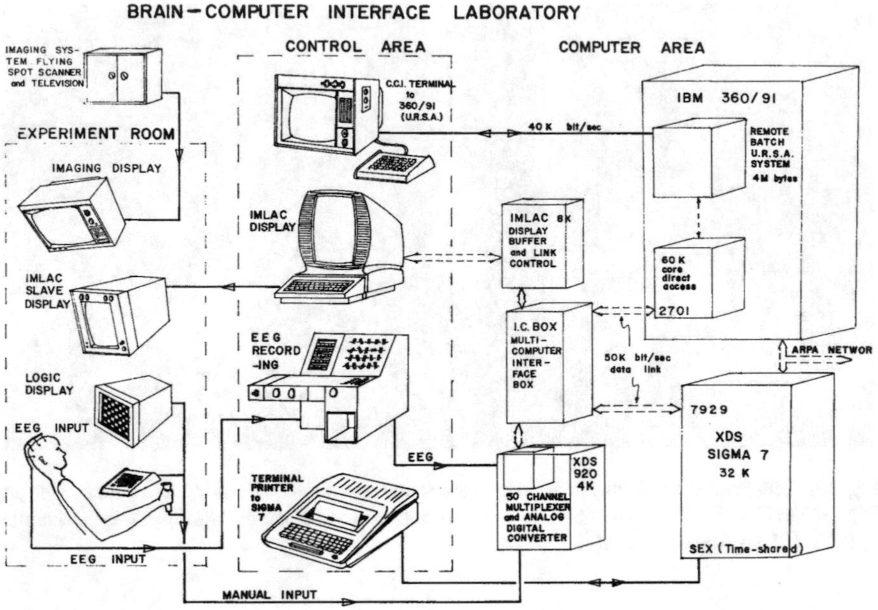

BRAIN—COMPUTER INTERFACE LABORATORY

FIGURE 2. General organization and computer architecture of the Brain-Computer Interface Laboratory at UCLA.

Fig. 2.7 Jacques Vidal, "General organization and computer architecture of the Brain-Computer Interface Laboratory at UCLA," 1973, printed in "Toward Direct Brain-Computer Communication." *Annual Review of Biophysics and Bioengineering* 2, no. 1 (1973): 157–80, 172

line of research continued. In this new direction of scholarship, Vidal and others were interested in creating a representation of a "virtual workspace" that would "evoke and emulate a mental representation" of a user (Vidal 1984; see Vidal 1983). It was the interface, that constituted the work.

2.8 Alpha's "Dark Ages" and the Re-emergence of Neurofeedback

In the 1970s, the promise of non-invasive BCI-EEG feedback (to train the brain for specific EEG-brainwave states by computer-assisted feedback) was hard to fulfill. The above-cited critique of the experimental methods and results of EEG-feedback persisted. In 1983, one general textbook on biofeedback stated that while alpha feedback had been the most dominant and popularly acclaimed form of biofeedback in the late 1960s, it had since "virtually dried up as a scientifically defensible clinical tool" (Basmajian 1983, cited in Schwartz et al. 2017). Indeed, in histories of (non-invasive) BCI and EEG-feedback research, the 1980s are described as the "dark age" or "dormant" phase of research (Evans 1998; Lotte et al. 2018).[13] The scientific demise of EEG-neurofeedback research in the 1970s and 1980s is often framed as a technological failure of processing speed, one that was eventually solved with faster computing, improved displays and better software (Schwartz et al. 2017). Around 1990, the start of the decade of the brain, saw a surge of new research in BCI, when technological advances allowed for feedback on 'a moment-to-moment' or "nearly-concurrent" basis (Thibault 2016). Because of the development of 'real' real-time technology, as artist David Rosenboom described in 1990, "ideas that were impractical when they were proposed many years ago are now practical" (Rosenboom 1990). Rosenboom's EEG-feedback artworks experienced a revival, in the early 1990s, as did the advertising of brain-training devices and techniques for "scientizing the Yogi-mind" (Dumit 1995, p. 351).

Hence, an important driver for the renewed potential of neurofeedback was the invention of faster and more detailed imaging technologies that could provide the high-paced, fine-grained, complex tempo-spatial measurements necessary to train the brain with feedback. Another important change, since the late 1980s, was the new possibility to record larger ensembles of neurons (Lenoir 2011; Lebedev and Nicolelis 2017). Together, these developments have been successful in creating, for example, basic communication tools for motor-impaired and "locked-in" patients. Today, the future of neurofeedback is linked to spatio-temporal innovation: faster computing speeds, the invention of MRI scanners with ever higher magnetic fields (seven Tesla, or more), the use of fNIRS or MEG, the possibility of measuring functional connectivity or the combination of these different technologies in multi-model imaging. Very recently, a 2018 systematic review of research on neurofeedback with fMRI (fMRI-nf) concludes that, while it is still unclear how fMRI-feedback approaches might lead to behavioral improvements in the clinical domain, it is certainly possible to change brain activity with fMRI-nf (Thibault et al. 2018).

Conversely, scientific evidence for the efficacy of various other forms of neurofeedback is contested, proof of its effectiveness has been controversial since the 1960s, and still is today. For example, in a recent systematic review, Robert Thibault and his co-researchers conclude that, while there is evidence that subjects partaking in EEG-neurofeedback experiments show changes both in brain activity patterns and in behavior, "the current literature does not support a direct connection between the specific feedback and the observed alterations" (Thibault et al. 2015, p. 196). In other words: EEG-neurofeedback has various (positive) effects, but these effects are—as far as research in the past fifty years shows—not evidently caused by the feedback about specific brain activities.[14] Instead, what is most prominent in this half decade or so is the development of powerful new computing methods, visualization techniques, display designs and new repertoires for (playful) feedback situations.

---

[13]"Dark ages" in (Evans 1998), "dormant" phase in (Lotte et al. 2018). EEG-feedback practices never stopped entirely in the 1980s, see (Sala 1999) for references to (what he calls the) "alpha-conditioning craze" that extended into the 1980s.

[14]In this context, Thibault et al. (2016) note neurofeedback's "multi-faceted nature:" participants may be influenced by non-neurofeedback-specific factors such as the tendency to comply to a demand, psychological facts, the overestimation of technology, positive effects of attention, etc.

Still today, as in the 1960s, the true fulfilment of successful neurofeedback is pro-jected to the (technological) future: with faster transmission speeds and higher imag-ing resolutions, ultimately, true efficacy of the training of the self through the brain is thought to be attainable. Michael Hagner and Cornelius Borck have dubbed this promissory logic in neuroscience a "proleptic structure," emphasizing that present-day practices in neuroscience are often based on promises that are not yet fulfilled, as a science that "anticipate[s] a future of comprehensive understanding" (Borch and Hagner 2006). As such, neurofeedback is a typical example of the future-oriented character of our current technoscientific age, and is based on the much older promise that the precise connection of mind and brain is within reach, what Borck has called "the phantasm of the imminent elucidation of the mind-brain" (Borck 2009). In this chapter, I have studied a characteristic element of this promissory domain, an asso-ciated emphasis, since the 1960s, on interface design and a particular performative style of EEG-feedback research.

2.9   Glass Brains and New Forms of Neuroscientific Life

Looking back to the history of EEG-feedback and early BCI-design helps to rec-ognize the continuing importance of an emphasis, in the neuroscientific field, on public demonstrations of new feedback interfaces. A characteristic example of this performative research culture comes from the team of BCI-creators (Tim Mullen and co-researchers) mentioned in the opening paragraphs of this chapter. A 2014 'Mozart & the Mind' festival in San Diego, in which several of the authors participated, is but one instance of an explosion of arts-meeting-neurosciences events characterized by new aesthetic, educational and spectacular formats for performing science. Highlight of this 2014 event was a live demo of a new visualization software called the *Glass Brain*, described as the "world's first interactive, real-time, high-resolution visual-ization of an active human brain designed specifically for virtual reality" (Intheon Community Projects 2019) (Fig. 2.8). On stage, neuroscientist Tim Mullen used VR-goggles to fly through the "live brain" of *Grateful Dead* drummer Mickey Hart (also on the podium) while the latter was playing a musical brain-training game, wear-ing a high-resolution EEG-cap (Greenemeier 2014). Projected on a large screen, the audience could see what was simultaneously happening in the neuroscientist's VR-world: a shimmering, translucent brain showed white flashing sparks (estimated information transfer between brain regions), bursts of color (activity in theta-, alpha-, beta-, gamma- EEG frequency bands) and golden threads (white matter tracts).

The iconic-looking *Glass Brain* visualization is not a clinical tool, but serves as a visual demonstration of the underlying technology's success: its algorithms and software pipeline are able to measure complex brain activity dynamics with EEG in real-time (Mullen et al. 2013; Mullen et al. 2015). With this capacity, such computational approaches could drive new forms of what is called 'closed-loop cog-nition' (experiments in which tasks are continuously updated based on the subject's neuro-cognitive state) that help to understand the fundamental mechanisms in neural

Fig. 2.8 Glass Brain
demonstration, San Diego
2014, courtesy of
Neuroscape, University of
California, San Francisco

networks, but can also be used in the development of brain training games ('closed-loop neuro-cognitive training,' Mishra and Gazzaley 2014, 2015). In 2013, the cover of *Nature* magazine announced the success of the brain training game 'neuroracer' in enhancing the cognitive control of older adults (Anguera et al. 2013). Hence, four decades after Barbara Brown's 'alpha racer' game and Jacques Vidal's *Spacewar* brain, these new computational methods are now welcomed for allowing "biofeedback on the next level" (Gazzaley in *Live Science*, Lewis 2014).

The sparkling, moving images of the real-time *Glass Brain* visualization strengthen the claim that closed-loop brain-training games are working on the physical, active organ in real-time—a brain that is finally truly accessible for enhancement (Fig. 2.9). Such persuasive visuals are necessary, because the brain-gaming field is a contested terrain. Scientists have to pursue active boundary-work to distinguish genuine, evidence-based (and perhaps future FDA-approved) neuro-games from commercial games that make inflated claims.[15] The vision of BCI as a closed-loop system that includes a fully accessible, transparent brain—a new type of interface—sets the horizon for this scientific research program. During a 2016 live demo of the *Glass Brain* at a *Fortune Magazine* conference, professor of neurology Adam Gazzaley, one of the scientists involved in the neuro-racer game and chief scientific advisor for a major digital medicine startup, projected a future in which we could "use this

---

[15]In 2014, The Stanford Centre for Longevity issued "A Consensus on the Brain Training Industry from the Scientific Community," stating that there is "no compelling scientific evidence" that brain games can reduce cognitive decline (Stanford Centre for Longevity 2014). The statement was countered by a second group of scientists, who argued "a substantial and growing body of evidence shows that certain cognitive training regimens can significantly improve cognitive function" (*Cognitive Training Data Response Letter*, 2014; see Span 216).

Fig. 2.9 *Glass Brain*
visualization, courtesy of
Neuroscape, University of
California, San Francisco

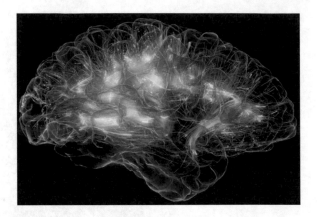

kind of technology to do real-time diagnostics. I could be flying through your brain, while you are in your virtual world [and I would be] pushing on different knobs to challenge you, see how your brain responds" (Fortune Magazine Conference 2016).

The performative and prospective dimension of various BCI-projects, exemplified by the *Glass Brain*, is typical of new structures of research that historian Steven Shapin has characterized as creating "pictures of possible worlds-to-come" and as engaged in "technoscientific and economic future making" (Shapin 2008, pp. 309, xv). Taking San Diego's venture-capital funded bio-tech sector as his main example, Shapin argues that practices of 'future making' are particularly important in a new era of industrial-scientific entrepreneurship with an intensified uncertainty about whether a project will yield results. In order to build (new forms of) trust and authority, entrepreneurial science frames research sites as creative playgrounds and places new emphasis on the charisma of the individual researchers. It is through the creative personalities of researchers that these fields can best embody and build visions of the future, accentuating elements of play and fun in research praxis. Such strategies also characterize BCI-research: the *Glass Brain* is a partly open source programming interface, one of various products and community projects for "plug-and-play neurotechnology" (Intheon 2019). In turn, the *Glass Brain's* computations have served as building blocks for the recently announced first internationally patented platform technology for a digital "personalized treatment experience" that may improve cognitive abilities (Akili Interactive 2018). The prospects of BCI are here phrased with great care and with the necessary official "forward looking" disclaimer, yet beyond these scientific papers and patent descriptions, the field of BCI is equally built by assemblages of images, animations, video lectures and demo's that engage in 'future making'.

Artists are an integral part of these practices of 'future making'. In San Francisco, the translational neuroscience center that co-developed the *Glass Brain* actively engages with artists who want to integrate these visualizations and technologies in their work. On various occasions, academically-trained entrepreneurs simultaneously work as neuroscientists and media artists, developing their augmentations of brain activity in tandem with artistic installations, demonstrating their work at hybrid academic-, public- and industry-oriented art-science events. The BCI-field is thus characteristic of what Shapin describes as new forms of "scientific life" in late technoscientific modernity, typified by the emergence of the scientist-entrepreneur and the scientist as performance artist, i.e. the rise of new norms and forms in managing and cultivating relations between science and society (Shapin 2008; see Jasanoff 2012; Bensaude-Vincent 2013; Bucchi et al. 2007). BCI-neurofeedback thrives in this hybrid world of clinical-artistic-commercial applications that emphasize the experience of feedback, the ingenuity or aesthetics of feedback design and the value of real-time interaction. Ultimately, historical observations help to better apprehend these contemporary, spectacular forms of 'neuroscientific life,' i.e. the structures of researching, experimenting, demonstrating, in place in research practices today.

2.10   Conclusion

In this chapter, I portrayed the emergence of a research culture in EEG-feedback in the period between Edmond Dewan's 1964 brain-controlled light switch to Vidal's 1977 paper on "Real-time detection of brain events in EEG." I have argued that the development of EEG-feedback in this period is characterized by a particular aesthetics of research practice: a strong emphasis on the performative demonstration of a feedback situation and the designed set-up that gives feedback, a prominence of interface over EEG-data—the idea that 'the interface is the work.' This aesthetics of research practice must be understood, I claim, in light of a longer established performative dimension of cybernetic feedback circuits, which transformed by the end of the 1960s into an interest in the 'circuited self' as part of a media ecological discourse.

Throughout my examples of such brain-media circuits, I have shown how artists and researchers stood in a continuous relation to popular reporting about feedback situations in newspapers and television broadcasts, which created a particular sphere of resonance for the emphasis on playful and spectacular demonstrations of interfaces. When computing was introduced in EEG-feedback around 1970, it brought notions of 'on-line' and 'real-time' into the circuit. These developments were not only understood as technological advancement through faster feedback, but they also brought an emphasis on the social potential of computing: self-insight, augmenting the self and connecting with others. My tracing of the genealogies of EEG-feedback and BCI with close attention to circuited selves and media ecologies, thus has helped to outline a particular performative research culture in the 1960s and 1970s.

In this chapter, the envisioned brain-computer interfaces, as well as the artistic experiments with brain-media circuits, are examples of what I call 'brainmedia'. With this term, I denote the assemblages of brains and media constituted by social-technological practices as well as imaginaries and visions, which together conjure new enactments of (for example, as here discussed) synchronicity, communication, wholeness, control, augmentation and awareness, that in turn impact our understandings of both brains and technological media. Brainmedia emerge through particular historical and situated practices, in which science and art are intermingled and jointly shape the directions in which a research field develops.

Brainmedia are political. Mentioned in this chapter are, for example, the ambiguous imaginaries of mind-control, self-circuiting and anti-authoritarianism, which become visible in tracing a longer genealogy of EEG-feedback and BCI. As such, this chapter provides historical background to contemporary discussions of the way current BCI-discourses frame the "neuro-technologized subject" and can emphasize autonomy and self-optimization over and above issues of neuro-governmentality (Schmitz 2012, 2016; Rose and Abi-Rached 2014; Brenninkmeijer and Zwart 2016). By tracing genealogies of EEG-feedback and BCI, this chapter has particularly contributed to a better grasp of the performative research cultures in the 1960s and 1970s, which also helps to see the persistence of performative elements in today's forms of 'neuroscientific life'. My expanded perspective on the 'art' of EEG-feedback and BCI shows how an aesthetics of research practice is shaped in hybrid, intersecting spheres of artistic, experimental and popular performing of knowledges. Underlining such interactions between heterogeneous sites of knowledge production in this genealogy of EEG-feedback demonstration, complements ongoing research into practices of 'performing knowledges' (Dupree and Franzel 2015).

Recently, on a podium at a 2018 TEDtalk-event in San Diego, neuroscientist Tim Mullen demonstrated a novel online and cell-phone-accessible interface, which allows users to view changing sensory measurements, including EEG-activity, in real-time.[16] At the event, Mullen sketched a future of neural interface technologies, about fifteen years from now, that would allow, for example, a teenager to improve learning by means of homework that adapts to cognitive states; the creation of VR-games based on group emotional states; and brain-health monitoring for astronauts on Mars. While Mullen acknowledged that such narratives, "may sound like science fiction," he emphasized that they are "all realistic applications of scientific research and development that's already happening today." When a demo of the real-time interface was presented on stage, for a very brief moment, the internet connection faltered. The colorful bars of the EEG-graph were at a stand-still. Yet, quickly, movement appeared, setting the scene for Mullen's vision of a future of neural interfaces merging with distributed computing to form a "sort of cognitive halo extending our minds beyond the limits of our bodies." "Today the mind is the new frontier," Mullen asserted, "and I believe we are at a point in time where advances in neuro-technology and computing and machine learning are converging on a singular point in history, beyond which lies a new era of discovery of who we are and what we can become." This point in history is, however, as I have shown in this chapter, part of a history of futures long in the making.

---

[16]TEDx Talks. *The Mind Is the New Frontier with Ubiquitous Neurotechnology | Tim Mullen | TEDxSanDiegoSalon*, published 2 November 2018, Accessed November 10, 2018. https://www.youtube.com/watch?v=ra6v0EvWclc.

I would like to thank the anonymous reviewers of this article for their comments, as well as Anton Nijholt, Jonna Brenninkmeijer, Peter Crown, David Gauthier, Richard Lowenberg, David Rosenboom and Michael Stevenson.

References

Adrian ED, Matthews BHC (1934) The Berger rhythm: potential changes from the occipital lobes in man. Brain 57:355–385

Akili Interactive (2018) Akili digital medicine technology platform granted multiple patents. Business Wire, April 10, 2018. https://www.businesswire.com/news/home/20180409006524/en/Akili-Digital-Medicine-Technology-Platform-Granted-Multiple

Anguera JA, Boccanfuso J, Rintoul JL, Al-Hashimi O, Faraji F, Janowich J, Kong E et al (2013) Video game training enhances cognitive control in older adults. Nature 501(7465):97–101. https://doi.org/10.1038/nature12486

Bardini T (2000) Bootstrapping: Douglas Engelbart, coevolution, and the origins of personal computing. Stanford University Press, Stanford, CA

Basmajian JV (1983) Biofeedback: principles and practice for clinicians. Williams & Wilkins, Baltimore, London

Bateson G (1971) The cybernetics of 'Self': a theory of alcoholism. Psychiatry J Study Interpers Process 34(1):1–18

Bernadette B-V (2013) Reconfiguring the public of science. In: Hubert-Curien J, Baranger P, Schiele B, Le Déaut J-Y, Bordenave A (eds) Science communication today international perspectives, issues and strategies. CNRS éd, Paris, pp 105–118

Benson D (1970) Neurone cluster grope. Radic Softw 1(2)

Benson E (2014) Environment between system and nature: Alan Sonfist and the art of the cybernetic environment. Communication 1 3(1):1–26. https://doi.org/10.7275/R5HT2M7T

Borck C (2018) Brainwaves: a cultural history of electroencephalography. Routledge, London & New York

Borck C (2009) Through the looking glass: past futures of brain research. Med Stud 1(4):329–338. https://doi.org/10.1007/s12376-009-0030-8

Brand S (1972) Space war, fanatic life and symbolic death among the computer bums. Roll Stone 7:50–58

Brenninkmeijer J (2016) Neurotechnologies of the self: mind, brain and subjectivity. Palgrave Macmillan UK, London

Brenninkmeijer J, Zwart H (2016) From 'Hard' neuro-tools to 'Soft' neuro-toys? Refocussing the neuro-enhancement debate. Neuroethics:1–12. https://doi.org/10.1007/s12152-016-9283-6

Brown BB (1970) Awareness of EEG-subjective activity relationships detected within a closed feedback system. Psychophysiology 7(3):451–464. https://doi.org/10.1111/j.1469-8986.1970.tb01771.x

Brown BB (1974) New mind new body; bio-feedback: new directions for the mind. Harper & Row, New York

Bryant WH (2006) Whole system, whole earth: the convergence of technology and ecology in twentieth century American culture. University of Iowa. https://www.researchgate.net/publication/33936896_Whole_system_whole_earth_the_convergence_of_technology_and_ecology_in_twentieth-century_american_culture

Bucchi M, Neresini F (2007) Science and public participation. In: Hackett EJ, Amsterdamska O, Lynch M, Wajcman J (eds) The handbook of science and technology studies, 3rd edn. MIT Press, Cambridge, MA, pp 449–472

Chu NNY (2015) Brain-computer interface technology and development: the emergence of imprecise brainwave headsets in the commercial world. IEEE Consum Electron Mag 4(3):34–41. https://doi.org/10.1109/MCE.2015.2421551

Clarke B, Hansen MBN (2009) Introduction: neocybernetic emergence. In: Clarke B, Hansene MBN (eds) Emergence and embodiment: new essays on second-order systems theory, Duke University Press Books, Durham, pp 3–25

Cleeland CS, Booker HE, Hosokawa K (1971) Alpha enhancement—due to feedback or nature of task? Psychophysiology 8(2):262

Cognitive Training Data Response Letter. Cognitive training data (blog), 2014. https://www.cognitivetrainingdata.org/the-controversy-does-brain-training-work/response-letter/

Collopy P (2015) The revolution will be videotaped: making a technology of consciousness in the long 1960s. University of Pennsylvania. https://repository.upenn.edu/edissertations/1665

Cramer F (2011) What is interface aesthetics, or what could it be (not)? In: Andersen CU, Pold SB (eds) Interface critism: aesthetics beyond buttons. Aarhus University Press, Aarhus, pp 117–129

Crowley K (2011) Feminism's new age: gender, appropriation, and the afterlife of essentialism. SUNY Press, Albany, NY

Daly J (1981) Close-coupled man/machine systems research (biocybernetics) completion report. Cybernetics Technology Division Program, DARPA

Dewan EM (1964) Communication by electroencephalography. Experiment at the Stanley Cobb Laboratories at Massachusetts General Hospital. Air Force Cambridge Research Laboratories, United States Air Force, pp 1–6

Drucker J (2011) Humanities approaches to interface theory. Cult Mach 12:1–20

Dumit J (1995) Brain-mind machines and American technological dream marketing. Towards an ethnography of cyborg envy. In: Gray CH (ed) The cyborg handbook. Routledge, New York, pp 347–361

Dunbar-Hester C (2010) Listening to cybernetics music, machines, and nervous systems, 1950–1980. Sci Technol Hum Values 35(1):113–139. https://doi.org/10.1177/0162243909337116

Dupree MH, Franzel SB (2015) Introduction: performing knowledge, 1750–1850. In: Dupree MH, Franzel SB (eds) Performing knowledge, 1750–1850. Walter de Gruyter GmbH & Co KG, Berlin & Boston, pp 1–24

Eaton ML (1971) Bio-music: biological feedback experimental music systems. ORCUS, Kansas City

Evans JR (1998) Reflections on neurotherapy: past, present and future. J Neurother 2(4):i–vi. https://doi.org/10.1300/J184v02n04_a

Ezios R (1971) Implications of physiological feedback training. Radic Softw 1(4):2–4

Fortune Magazine Conference (2016) How video games could help treat brain disorders. https://www.youtube.com/watch?v=3eXNEXwVIwQ

Foucault M (1988) Technologies of the self: a seminar with Michel Foucault. In: Martin LH, Gutman H, Hutton PH. University of Massachusetts Press, Amherst

Getting wired. Radic Softw 1(4):2 (1971)

Greenemeier L (2014) 'Glass Brain' offers tours of the space between your ears. Scientific American, September 29, 2014. https://www.scientificamerican.com/article/glass-brain-offers-tours-of-the-space-between-your-ears/

Grossberg JM (1972) Brain wave feedback experiments and the concept of mental mechanisms. J Behav Ther Exp Psychiatry 3(4):245–251. https://doi.org/10.1016/0005-7916(72)90043-2

Grudin J (2005) Three faces of human-computer interaction. IEEE Ann Hist Comput 27(4):46–62

Hadler F, Irrgang D (2015) Instant sensemaking, immersion and invisibility. Notes on the genealogy of interface paradigms. Punctum 7–25

Hagner M, Borck C (2006) Brave neuro worlds. In: Hagner M (ed) Der Geist bei der Arbeit: historische Untersuchungen zur Hirnforschung. Wallstein, Göttingen, pp 7–34

Hartzell E, Sobell N (2001) Sculpting in time and space: interactive work. Leonardo 34(2):101–107. https://doi.org/10.1162/002409401750184636

Hayles NK (1994) Boundary disputes: homeostasis, reflexivity, and the foundations of cybernetics. Configurations 2(3):441–467. https://doi.org/10.1353/con.1994.0038

Hookway B (2014) Interface. MIT Press, Cambridge

Intheon community projects. https://intheon.io/projects. Accessed 26 Jan 2019

Jasanoff S (2012) Science and public reason. Oxon & New York, Routledge

Joselit D (2007) Feedback: televison against democracy. Mit Press, Cambridge, MA; London

Joseph BW (2011) Biomusic. Grey Room 45:128–150. https://doi.org/10.1162/GREY_a_00053

Kahn D (2013) Earth sound earth signal. Energies and earth magnitude in the arts. University of California Press, Berkeley and Los Angeles

Kaizen W (2008) Steps to an ecology of communication: radical software, Dan Graham, and the legacy of Gregory Bateson. Art J 67(3):86–106. https://doi.org/10.1080/00043249.2008.10791316

Kantor RE, Brown D (1970) On-line computer augmentation of bio-feedback processes. Int J Bio-Med Comput 1(4):265–75. https://doi.org/10.1016/0020-7101(70)90002-4

Kamiya J (1968) Conscious control of brain waves. Psychol Today 1:56–60

Kirsch R (1974) Biofeedback: in the beginning was alpha. Los Angeles Times, August 18, 1974

Kornmuller A (1945) Signalisierung Der Langsamen Wellen Des EEG Im Sauerstoffmangel, November 15, 1945. MPG-A III/16/41

Lebedev MA, Nicolelis MAL (2017) Brain-machine interfaces: from basic science to neuroprostheses and neurorehabilitation. Physiol Rev 97(2):767–837. https://doi.org/10.1152/physrev.00027.2016

Lenoir T (2011) Neurofutures: brain-machine interfaces and collective minds. Open Humanites Press. http://www.livingbooksaboutlife.org/books/Neurofutures

Lewis T, Writer S | March 10, and 2014 07:37 pm ET. Virtual reality system lets you explore your brain in real-time. Live Science. https://www.livescience.com/44000-virtual-reality-system-reveals-brain.html. Accessed 28 Jan 2019

Lowenberg R, Crown P (1971) Environetic synthesis. Techno-Sensory Interface Projects, January 31, 1971. www.vasulka.org/archive/Contributors/PeterCrown/Teluride/EnvironeticSynthesis.pdf

Luce G, Peper E (1971) Mind over body, mind over mind. New York Times, September 12, 1971

Lynch JJ, Paskewitz DA (1971) On the mechanisms of the feedback control of human brain-wave activity. J Nerv Mental Dis 153:205–17. https://doi.org/10.1097/00005053-197109000-00005

Lynch JJ, Paskewitz DA (1979) On the mechanisms of the feedback control of human brain-wave activity. In: Peper E, Ancoli S, Quinn M (eds) Mind/body integration: essential readings in biofeedback. Springer US, Boston, MA, pp 325–340. https://doi.org/10.1007/978-1-4613-2898-8_28

Markoff J (2002) A long time ago, in a lab far away ... The New York Times, February 28, 2002, sec. Technology. https://www.nytimes.com/2002/02/28/technology/a-long-time-ago-in-a-lab-far-away.html

Markoff J (2006) What the Dormouse said: how the sixties counterculture shaped the personal computer industry. Penguin Books, New York

Martin J (1973) Design of man-computer dialogues. Prentice-Hall, Englewood Cliffs, N.J.

Martin J (1967) Design of real-time computer systems. Prentice-Hall, Englewood Cliffs, N.J.

Maskeliunas R, Damasevicius R, Martisius I, Vasiljevas M (2016) Consumer-grade EEG devices: are they usable for control tasks? PeerJ 4. https://doi.org/10.7717/peerj.1746

McGee M (2005) Self-Help Inc: makeover culture in American life. Oxford University Press, Oxford

Mishra J, Gazzaley A (2014) Closed-loop rehabilitation of age-related cognitive disorders. Semin Neurol 34(5):584–590. https://doi.org/10.1055/s-0034-1396011

Mishra J, Gazzaley A (2015) Closed-loop cognition: the next frontier arrives. Trends Cogn Sci 19(5):242–243. https://doi.org/10.1016/j.tics.2015.03.008

Mozart & the mind 2014. Accessed October 13, 2018. http://www.antillipsi.net/art-1/matm/matm2014. Accessed 22 Oct 2018

Mullen T, Kothe C, Chi YM, Ojeda A, Kerth T, Makeig S, Cauwenberghs G, Jung T (2013) Real-time modeling and 3D visualization of source dynamics and connectivity using wearable EEG.

In 2013 35th annual international conference of the IEEE engineering in medicine and biology society (EMBC), pp 2184–2187. https://doi.org/10.1109/EMBC.2013.6609968

Mullen T, Khalil A, Ward T, Iversen J, Leslie G, Warp R, Whitman M et al (2015) MindMusic: playful and social installations at the interface between music and the brain. In Nijholt A (ed) More playful user interfaces. Gaming media and social effects. Springer Singapore, pp 197–229. http://link.springer.com/chapter/10.1007/978-981-287-546-4_9

Nam CS, Nijholt A, Lotte F (2018) Introduction: evolution of brain-computer interfaces. In: Nam CS, Nijholt A, Lotte F (eds) Brain-computer interfaces handbook: technological and theoretical advances. CRC Press is an imprint of the Taylor & Francis Group, Boca Raton; London; New York, an Informa business

Neuroscapelab (2019) Glass brain flythrough—Gazzaleylab/SCCN/Neuroscapelab. https://www.youtube.com/watch?v=dAIQeTeMJ-I. Accessed 28 Jan 2019

Nijholt A (2015) Competing and collaborating brains: multi-brain computer interfacing. Brain-computer interfaces: current trends and applications, pp 313–335. https://doi.org/10.1007/978-3-319-10978-7_12

Pickering A (2010) The cybernetic brain. University of Chicago Press, Chicago. http://site.ebrary.com/id/10386301

Robbins J (2008) A symphony in the brain: the evolution of the new brain wave biofeedback. Grove Press

Rose N, Abi-Rached J (214) Governing through the brain: neuropolitics, neuroscience and subjectivity. Camb J Anthropol 32(1):3–23. https://doi.org/10.3167/ca.2014.320102

Rosenboom D (1983) Artificial intelligence and art education (1970 & 1981) (Revised 1983), in collected articles 1968–1982, pp 30–34. https://www.researchgate.net/publication/265116281_ARTIFICIAL_INTELLIGENCE_AND_ART_EDUCATION. Accessed 13 Dec 2018

Rosenboom D (1996) B.C.-A.D. and two lines: two ways of making music while exploring instability in tribute to Salvatore Martirano. Perspect New Music. https://www.thefreelibrary.com/B.C.-A.D.+and+two+lines%3A+two+ways+of+making+music+while+exploring...-a019554504

Rosenboom D (1972) Method for producing sounds or light flashes with alpha brain waves for artistic purposes. Leonardo 5(2):141–145. https://doi.org/10.2307/1572548

Rosenboom D (1976) Biofeedback and the arts: results of early experiments. A.R.C. Publications, Vancouver, B.C

Rosenboom D (1990) The performing brain. Comput Music J 14(1):48–66. https://doi.org/10.2307/3680116

Rosenzweig R (1998) Wizards, bureaucrats, warriors, and hackers: writing the history of the internet. Am Hist Rev 103(5):1530–1552. https://doi.org/10.2307/2649970

Sala SD (ed) Pseudoscience and the brain: tuner and tonics for aspiring superhumans. In: Mind myths: exploring popular assumptions about the mind and brain. Wiley, New York, pp 60–82

Schlimm D (2008) Learning from the existence of models: on psychic machines, tortoises, and computer simulations. Synthese 169(3):521–538. https://doi.org/10.1007/s11229-008-9432-5

Schmitz S (2012) The neurotechnological cerebral subject: persistence of implicit and explicit gender norms in a network of change. Neuroethics 5(3):261–274. https://doi.org/10.1007/s12152-011-9129-1

Schmitz S (2016) The communicative phenomenon of brain-bomputer interfaces. In: Pitts-Taylor Victoria (ed) Mattering: feminism, science, and materialism. NYU Press, New York, pp 140–157

Schwartz MS, Collura TF, Kamiya J, Schwartz NM (2017) The history and definitions of biofeedback and applied psychophysiology. In: Schwartz MS, Andrasik F (eds) Biofeedback, fourth edition: a practitioner's guide. Guilford Publications, New York & London, 3–23

Shapin S (2008) The scientific life: a moral history of a late modern vocation. University of Chicago Press, Chicago, Illinois

Sobell N, Stermitz E (2007) Interview with Nina Sobell, August 21, 2007. http://archive.rhizome.org/digest/?msg=00298

Span P (2017) F.T.C.'s lumosity penalty doesn't end brain training debate. The New York Times, December 21, 2017, sec. Health. https://www.nytimes.com/2016/01/19/health/ftcs-lumosity-penalty-doesnt-end-brain-training-debate.html

Stanford Centre for Longevity (2014) A consensus on the brain training industry from the scientific community (full)—stanford center on longevity. http://longevity.stanford.edu/a-consensus-on-the-brain-training-industry-from-the-scientific-community-2/

Stoyva J, Kamiya J (1968) Electrophysiological studies of dreaming as the prototype of a new strategy in the study of consciousness. Psychol Rev 75(3):192–205

TEDx Talks (2018) The mind is the new frontier with ubiquitous neurotechnology | Tim Mullen | TEDxSanDiegoSalon. https://www.youtube.com/watch?v=ra6v0EvWclc. Accessed 10 Nov 2018

Teitelbaum R (1976) In tune: some eearly experiments in biofeedback music (1966–74). In: Rosenboom D (ed) Biofeedback and the arts: results of early experiments. A.R.C. Publications, Vancouver, B.C, pp 35–65

Thibault RT, Lifshitz M, Birbaumer N, Raz A (2015) Neurofeedback, self-regulation, and brain imaging: clinical science and fad in the service of mental disorders. Psychother Psychosom 84(4):193–207. https://doi.org/10.1159/000371714

Thibault RT, Lifshitz M, Raz A (2016) The self-regulating brain and neurofeedback: experimental science and clinical promise. Cortex, What's your poison? Neurobehavioural consequences of exposure to industrial, agricultural and environmental chemicals 74:247–261. https://doi.org/10.1016/j.cortex.2015.10.024

Thibault RT, MacPherson A, Lifshitz M, Roth RR, Raz A (2018) Neurofeedback with FMRI: a critical systematic review. NeuroImage 172:786–807. https://doi.org/10.1016/j.neuroimage.2017.12.071

Turner F (2006) From counterculture to cyberculture: stewart brand, the whole earth network, and the rise of digital utopianism. University of Chicago Press, Chicago, London

Vasulka W, Lowenberg R (1970) Environetic synthesizer. Radic Softw 1(2)

Vidal JJ (1975) Neurocybernetics and man-machine communication. In: Proceedings of the 1975 international conference on cybernetics and society. San Francisco, California, September 23–25, 1975. Institute of Electrical and Electronic Engineers, New York, pp 421–422

Vidal JJ (1977) Real-time detection of brain events in EEG. Proc IEEE 65(5):633–641. https://doi.org/10.1109/PROC.1977.10542

Vidal JJ (1983) Silicon brains: whither neuromimetic computer architectures. In: Proceedings IEEE international conference on computer design: VLSI in computers (ICCD '83). Port Chester, NY, pp 17–20

Vidal JJ (1973) Toward direct brain-computer communication. Annu Rev Biophys Bioeng 2(1):157–180. https://doi.org/10.1146/annurev.bb.02.060173.001105

Vidal JJ, Miller LH, Devillez G (1984) Adaptive user interfaces for distributed information management. In: Human-computer interaction: proceedings of the first U.S.A.-Japan conference on human-computer interaction, Honolulu, Hawaii, August 18–20, 1984. Amsterdam: Elsevier, 99 77–81

Walter WG, Shipton HW (1951) A new toposcopic display system. Electroencephalogr Clin Neurophysiol 3(3):281–292. https://doi.org/10.1016/0013-4694(51)90074-0

Walter WG (1953) The living brain. Norton, New York

Wiener N (1961) Brain waves and self-organizing systems. In: Communication and control, 2nd ed. MIT Press, Cambridge, MA, pp 181–203

Chapter 3
Brain-Computer Interfaces in Contemporary Art: A State of the Art and Taxonomy

Mirjana Prpa and Philippe Pasquier

Abstract In this chapter, we present a state of the art on Brain-Computer Interface (BCI) use in contemporary art. We analyzed sixty-one artworks that employ BCI dating from 1965 to 2018, and present a taxonomy with five categories guiding the discussion of specific BCI artworks: input, mapping, output, format, and the presence of an audience. Moreover, we briefly present and discuss key points about BCI devices used in some of the artworks that are available on the market. Finally, we present insights from nineteen artists that we surveyed about their BCI art practices, experiences with BCI devices and peculiarities of working with brain activity as a resource for art creation. We then conclude with our summary of challenges and potentials for BCI art in the future.

Keywords State of the art · BCI · EEG · Taxonomy · Contemporary art

3.1 Introduction

Revealing the intricacies of the human brain and its functioning is a source of intrigue and a subject of study for various disciplines with the same goal: to understand how we behave and experience the world. One of these disciplines, that of art, has been providing a unique perspective on understanding the human brain. Through their practices, artists' contribution to this understanding requires rigorous involvement in the process of discovery: "*...the artist is in a sense, a neuroscientist, exploring the potentials and capacities of the brain, though with different tools... How such creations can arouse aesthetic experiences can only be fully understood in neural terms.*" (Shimamura and Palmer 2012).

M. Prpa (✉) · P. Pasquier
School of Interactive Arts + Technology, Simon Fraser University,
Surrey, Canada
e-mail: mprpa@sfu.ca

P. Pasquier
e-mail: pasquier@sfu.ca

© Springer Nature Switzerland AG 2019
A. Nijholt (ed.), *Brain Art*,
https://doi.org/10.1007/978-3-030-14323-7_3

Although the modes of artistic exploration of the brain can take upon various forms (such as metaphorical), in this chapter, we are concerned with the utilization of neurophysiological brain data through artistic processes and into a creative output. For this inquiry, we identified two historically significant events: the first took place in 1924 when German psychiatrist Hans Berger recorded the electrical activity of the human brain for the first time in history. The recording of *Berger's wave* or what is known today as *"Alpha rhythm"* marked the beginning of electroencephalography (EEG), a neuroimaging technique that has since been utilized in the context of art. The second event, following Berger's work, was when two leading physiologists from Cambridge's Physiological Lab, Edgar Douglas Adrian, and Bryan Matthews, mapped Alpha waves into audio signals in 1934 (Adrian et al. 1934; Rosenboom and Number 1990). While the first event made the utilization of brainwaves possible, the second event naively marked the beginning of creative explorations of brain activity that advanced outside of science labs into the world of contemporary art.

In the early days of artistic experimentation with brain sensing, due to the complexity of early EEG apparatus, collaborations between scientists and artists were common. Alvin Lucier was initially introduced to "brain-music" by his friend, physicist Edmond Dewan. With the assistance of Dewan and support from John Cage, Lucier performed *Music for Solo Performer* at the Rose Art Museum (Waltham, Massachusetts) in 1965 which constitutes the first recorded brainwave music performance. Moreover, Lucier's sonification of Alpha waves laid the foundation for what we refer to in this chapter as **brain-computer-interface art (BCI art)**.

The term brain-computer interface (BCI) was coined by Jacques Vidal, UCLA's[1] professor and pioneer in this field (Vidal 1977, 1973). BCI is a system that senses and utilizes brain activity in one-way communication from a brain to a computer. BCI definitions vary though, depending on how BCI is utilized. For example, (Wolpaw and Wolpaw 2012) define BCI as *"a system that measures central nervous system (CNS) activity and converts it into artificial output that replaces, restores, enhances, supplements, or improves natural CNS output, and thereby changes the ongoing interactions between the CNS and its external or internal environment"*.[2] However, Wolpaw and Wolpaw's definition describes one approach to utilizing BCI (active BCI, Sect. 3.2.2.2) that Zander et al. (2010) recognize as *Direct BCI*, in which mental activity is consciously controlled and directed in order to change the output of the system. The same authors also juxtapose *Direct BCI* with *Indirect BCI*, as the latter collects and utilizes passive, spontaneous brain activity that is not consciously controlled (this will be expanded on in Sect. 3.2.2.2).

While early BCI devices emerged within the context of medical research, recent interest in ubiquitous computing, wearable technologies, body interfaces, affective computing, and a movement towards the "quantified self" emphasize the potential impact that commercial BCI devices could have on the market. Since the first International Meeting on BCI in New York in 1999 (Wolpaw et al. 2000), the expansion

---

[1] University of California, Los Angeles.

[2] An example of this definition is a participant with impaired motor neurons who utilizes BCI input to control their wheelchair.

of BCI devices on the market resulted in a large number of open source as well as proprietary devices that are non-invasive, affordable and user-friendly. Following these technological advances in BCI technologies and their diverse uses beyond laboratories and into the wild of the consumer market, the corpus of BCI art has grown extensively. However, the lack of a systematic overview of ideas, concepts, implemented approaches, and typologies prevents us from a comprehensive understanding of the BCI art landscape.

To that end, with this chapter, we aim to contribute to understanding the complex landscape of BCI art. Our research process was as follows: first, we surveyed EEG-based BCI devices with a focus on EEG approaches and related control paradigms. Then we analyzed 61 BCI artworks (see Table 3.3 in Sect. 3.7) based upon which we created a taxonomy (see Fig. 3.1b in Sect. 3.2.1 and Table 3.4 in Sect. 3.7) that we present here. Following the logic of our research process, we begin this chapter by introducing the field of brain-computer interface and laying out the landscape of EEG-based BCI devices and types of brain data. Second, we group, combine, analyze, and categorize the work that has been done in BCI art so far. Within each taxonomy category and their subcategories, we provide background knowledge and concepts necessary for understanding the nuances of that category, illustrated with examples from BCI art. A comprehensive list and video/images of BCI artworks that we analyzed can be found in the online database that we created at https://bci-art.tumblr.com/. Finally, in addition to the taxonomy, we discuss challenges and potentials of the exploitation of brain activity in art, based on the insights gained through our practice, analyzed examples, and direct correspondence with nineteen authors. Our aim is to provide a clear framework as guidance for artists and researchers in all future creation and discussion of BCI artworks.

3.2 Categories for BCI Art Analysis

In this section, we present the complex landscape of BCI art (Fig. 3.1) by looking at the characteristics of EEG-based BCI devices used in an art context (Fig. 3.1a), and BCI artworks (Fig. 3.1b). First, in Sect. 3.2.1 we present the main characteristics of EEG-based BCI devices (Table 3.1). Then we introduce 61 artworks starting from the mid-1960s until 2018 (Table 3.3) through the categories of the **Taxonomy of BCI art** (Table 3.4). The proposed taxonomy consists of 5 main dimensions that guided our comparison and analysis of the artworks. In *Input dimension*—Sect. 3.2.2 —we discuss different types of brain data, detailing EEG classification approaches, control paradigms, timeliness of input, and finally we discuss modality of BCI artworks because some of the analyzed artworks combine EEG data with other types of input data (heart rate, electrodermal activity, etc.). Then in Sect. 3.2.3—*mapping*— we discuss the different ways that input is transposed to output in BCI artworks. This is followed by a discussion of the *Output dimension* in which we present a variety of outputs that BCI artworks have, including visual, sound, audio-visual, moving images, immersive, and control of a physical object (Sect. 3.2.4). Output is

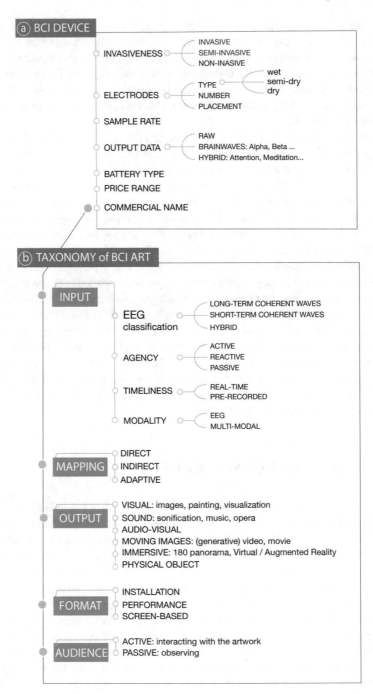

Fig. 3.1 Taxonomy of BCI devices (**a**) and BCI artworks (**b**). Image by the authors

closely related to *presentation format* that we discuss in Sect. 3.2.5, which is then followed by a discussion of the presence and the role of the audience in BCI artworks (Sect. 3.2.6).[3] The descriptions of the taxonomy categories are illustrated with some artwork examples and a brief description of their features. More details about the artworks can be found by following the number indicated in $\widehat{x}$ in the survey Tables 3.3 and 3.4 (Sect. 3.7).

3.2.1 EEG-Based BCI Devices

Brains are complex systems within which dynamic electrochemical processes take place. Neuroimaging (brain-imaging) techniques provide insights into *structural* and *functional* properties of the nervous system. While structural imaging allows for a better understanding of brain structures, functional imaging provides recordings of the activity across different brain areas. The practical application of brain-imaging in artworks discussed in this chapter is concerned with electroencephalography (EEG), an approach to the understanding of brain functioning through measuring electrical activity in the brain by multiple electrodes that are placed on the skull's surface. Available today are various other techniques for neuroimaging, such as: *magnetoencephalography*—records magnetic fields produced by electrical currents occurring in the brain (Panoulas et al. 2010), *functional Near-Infrared Spectroscopy*—measures hemodynamic (the flow of blood in the brain) responses associated with neuron behavior (Coyle et al. 2007), *Event-related optical signal*—measures changes in the optical properties of active areas of the cerebral cortex (Nam et al. 2018). These techniques provide higher spatiotemporal resolution of recordings compared to EEG. However, the majority of consumer-grade BCI devices utilize EEG only. Just recently there has been a push towards hybrid BCIs that combine EEG (high temporal, low spatial accuracy) and fNIR (low temporal, high spatial accuracy) (Naseer and Hong 2015; von Luhmann and Muller 2017), however, our query yielded one artwork that utilizes hybrid BCI approach. We are certain that we are very close to embracing hybrid BCIs, or even solely fNIR-based BCI (NIRSIT 2018) in the art field. The push towards the development and proliferation of ergonomic and aesthetically pleasing headsets spanning beyond EEG into more precise (higher spatial-temporal resolution), reliable, and wireless, headsets opens many possibilities for art applications in the future.

EEG-Based BCI Devices Used in Art Contexts are Non-invasive—From the early days of Berger recording brain activity by inserting electrodes into a patient's skull until today, the advancement of BCIs and underlying technologies is undeniable. The devices available today are capable of detecting electrical signals of the smallest magnitude from the electrodes placed on the surface of the skull. Compared to Berger's rudimentary and invasive approach, the degree of invasiveness of BCI devices on the participant has decreased significantly. Overall, regarding the inva-

---

[3]In this chapter we use word "participant" to differentiate between an audience at large and a person—a participant—whose EEG data is utilized in an artwork.

siveness, BCI devices are classified into three classes: invasive, partially invasive, and non-invasive (Nicolas-Alonso and Gomez-Gil 2012). Both *Invasive BCI* and *Semi-invasive BCI* require the surgical placement of microelectrodes inside gray matter to record brain activity and are used exclusively for medical applications. While we have no knowledge of these devices being widely employed in an artistic context at the time of writing, interest in a cyborg movement (Harbisson and Ribas 2010) has offered some perspectives on how these interfaces could be used in the future. Neil Harbisson, a color blind artist, became the first cyborg known for an invasive implant in his skull—an eyeborg antenna—that translates colors to sound to overcome color blindness (Eyeborg 2019). In a performance set in two locations in New York, Harbisson "perceives" the colors from a canvas painted by volunteers on Times Square via Skype connection and projections at his location. Without looking at the projection, his implanted antenna translates the projected colors into sound frequencies that he then paints on a canvas (Pearlman 2014, 2015). To our knowledge, the use of invasive brain implants is tied only to the cyborg art movement and the work of Haribsson, compared to non-invasive BCI devices that are widely employed in an art context and therefore will be the focus of this chapter.

The use of *non-invasive* BCI spans beyond medical into various everyday applications, from gaming, meditation, to utilization in art. One of the main differences among consumer EEG-based BCI devices is in the type of electrodes: wet, semi-dry, and dry. Wet electrodes require the application of a gel to secure the connectivity between the skull and the electrode. In the past, all non-invasive EEG-based BCIs used wet electrodes. However, due to the inconvenience of the gel residues, devices with wet electrodes are now used mainly in a medical context. *Semi-dry* electrodes partially overcome the residue problem by replacing the conductive gel with a saline solution. However, the saline solution on the electrodes' felt pads tends to dry quickly, so these electrodes need moistening more often than gel-based electrodes. Compared to gel-based electrodes that can hold high conductivity for up to eight hours, semi-dry compromise the endurance for comfort. *Dry* electrodes require reduced set-up time and no need for gel/paste application. However, this type of sensor requires firm pressure on the head. Devices with dry electrodes must penetrate through hair and achieve solid scalp contact which is often experienced by the participants as uncomfortable. Finally, focus and high expectations are on a new generation of dry electrodes (Lin et al. 2011) moisturized by human perspiration (for example, the hydrophilic polymer electrodes built in devices such as Emotiv's Insight (Emotiv n.d.)). This type of dry electrodes do not require firm pressure on the skull. However, their price is higher than the price of gel-based electrodes.

Range of EEG-Based BCI Devices on the Market—Table 3.1 presents the range of BCI devices available on the market at the time of writing. Since the end of the 1990s, the number of low-cost EEG BCI devices on the market has rapidly increased, resulting in head-mounted devices such as Emotiv EPOC and Emotiv Insight (n.d.), Muse (n.d), and NeuroSky Mindwave (n.d.-b). These devices vary in the type, number, and placement of electrodes, output signal, sample rate, as well as price (Table 3.1). It is expected that the number of head-mounted BCI devices will continue to increase, however the most recent direction for BCI is towards in-ear

Table 3.1 EEG-BCI devices – continued from previous page

Device	Electrode #: type	Electrode placement	Sample rate/resolution	Output data	Battery/run time	Price (USD)
ABM X10 (B-Alert X series mobile EEG 2018)	9+1*: dry	F3, Fz, F4, C3, Cz, C4, P3, POz, P4	256 Hz/16-bit	Delta, Theta, Alpha, Beta, Gamma, and High Gamma	Lithium Ion/11 h	—
ABM X24 (B-Alert X series mobile EEG 2018)	20+4*: dry	Fz, Fp1, Fp2, F3, F4, F7, F8, Cz, C3, C4, Pz, P3, P4, POz, T3, T4, T5, T6, O1, O2	256 Hz/16-bit	Delta, Theta, Alpha, Beta, Gamma, and High Gamma	Lithium Ion/6 h	—
Cognionics Quick-20 (Quick20 n.d.)	8/20 +2*: dry	Fz, Fp1, Fp2, F3, F4, F7, F8, Cz, C3, C4, Pz, P3, P4, P7, P8, T3, O1, O2	0–131/262/524 Hz/24-bit	Raw data, 3-axis accelerometer	Lithium-ion/up to 12 h	6,000–15,000
Cognionics Quick-30 (Quick30 n.d.)	30 +2*: dry	Af3, Af4, Fz, Fp1, Fp2, F3, F4, F7, Fc5, Fc6, F8, Cz, C3, C4, Cp5, Cp6, Pz, Po7, Po3, Po4, Po8, P3, P4, P7, P8, T3, O1, O2	0–131/262/524 Hz/24-bit	Raw data, 3-axis accelerometer	Lithium-ion/up to 12 h	22,000
Cognionics Mobile-128 (Quick30 n.d.)	64/128 +2*: wet	128 electrodes, 10–20 placement	0–131/262 Hz/24-bit	Raw data, 6-axis IMU (Acc+Gyro)	Lithium-ion/up to 8 h	38,000–50,000
Emotiv EPOC (Emotiv n.d.)	14+2*: semi-dry	10–20: AF3, F7, F3, FC5, T7, P7, O1, O2, P8, T8, FC6, F4, F8, AF42: P3*,P4*	128 or 256 Hz/ 14-bit per channel	Raw data, gyroscope, accelerometer, magnetometer (with EmotivPro licence); MyEmotiv app: focus, stress, excitement, relaxation, interest and engagement	Lithium Polymer battery 640mAh/12 h	800
Emotiv Insight (Emotiv n.d.)	5 + 2*: dry	10–20:AF3, AF4, T7, T8, Pz: P3*, P4*	128 Hz/: 14-bit per channel	Raw data, gyroscope, accelerometer, magnetometer (EmotivPro licence); MyEmotiv app: focus, stress, excitement, relaxation, interest and engagement	Lithium Polymer battery 480mAh/4 h	300

(continued)

Table 3.1 (continued)

Device	Electrode # : type	Electrode placement	Sample rate/resolution	Output data	Battery/run time	Price (USD)
IMEC (wireless 2018)	8: dry	—	512 Hz/—	Raw data (0.5–100 Hz)	—	—
Interaxon Muse (Muse n.d.)	4+1*: dry	10–20: TP9, AF7, AF8, TP10: FPz*	256 Hz/12-bit per channel	Raw data, calm score, accelerometer, gyroscope	Lithium Ion/5 h	250
MyndBand EEG Brainwave Headset (MindPlay n.d.)	1+2*: dry	10–20: FP1	512 Hz/—	Raw data (3–100 Hz), Attention, Meditation, Eye blink	Lithium Ion/10 h	300–600
Neuro- electrics Enobio ENOBIO (Neuroelectrics n.d.)	8/20/32: wet/dry	10–10: a cap with 39 possible positions	500 Hz /24-bit per channel	Raw data	—/up to 16 h	4,000–17,000
Neurosky MindWave Mobile (NeuroSky Mind-Wave n.d.-a)	4: dry	10–20: TP9, FP1, FP2, TP10	512 Hz/12-bit per channel	Attention, Meditation, Eyeblinks, Brainwave Bands, raw Output	1 AAA/8 h	100
Neurosky MindSet (NeuroSky n.d.)	1: dry	FP1	128Hz or 512Hz/—	Raw data (0–50Hz), Attention, Relaxation, Blink detection	Lithium Ion/—	Obsolete
OpenBCI (OpenBCI n.d.)	8/16: dry	10–20: (for 8) Fp1, Fp2, C3, C4, P7, P8, O1, O2: (16): add F7, F3, F8, F4, T7, T8, P3, P4	—	Raw data	Lithium Ion 500 mAh/—	Open source

EEG BCI devices (Looney et al. 2012; Mikkelsen et al. 2015; Ear EEG demo 2018; Ear EEG project 2018).

3.2.2 Input of BCI Artworks

In this section, we cover four subcategories of the input dimension: EEG classification approaches, agency paradigms, timeliness of the input, and modality of input, illustrating them with BCI art examples.

3.2.2.1 Input: EEG Classification Approaches

EEG is a functional neuroimaging technique for recording the electric component of the brain's electrochemical processes. EEG captures "neural oscillations" —a synced electrical activity of the clusters of neurons across the brain that are constantly firing electrical discharges. While a large number of the neurons fire simultaneously across the brain, the activation of the neuron clusters in particular regions of the brain indicates specific actions or processes. For example, brainwaves associated with cognitive processing are most prominent in the occipital region of the brain (back and lower part of the skull). To capture brain activity across various brain regions, one of the most widely accepted approaches to electrode placement on the skull is the *10–20 International System of Electrode Placement* (Silva and Niedermeyer 2012) developed by Dr. Herbert Jasper in the 1950s (Fig.3.2) (Szafir 2010). For higher density electrode setting, the 10–10 system has been used for placement of up to 81 electrodes, and beyond that the 10–5 system is used for placing up to 320 electrodes (Jurcak et al. 2007).

Recorded brainwaves are classified by their frequencies, amplitudes, location, and shape (Kumar and Bhuvaneswari 2012). Regarding frequency, spontaneous neural activity shows fast cortical potentials (FCP) that range from 0.5 Hz to 100 Hz (Moss 2003). Raw, unprocessed data of electrical activity of the brain exposes background noise which is mixed with brainwaves. Therefore, to understand the relationship between brainwaves and the presented stimulus or cognitive processes better, two distinct approaches are discussed in the literature. The first approach is the recording and analysis of *Long-Term Coherent Waves* (LTCW), and second, *Short-Term Transient Waves* (STCW) (Rosenboom and Number 1990). The third approach, **Hybrid**, emerged due to the progress in machine learning and artificial intelligence, and builds upon LTCW and STCW, using all possible data combinations to train artificial models for high-level prediction.

Long-Term Coherent Waves (LTCW)—In this approach, also known as **neuro-feedback**, captured EEG activity is classified based on brainwave frequencies in the range from 1 to 30 Hz. According to some authors, different brainwaves are more prominent in some parts of the brain than in others, and the probability of capturing a particular brainwave can be increased by positioning electrodes in the regions of the brain associated with it. For capturing slow brainwaves (0.5–2 Hz), the electrodes

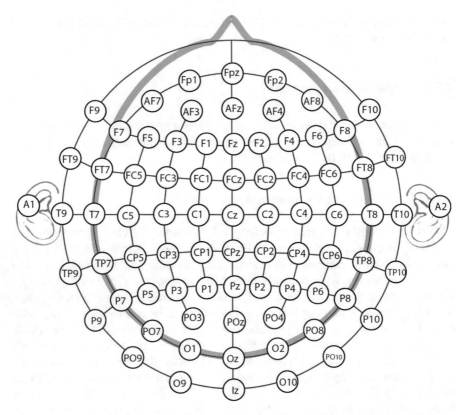

Fig. 3.2 Electrode Placement according to the International 10–20 System. Letters correspond to the lobes –F(rontal), T(emporal), P(arietal), and O(ccipital). C stands for Central position. Image by the authors

should be placed in F4-A1 positions (see Fig. 3.2), for brainwaves in the frequencies between 11 and 16 Hz in C4-A1 position, and finally, for Alpha wave (8–13 Hz) in 02-A1 positions (Morley et al. 2013). However, further classification of brainwave

Table 3.2 EEG waves, their frequencies and features

Name	Frequency range (Hz)	Associated features
Delta	0.5–4	Fatigue, sleep, severe slowing of mental processes, possible to occur in meditation by very experienced practitioners able to maintain consciousness in delta state
Theta	4–7	Deep meditation, reduced consciousness, hypnosis, attention lapses, slowed processing, stage 1 of sleep, memory consolidation
Alpha	8–14	Relaxed wakefulness, readiness, inactive cognitive processing, most prominent during meditation
Slow Beta	15–20	Intense focus, cognitive enhancement
Medium Beta	20–30	Anxiety, distractibility
Fast Beta (Gamma)	30–70	Hyper-alertness, processing of various attended stimuli (tactile, visual, auditory), stress

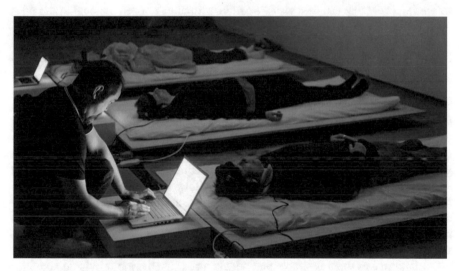

Fig. 3.3 George P. Khut and James P. Brown, *Alpha Lab*, 2013. *Alpha Lab. 2013. George P. Khut, James P. Brown*. With the permission of George P. Khut

frequencies into brainwave bands is a subject of disagreement. While in some literature brainwaves can be found divided into five bands: Alpha, Beta, Gamma, Delta, and Theta, the majority follows the guidelines provided by the *International Federation of Electrophysiology and Clinical Neurophysiology* (Steriade et al. 1990). Based on this classification, brainwaves are classified into six bands with associated features, as presented in Table 3.2.

In our survey, the majority of the artworks (43/61) utilized LTCW. However, the documentation of only fifteen artworks specified which brainwaves were utilized. For example, Khut's **Alpha Lab** (37) is built upon the activity of the Alpha brainwave. This installation (Fig. 3.3) invites its audience to explore their consciousness through an immersive soundscape generated in real-time by Alpha brainwave activity. The installation takes place in a dark chamber in which three participants lay comfortably. Each participant wears headphones and a BCI device that reads the levels of Alpha waves and translates them into a soundscape. While there is no "desired result," the experience takes the form of lucid dreaming supported by a soundscape that reacts to fluctuations in Alpha waves, which naturally occur during meditation or just before falling asleep.

Short-Term Transient Waves (STTW)—SSTWs are the brain's response to sensory, cognitive or motor stimuli, and are also known as *slow cortical potentials* (SCP). SCP last between 300 ms to several seconds (Psychophysiological 2000) and are observable as shifts in cortical electrical activity after the stimulus. These event-related potentials (ERP) are time-locked EEG activity which means that they occur (only temporarily) after a specific time following the sensory stimuli or cognitive processes. For example, P300 stands for an ERP that occurs around 300 ms after the triggering event that can be a visual or audio stimulus, or even a thought (Panoulas et al. 2010).

Besides ERP, the other approach to input EEG classification builds upon Steady-State Evoked Potentials (SSEP) that are elicited by the repetitive external stimulus. SSEP can be visual (Steady-State Visual Evoked Potentials-SSVEP), auditory (Steady-State Auditory Evoked Potentials-SSAEP) or tactile (Steady-State Somatosensory Evoked Potentials-SSSEPs). The premise behind these methods is that external stimuli at specific frequencies can stimulate brain activity. For example, Steady-State Visual Evoked Potentials are visually induced brain responses at frequencies ranging from 3.5 to 75 Hz. When the retina is visually stimulated, the brain generates an electrical response at the same frequency as the frequency of visual stimuli. A wide range of SSVEP frequencies allows for a wide range of utilization of this paradigm in creative endeavors. Exposing the audience to visual stimulation of a particular frequency at the same time opens a design possibility to utilize as many input points as there are audience members, whose now altered brainwaves are synchronized. Moreover, SSVEP's relative immunity to the artifacts (e.g., muscle potentials) makes them desirable and widely used. Lastly, motor-related activities can be captured in the brain as a *sensorimotor rhythm* (SMR) or μ-rhythm. SMR is a recording of brain activity in ranges between 12 and 15 Hz over sensory-motor areas on the skull during a motor task (movement) or even motor imagery (imagined movement) (Thompson and Thompson 2003).

In our analysis, we came across seven artworks that employ various STTWs. One of the artworks, **The Gender Generator** (58) by Josh Urban Davis, utilizes a modified P300 paradigm (n250 ERP) in an exploration of gender expression and dysphoria. First, flashing characters on a screen (Fig. 3.4) are presented to the participant who is then prompted with the question "Which Is You?" Second, after the question the participant makes a mental selection of the character and they count the appearance of the same character as it repeats in random order several times on the screen. This

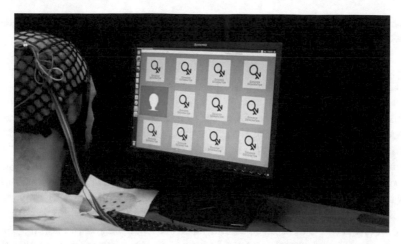

Fig. 3.4 Josh Urban Davis, *The Gender Generator*, 2017. *The Gender Generator. 2017.* Josh Urban Davis. A video still from a technical evaluation of the Gender Generator. With the permission of Josh Urban Davis

procedure is then repeated for various questions about one's physical appearance: body type, hair, etc. until the "complete" representation of the gender of a person is displayed on the screen.

Finally, the third and most recent approach allowed by the progress in machine learning is what we termed here: **hybrid**. In the hybrid approach, the device's proprietary software employs machine learning models to captured EEG data in order to detect complex categories of affective or cognitive functioning. One example of such hybrid classification is Emotive's *MyEmotive suite* (previously known as EPOC Affectiv Suite) that allows participants to measure six cognitive metrics: interest, excitement, relaxation, engagement, stress, and focus (Emotiv n.d.). Other headsets provide different categories, such as Interaxon's Muse that outputs levels of meditation only or NeuroSky that provides scores for: attention, meditation, blink detection, mental effort (engagement), familiarity, appreciation (enjoyment), cognitive preparedness, creativity, alertness, and emotional spectrum (intensity, and pleasantness) (NeuroSky Algorithms n.d.). While we speculate that these algorithms employ machine learning models on complex EEG data, none of these commercial software provide insights into how these levels are measured or extracted from the raw data, and these procedures are therefore subject to speculation and ambiguity.

In our survey, we identified nine artworks that employed the hybrid approach in EEG classification. All of these artworks utilize device proprietary software to extract participants' states and employ the information in various outputs. For example, Ramchurn's brain-controlled movie—**The Moment** (61) utilizes the participant's attention levels to alter the narrative of the movie. When the participant's attention levels drop, the movie changes from the initial narrative to show scenes from a secondary narrative. The movie then unfolds dynamically, driven by the changes in attention levels, leaving room for "101 trillion ways to view the content" (Ramchurn 2018).

3.2.2.2 Input: BCI Agency Paradigms

In this section, we describe three BCI agency paradigms and present a few examples for each. For this input category, we prefer to use the word "agency" over "control" for a few reasons. First, in the context of BCI artworks, the participant-artwork interaction is not always built upon the control of creative output. Often, a participant's brain activity is utilized in the creative output without the participant's awareness of their explicit "control" over the artwork. Second, the agency in our taxonomy is concerned with the degree of impact that the participant's brain activity has within the artwork, revealing "the capacity, condition, or state of acting or of exerting power" (Agency n.d.). In this sense, an artist has creative control over the final output, by choosing interaction paradigms that restrict or support the degree of the impact that the participant's brain activity can have on the creative output. A somewhat different perspective on creative control is presented by Wadeson et al. who discuss four categories of a participant's *creative control*: passive, selective, direct, and collaborative (Wadeson et al. 2015). However, we find that the first three categories of Wadeson et al.'s classification relate to input agency and mapping, while the fourth, collaborative, refers to the number of participants in the artwork.

While we find the use of the word "agency" more suitable in our classification, we borrowed two categories from existing BCI input paradigm literature: *active*, and *passive*. The third category: *reactive* was borrowed from a classification by (Zander et al. 2010). While active and reactive inputs require a participant to train the system, passive does not require any training (Zander et al. 2010).

Active Input Agency—The initial development of BCI as assistive technology allowed people with sensory-motor or cognitive impairments to perform actions that were otherwise inaccessible (Millán et al. 2010). For example, BCI-controlled wheelchairs allow people with motor difficulties to move in physical space. Similarly, BCI-controlled cursors and pointers on screen enable impaired participants to use computers, communicate, and participate in activities that were otherwise inaccessible (Lebedev and Nicolelis 2006). In both examples, the participant has active control over the performed task; however, the drawback of this approach is the long and demanding training process that the participant must undertake before performing a task. Another drawback is mental fatigue that occurs after a certain period of BCI usage.

In our analysis, we identified twelve artworks that utilize active input agency. The participant's active control is used to control the behavior or physical characteristics of the artwork. Duenyas' installation **Ascent** (25) is an example of active input agency. This brain-controlled levitation performance defies gravity, as the participant, suspended in a flying harness, starts to meditate (see Fig. 3.5). The higher

Fig. 3.5 Yehuda Duenyas, *Ascent*, 2011. *Ascent. 2011.* Yehuda Duenyas. Image credit: Andrew Federman.With the permission of Yehuda Duenyas

the meditation levels are, the higher the participant ascends. Also, meditation levels control the sound and light of the installation.

Reactive Input Agency—employs brain activity that is altered by an external stimulus. The participant is simply attending to the stimulus. The participant's short-term transient waves then reveal the presence of the stimulus with an onset time (using, for example, P300 paradigm) and those fluctuations in brainwaves are then employed as a reactive input control (Zander and Kothe 2011).

Reactive input agency is often found in artworks that employ some form of short-term transient waves (see Sect. 3.2.2.1), such as **The Multimodal Brain Orchestra** (19). The orchestra members equipped with BCIs, attend to a range of flashing stimuli (in this case, to a visual representation of music excerpts). They change the piece by making a mental selection of one of the flashing stimuli and then count the number of its occurrences (similar to the interaction in *The Gender Generator* (58)). Another example, Batoh's live performance **Brain Pulse Music** (31), illustrates how reactive agency input can be utilized in a stage performance to create a relationship between the participant and the performer. In this collaborative piece between the artist and the

Fig. 3.6 Masaki Batoh, *Brain Pulse Music*, 2012. A video still from performance. Image credit: Masaki Batoh. With the permission of Masaki Batoh

audience member (see Fig. 3.6), the participant wears goggles, and a custom-made headset with EEG electrodes and flickering LED lights which affect the participant's brain activity (VEP). The participant's brain data is then sent to Batoh who maps it to sound in his on-stage performance. While this is an appealing approach to influencing brainwaves via external stimulus, it remains unclear what impact these LED lights have on the brain activity if at all, and whether the participant is instructed on how to attend to the LED lights.

Passive Input Agency—Opposite to active, passive input does not require the participants to perform any particular task to change or influence their brain activity explicitly. Referring to the shortcoming of active BCI, some authors advocate for the development and use of passive BCI. Passive BCI has been advocated as an adequate technology for open monitoring of ongoing processes in the brain that are not always easy to otherwise capture and translate. To that end, (George and Lècuyer 2010) presented a few applications of passive BCI: adaptive automation (when the participant's engagement levels decrease, the system takes control over driving), multimedia classification, video games (control of aesthetics and game mechanics based on the participant's engagement), and error detection.

Artworks that employ passive control rely on the changes and fluctuations in either brainwaves (Long Term Coherent Waves) or the participant's states (Hybrid classification) that are utilized in the artwork. **The Magic of Mutual Gaze** (29) is an installation/performance piece for two participants who are seated across from each other. While the participants are directing their gaze towards each other (Fig. 3.7), their brainwaves are captured and analyzed for synchronicity. The synchronous functioning of two brains generate visuals that show the connection. In this case, the participants are not instructed what to do, and the experience emerges from the moment-to-moment synchronicity of their brains oscillating at the same frequencies.

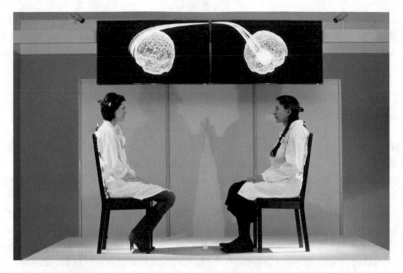

Fig. 3.7 Suzzane Dikker, Marina Abramovic, Matthias Oostrik, Jason Zevin, *The Magic of Mutual Gaze*, 2011–2014. *Measuring the Magic of Mutual Gaze. 2011*. Marina Abramovic, Suzanne Dikker, Matthias Oostrik and participants of the Annual Water mill Art and Science: Insights into Consciousness Workshop. Photo by Maxim Lubimov, Garage Center for Contemporary Culture. With the permission of Abramovic LLC

Finally, in our survey we faced a situation in which one artwork can be defined as active or passive, depending on the information disclosed to the participant. For example, in our taxonomy we list a number of artworks with passive input agency, such as: **Mind Pool** (21), **Solaris** (47), and **UFO wave** (17). What is common for the artworks with passive agency is that the nuances and details behind the interaction are usually undisclosed to the participant. However, once the participant becomes aware, for example, that their meditation levels have a particular impact on the artwork, they might purposefully try to alter their brain activity by focusing on the practice of meditation. In that case, passive BCI is utilized as an active BCI as long as the participant is engaged in performing actions that alter brain activity and therefore, the final output. Similarly, we classify **PrayStation** (30) and **Eunoia** (38) as active, but if their participants do not perform the required task of meditating/praying, the piece becomes passive.

3.2.2.3 Input: Timeliness of Input Data

Timeliness of input data refers to the time when data is captured. Our analysis encompasses artworks that employ real-time EEG data capture and mapping into the artwork. However, we came across two pieces that utilize pre-recorded data, and we include them in the Taxonomy. Casey's *Dream Zone* (33) is a generative video showing patterns and mandalas that respond to changes in pre-recorded data. The artist records the participants' brain activity while meditating "on the morphing hexagon kaleidoscope", which is then used to generate the video, with the hope that such imagery will stimulate viewers' Theta wave activity, associated with the profound states of consciousness otherwise normally reached only through meditation.

3.2.2.4   Input: Modality

The majority of the artworks reviewed in this chapter are mono-modal (54/61) in that they employ EEG-data only. Multimodality stands for an approach in which EEG data is combined with other physiological data such as EKG (electrocardiography), EMG (electromyography), or GSR (galvanic skin response). One of the multimodal projects analyzed here is **Naos** (see (18) in Table 3.4), an installation and platform for "sensing" the participant. Built upon the *Biometric Tendency Recognition and Classification System* (Castellanos et al. 2008), this system presents the participant with visual stimuli carrying affective content. Based on the physiological response of the participant (EEG, EMG, GSR) the system determines in real-time what next image should be displayed. This process creates an affective loop between the participant and the system. The ultimate goal of the system, according to the authors, is to reach "equilibrium" in which the image's expected physiological response, and the participant's actual response and classification are the same.

For further analysis of multi-modal artworks, it is critical that we delve into a comprehensive understanding of how data of other input modalities are used in these artworks. However, we came across an obstacle: a lack of documentation regarding how different data contributed to the overall experience of the artworks, besides EEG data. This is one of the few limitations that is mentioned in the Discussion (Sect. 3.4).

3.2.3   Mapping Strategies

The analysis of how EEG data is mapped to the parameters and interaction nuances of the artworks revealed a severe challenge similar to the one above, that is a lack of documentation about mapping details. Most of the artwork documentation we came across did not disclose mapping details, making the analysis of it difficult without speculation. However, we identify three possible mapping situations: direct, indirect, and adaptive.

Direct Mapping is the simplest of the three, in which the input EEG data is always mapped to the same parameters of the artwork and the output is somewhat predictable. For example, **The SubConch** (20) is an installation consisting of a lit conch sculpture in which direct mapping is realized by calibrating the lighting levels to brain activity. The participant's brainwaves are mapped to the sounds and control the brightness of the light through a passive agency (Fig. 3.8). Therefore, the participant passively creates the audio-visual installation by utilizing the direct mapping between brain activity on one side, and the sounds and lighting levels on the other, resulting in a somewhat predictable outcome.

Indirect Mapping is found in artworks that map EEG data to one set of parameters and then influence the values of another set of parameters. An example of indirect mapping is Ulrike Gabriel's **Terrain 01** (04), a piece that reveals the artist's intention to show a failure in our attempts to keep the role of mere observers. In **Terrain 01**, Gabriel puts the participant in the position of a robot's "brain" that controls their behavior. A few tiny robots that resemble roaches, with photovoltaic cells and proximity electrodes attached to their backs, are placed on an oval plate. The participant's Alpha waves indirectly control the robots by regulating the lightning in the installation; the more relaxed the participant is, their Alpha waves would be more prominent, which finally results in the lights shining brighter, giving the robots more energy for moving.

Finally, **adaptive mapping** arose from artificial intelligence and models capable of listening and changing how and to what EEG data is mapped, following the programmed logic. This type of mapping could contribute to the ever-changing nature of the piece (anywhere between random and predictable), or could adjust to the participant-specific EEG activity. In the latter, the artwork with adaptive mapping could "listen" to the participant and gradually lead the interaction, keeping participant engagement levels at the optimum for flow experience (Nakamura and Csikszentmihalyi 2014). One of the artworks with adaptive mapping is **Naos** (18), previously described in Sect. 3.2.2.4.

Fig. 3.8 Mats J. Sivertsen, *The SubConch*, 2009. With the permission of Mats J. Sivertsen

3.2.4 Diversity of the Output Types of BCI Art

We classify all analyzed artworks into six categories regarding their type of output, as per Fig. 3.1b. First we discuss artworks with *visual output* such as BCI images, painting, and visualization in Sect. 3.2.4.1. This is followed by a discussion of *sound output* of BCI artworks spanning from sonification, orchestral compositions, and opera (Sect 3.2.4.2). Then we discuss artworks with *audio-visual output* in Sect. 3.2.4.3. Following the discussion on visuals, we continue by presenting more recent work in *moving images*, discussing BCI-based generative video artwork and a BCI movie in Sect. 3.2.4.4. Then we expand the discussion to encompass *immersive*, computer-generated environments (Virtual and Augmented Reality) and head-mounted 180 panorama in Sect. 3.2.4.5. Finally, we conclude the output section by discussing built BCI-based physical objects, installations, and instruments in Sect. 3.2.4.6.

3.2.4.1 Visual-Based Output of BCI Artworks

The examples that follow are classified into the visual category for two reasons. First, the media used in these artworks convey visual information. Second, the artworks are not context dependent, they do not occupy the space beyond a canvas or a screen, and do not create a sense of spatial immersion (such as in the case of immersive virtual environments presented on head-mounted displays). Thematically, it appears the majority of the artworks in this category are centered around searching for an answer to how we visually represent something that is invisible to our eyes; What are our thoughts like, and do they have a shape or a color?

In attempts to demystify the brain and find answers to these questions, many artists capture brain activity and translate brainwaves into paintings and digital prints. **The Shapes of Thought** (12) is a visual representation of EEG recorded during the participants' evocation of traumatic events. While participants alter between hypnotic and sleeping state, the system captures participants' brain activity and generates complex 3D meshes in real-time. These 3D forms are then printed as images and presented as a collection of traumatic experiences. Similarly, **Brain Art: Abstract Visualization of Sleeping Brain** (28) utilizes pre-recorded instead of real-time data of the brain during sleep. An interesting departure from printed images are systems that allow an audience to create EEG-driven digital paintings like **Cerebral Interaction and Painting** (36), or the commercial application **Braintone art** (Braintone 2019).

While the artworks above visualize brain activity of one participant at a time, one of the pioneers of BCI art, Nina Sobell, explores the synchronicity and non-verbal communication between two participants. In her **BrainWave Drawings** (02), a real-time video portrait of two participants is augmented by the drawing of a Lissajous curve on the screen when their brain activity is synced (Fig. 3.9). As Sobell shares "*a circular configuration or Lissajous figure forms on an oscilloscope, when both are emitting the same brainwave frequency simultaneously. The pattern distorts horizontally or vertically, indicating a person is plugged into the X-axis and which*

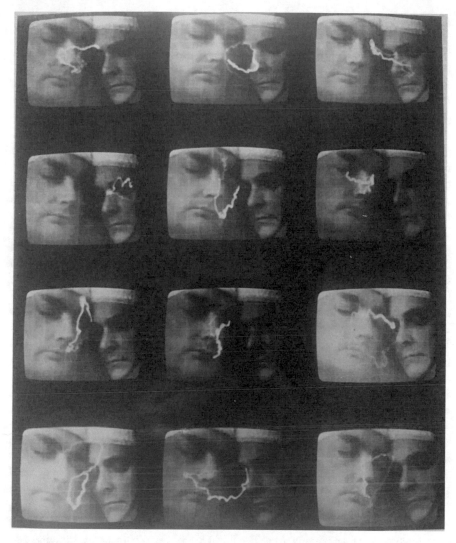

Fig. 3.9 Nina Sobell, *BrainWave Drawing*, 1973–2008. *BrainWave Drawing. 1973-2008*. Nina Sobell. With the permission of Nina Sobell

person is on the Y-axis. The people have been informed which axis, X or Y, they have been plugged into. So, when the pattern distorts horizontally (x-axis) or vertically (Y-axis) they can see immediately who is in the process of diverging."

3.2.4.2 Sonic Output: Brain Sonification, Music and Opera

Cerebral Music, a sonification of the brainwaves performed during a radio interview in 1961 by Grey Walter, has been speculated to be the first brain music (Haill n.d.). However, the lack of recordings of that event overshadows that claim. The first recorded performance of brain music is Alvin Lucier's **Music for Solo Performer** (01) from 1965. In this piece, Lucier, who was introduced to EEG by his friend, the scientist Edmond Dewan, used fairly simple equipment that consisted of one to three EEG electrodes placed on his forehead while performing. For the premiere on May 5th, 1965, he sent amplified Alpha waves to "16 loud speaker-percussion pairs deployed around the museum" (Straebel and Thoben 2014). These amplified Alpha waves required continuous distribution and redirection to the instruments in the room, and for this reason, Lucier was not the only one to perform that night. Lucier's assistant was John Cage, who took part in creating the piece as "an invisible performer, who raised and lowered the stereo amplifiers' volume controls, channeling the Alpha signal to various instruments around the room." (Straebel and Thoben 2014)

The interest in EEG sonification performances in contemporary music has not swayed since 1965. Some contemporary artists perform solo while others engage the audience on stage (like the previously mentioned Batoh's *Brain Pulse Music*). In solo performances, self-reflection through the sonification of brainwaves seems to be a reoccurring theme. In **Sitting.Breathing.Beating.NOT Thinking** (16) Adam Overton maps not only his brainwaves but changes in heartbeat and breathing rate to influence sound while performing *a meditative brain concert*. In this piece, Overton explores different mappings of the input, creating a unique performance each day for 7 days. As described by the author, the projected sound is generated by the software that plays data files as sound files, resulting in a purely digital, noise-like sound achieved in a process known as "data-bending."

Next, beyond sonification is brain-controlled music pioneered by David Rosenboom. As defined by Rosenboom, his piece **On Being Invisible** (03) is an "attention–dependent sonic environment". The sonic environment is generated by a brain-controlled set of electronic sound modules obtained from several inputs: small instruments, voice, and brainwaves. The brain signals are analyzed by applying pattern recognition to the brainwave frequencies. When a match between a new and one of the previous frequencies is found, the rhythm and the sound are affected by the same set of the rules previously applied to the matching pattern. In 1994, inspired by progress in physics, brain-science, and cosmology, Rosenboom returned to some of the ideas of **On Being Invisible** to realize them in a new piece, the self-organized opera **On Being Invisible II (Hypatia Speaks to Jefferson in a Dream)** (06). Even though some of the early technical solutions in Rosenboom's pieces were limited to the technology available at the time, his work has had a strong influence on contemporary practice. Beside Rosenboom, Richard Teitelbaum is yet another pioneer in the sonification of brainwaves, as seen in his work with an improvisational group –*Musica Elettronica Viva* (Holmes 2016).

The artworks presented so far relied on LTCW (neurofeedback) paradigm and direct EEG mapping of brain activity to sounds. A different approach is found in **The Multimodal Brain Orchestra** ⑲ that is performed by a quartet, a multimodal interactive system, and a conductor. In this concert, music is generated from a previously recorded tape. Quartet members voluntarily create a performance through two different stimulation approaches used to trigger sound events:

- P300 speller paradigm: in a matrix of 6×6 symbols, a symbol, a column or a row of symbols flashes. To trigger discrete sounds in real time, an orchestra member focuses on the flashing symbol and counts the number of times it flashed,
- Steady-State Visual Evoked potential: Four different light sources flicker at different frequencies and provoke the retina that causes the brain to generate activity at the same frequency triggered by the flickering light (see Sect. 3.2.2.1).

Both of these BCI approaches require a training period for the participants/ performers. The conductor directs the piece by giving cues to the performers, after which the performers focus on a specific row or column to ignite the desired brain activity, and consequently play the desired scores. Unlike the performance mentioned above, Eduardo Miranda's **Activating Memory** ㊹ does not have one central figure/conductor to direct the performance. Instead, the orchestra consists of a string quartet and the Brain-Computer Music Interface (BCMI) quartet. Each of the four performers in this BCMI quartet wears a cap with attached EEG electrodes and are seated in front of a screen. Four possible scores are displayed on the screen to each BCMI member out of which they choose only one at a time by gazing at it. The whole process relies on the approach of visually evoked potentials (VEP) and measured brain activity in the visual cortex, similar to the approach used in **Multimodal Brain Orchestra**. After the selection is made, one of the four string performers receives the score and performs it. In this case, all of the performers with EEG-caps are the creators of the collaborative piece in real time. For further reading about BCI and music, Eduardo Miranda and Julien Castet's book "Guide to Brain-Computer Music Interfacing" (Miranda and Castet 2014),,and Rosenboom's "Extended Musical Interface Human Nervous System: Assessment and Prospectus" (Rosenboom and Number 1990) are significant resources.

Finally, our analysis includes one opera. *Noor* ㊻ is an opera performance concerned with the theme of surveillance. The performer's affective states are obtained from their brain activity. Then those affective states such as excitement, interest, meditation, and frustration, are mapped to one of the four databases containing prerecorded sound, libretto, and videos. Through real-time feedback between changes in the performer's affective states and corresponding audio-visuals, the performer controls the libretto, music, and videos and creates the multi-media opera.

Fig. 3.10 Dmitry Morozov ::Vtol::, *eeg_deer*, 2014. *eeg-deer. 2014*. Dmitry Morozov.With the permission of Dmitry Morozov

3.2.4.3   Audio-Visual Output of BCI Artworks

In this section, we discuss two formats of audio-visual BCI artworks: BCI audio-visual installations, and BCI audio-visual performances.

Audio-Visual Installations—In "**Behind Your Eyes, Between Your Ears**" (54), the participants, one at the time, explore the states between "thinking and being" while their Alpha wave activity is mapped to interactive soundscape and visuals. Visuals are then projected on each participant's face, creating a dreamy portrait for the audience to enjoy (Khut 2015). Another example of audio-visual installation is **State.Scape** (51), a virtual environment exposing a flock of birds whose behavior depends on the participant's excitement, engagement, and meditation levels as obtained from Emotiv's *Affectiv Suite*. Changes in the EEG data controls the flock's position, birds' speed, and their number. Apart from controlling the flock properties, EEG states are mapped to control the volumes of different audio tracks, creating a dynamic atmosphere that changes in real-time. With this piece, the authors aim to create an intimate experience in an enclosed space that allows for self-reflection and ultimately, meditation. Immersive virtual environments presented on head-mounted displays are discussed in Sect. 3.2.4.5.

Audio-Visual Performances—In audio-visual performances, the agency and presence of a performer can vary significantly from one piece to other. For instance, performance can be merely brainwave-generated music and visuals projected on the screen in which the performer's presence is minimal, such as in Dmitry Morozov's **eeg_deer** (46) (Fig. 3.10).

On the contrary, in Novello's performance titled **Fragmentation,** (49) the performer's presence on the stage has a crucial part in creating the experience. The

Fig. 3.11 Alberto Novello, *Fragmentation*, 2014. *Fragmentation. 2014*. Alberto Novello. Image credits: Erin McKinney. With the permission of Alberto Novello

experience starts with the performer—a Butoh dancer—sleeping on the stage. Once awake, the performer practices a concentration task and finally jogs while their brainwaves generate the soundscape and the visuals projected on the wall behind the artist (Fig. 3.11). The performer's EEG controls an avatar in a virtual 3D maze project on the wall, that then controls the sounds and visuals as the avatar moves. In this complex piece, Novello challenges himself as a performer from "outside" and "inside" of his body, to create a performance through particular mental and physical tasks. Another piece with a strong presence of a performer is **The Escalation of Mind** (32). An artist seated on the stage is reading Herman Hesse's "The Glass Bead Game" while his facial expressions and brain activity control audio-visual sequences and their duration.

3.2.4.4 Moving Images as Output

In our analysis, we included three kinds of moving images: live video footage, brain-controlled movie, and screen-based virtual environments. The **Chromatographic Orchestra** by Ursula Damm (40) is an interactive BCI-controlled live video footage. In Damm's work, the participant's neural activity manipulates the software that, as a result of the interaction, defines the degree of abstraction of the displayed video from nearby cameras. Two other examples in this category are *The Moment* (61), a brain-controlled movie, and *Dream Zone* (33), a generative video piece; both are described previously.

3.2.4.5   Immersive Output of BCI Artworks

Immersive Virtual Environments—Our research of virtual reality artworks pre-sented exclusively on head-mounted displays (HMD) resulted in four artworks. **The Hidden Rooms** (24) is a panorama (180° image) from 2011 presented on head-mounted display. This piece, according to the author, represents the metaphor for the unconscious side of the brain. In this piece, the participant equipped with the *QBIC Belt Integrated Computer* (Amft et al. 2004) wanders through the panoramic envi-ronment defined by the author as "a brain-controlled panoramic experience using photography and spatiality".

The **Einstein's Brain Project** by Alan Dunning and Paul Woodrow resulted in a rich corpus of artworks presented on HMD that very often expand beyond HMD into a physical space. The overarching theme of the project examined "the idea of the world as a construct sustained through neurological processes contained within the brain" (Dunning et al. 2001). In **The Errant Eye** (11) the authors explore percep-tion, consciousness and the constructs of reality in the virtual reality medium while focusing on the brain as the main operator in handling this process. The participant, immersed in a virtual environment through HMD, and equipped with the *Interactive Brain Wave Visual Analyzer* [4] and a gesture recognition glove, explores the virtual environment. This environment is not stable; it changes according to the changes in the participant's EEG activity, distorting the images of "reality". A discrepancy in the images of the world as it is and the world as it is perceived (manipulated by EEG) creates a thought-provoking space for negotiation and exploration.

Immersive BCI Virtual Worlds Projected in Physical Space—Expanding beyond virtual environments, **The Mnemonic Body** (07), brings together virtual and physical space. The installation is composed of a life-sized mannequin of the human body equipped with electrodes. The participant interacts with it by touching, stroking, or breathing on the body covered with thermochromic paint that changes color when touched. An image of a field of stars is projected on four walls around the mannequin. The participant wears the *Heads Up Display* (HUD), a head-mounted display with attached electrodes for EEG recording, and haptic gloves. As described by the authors, the installation depends on the participant's affective states: calming states trigger a projection of fluid, slow-paced and smooth environments, whereas discomfort results in more startled, fast-paced environments. Similar work by the same authors are: **The Madhouse** (8), **Derive** (10), and **Pandaemonium** (09).

Lastly, **Conductar** (45) is somewhere between a virtual environment and aug-mented reality. Inspired by **Derive** (10), this audio-visual application is location dependent: it depends on the GPS location in the physical environment. Audio and a generated world visible on the screen of a mobile phone (Fig. 3.12) are generated through movement and EEG data as the participant is moving and exploring the city of Asheville (USA).

---

[4]http://www.ibva.co.uk/.

Fig. 3.12 Jeff Crouse, Gary Gunn, Aramique, *Conductar*, 2014. *Conductar. 2014*. Jeff Crouse, Gary Gunn, Aramique. With the permission of Aramique

3.2.4.6 Installations of Physical Objects

In this section, we present artworks that employ brain activity to manipulate properties or states of physical objects directly. Most of these artworks are installations and designed for a single participant. One noticeable similarity among these installations is in the position of the participant, who is usually centrally positioned in the installation, or in a position that allows for easy monitoring of changes caused by their brainwaves.

In **Mind Pool** (21) the participant's brainwaves are reflected in ferrofluid in the form of concentric circles, accompanied by sound. Brainwave frequencies trigger the electromagnets positioned under the surface of the dish filled with ferrofluid. Depending on the most prominent brainwaves, different electromagnets are activated which change the appearance of the circles on the surface. A similar project to **Mind Pool** is **Solaris** (47). While **Mind Pool** has more of a meditative character to it, **Solaris** creates a darker, experientially more stimulating experience through the choice of colors and sounds. Similarly, Lisa Park's **Eunoia II** (43) (Fig. 3.13) expands on her previous work **Eunoia** (38) by adding more physical elements—dishes installed on top of the dishes, half filled with water, represents one particular emotion. Once the real-time analysis of the participant's brainwaves (via proprietary software) reveals the participant's current emotion, the system generates a sound corresponding to the emotion which then causes the water to resonate in concentric circles on the surface. In both of these pieces, introspection and reflection through physical objects are apparent whereas the mapping is undisclosed and ambiguous.

While the projects mentioned above are single-participant, our survey encompasses a few multi-participant installations in the category of controlled objects such as Mariko Mori's **UFO wave** (17). Three participants enter a futuristic oval sculpture and lay on one of the pods while wearing EEG electrodes on their foreheads. The spherical ceiling projects six abstract shapes/blobs that represent the left and right lobes of the participants' brains. Shapes and colors of the blobs change based on the participants' Alpha, Beta and Theta levels. As explained by the author, the intention is to evoke "a deeper consciousness in which the self and the universe become interconnected."

Fig. 3.13 Lisa Park *Eunoia II*, 2014. *Eunoia II. 2014*. Lisa Park. With the permission of Lisa Park

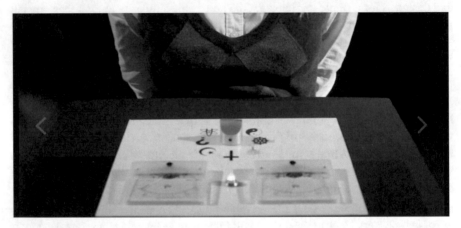

Fig. 3.14 Justin Love, Philippe Pasquier, *Praystation*, 2012. *Praystation. 2012*. Justin Love, Philippe Pasquier. A video still. With the permission of Philippe Pasquier

Barriere ⑤, another piece by Ulrike Gabriel, employs thirty robots controlled by two participants on each side of a five-meter-long tray. The sync between the activity of the participants' brains controls the level of the lightning. More light on the tray results in robots moving freely across the whole tray. In case of inconsistent and mismatching brainwave patterns, the tray is partially lit, which makes the robots "negotiate" their movement. Gabriel uses robots as a medium for displaying participants' inner states and the synergy between them.

The installations we analyzed often unfold around one central object, an instrument for controlling other elements of the installation. **PrayStation** ㉚ unfolds around a custom built instrument (prayer dial) (see Fig. 3.14). This piece is a commentary on "technology-as-placebo", combining ideas of religion and human thoughts in a unique and tangible experience. In this installation, the participant picks one out of the eight most popular religions on a custom designed prayer dial to pray to. After the choice is made, the system analyzes EEG data and associates it with a prayer and meditative states which then trigger the system to release virtual agents to create visual feedback on the canvas in front of the participant. Other examples of BCI-controlled instruments are: Dmitry Morozov's **Turbo-gusli** (48), a customized traditional instrument played by a participant's brainwaves; Jamie Gillett's **Neuro-Harp**, and Greg Kress' **The Brain Noise Machine** (22).

Lastly, the power of the mind to change the appearance or position of physical objects was the purpose of early applications of BCI in restoring movement limitations. Jody Xiong in their piece **Mind Art** (50), addresses the body's limitations through a series of paintings created by people with motor disabilities. The installation consists of four large canvases attached to form a box shape. A balloon, filled with a color that is picked by the participant, is placed in the center of the box, connected to detonators that are activated by the brainwaves of the participant. The explosion of the balloon results in the abstract paintings on canvases. Even though the idea of creating with the mind is not new (for example, see **Cerebral Painting** (36)), **Mind Art** expanded the 2-D canvas into space, transforming the intimate act of creating mind-painting into a collective event. Even though the final output is a painting, due to its spatial display, we included it in this section rather than in the visual BCI art category.

3.2.5   Presentation Format of BCI Artworks

So far in the descriptions of the artworks above, we have mentioned three *presentation formats*. *Screen-based* BCI artworks encompass mobile or desktop applications. The other two formats, *installation* and *performance*, are similar in that the artworks in both categories need human input, either real-time or recorded for the complete presentation. What distinguishes these two formats is that in performances the author/performer(s) generate(s) the output while the audience is in the role of passive observer. Performances, compared to installations, are usually rehearsed in advance and articulated in artistic expression. This is because the artist, the creator of the piece, takes part in it as a performer. However, while installations have open ends for their users/audience members to explore, sometimes with guidance but more often without, performances are more deterministic in what and how the author/performer wants to show. In most cases, performers know precisely how to use the device or how to trigger specific brainwave patterns to achieve a somewhat predicted result which is then consciously utilized or avoided.

EEG KISS (55) by Lancel/Maat (Figs. 3.15 and 3.16) is an artwork presented in both formats: as an installation and as a performance as well. It explores the act and intimacy of kissing through real-time collected EEG data—"a portrait of a kiss" that

Fig. 3.15 Karen Lancel, Herman Maat, *EEG KISS*, 2016. *EEG KISS*. 2016. Karen Lancel, Herman Maat. With the permission of Karen Lancel

Fig. 3.16 Karen Lancel, Herman Maat, *EEG KISS*, 2016. EEG KISS. 2016. Karen Lancel, Herman Maat. With the permission of Karen Lancel

generates the audio and visuals. Participants are invited to de-mystify E.E.G data through their own sense-making processes and to take part in co-creation by evoking their own experiences of kissing.

3.2.6   Audience of BCI Artworks

This category in our taxonomy describes two roles that the audience can take. The first role is of **an active audience member** who wears BCI equipment and whose brainwaves are actively fed into the artwork. The second role is of **a passive observer**, a spectator of the performance or installation.

Regarding the number of active audience members, the majority of BCI artworks allow only one person at a time to interact with the artwork. However, some artworks utilize two or more inputs, and this exploration started early on. According to Nijholt (2015), one of the first multi-brain artworks was *Alpha Garden* by Jacqueline Humbert in 1973. Among more recent artworks, our survey includes fifteen that utilized the brain activity of a minimum two and maximum of 48 participants. **Mood Mixer** (26) utilizes input from two audience members, reading their relaxation and sustained attention levels to create an audio-visual experience. **DECONcert series** (27) by Steve Mann, James Fung and Ariel Garten utilized the brainwaves of 48 visitors to create the sonic environment. Collective brain activity is analyzed and used to change real-time sonification through a continuous feedback loop between the sonic environment and the participants' brainwaves.

Regarding artworks with the presence of an audience as spectators (passive audience), most are open to a larger audience. However, some installations aim to create an intimate ambiance for those interacting with the artwork and limit the number of participants who can be present at the same time. For example, in Khut's **Alpha Lab**, (37) only those who wear BCI devices are part of the experience. In the case of screen-based applications, the presence of the audience depends on the context in which the artwork is experienced (home vs. gallery).

We conclude here our analysis of the artworks. We aimed to provide these descriptions of the selected artworks to serve as examples in the presented taxonomy. Moreover, we hope that this work will ignite the discussion and help identify a larger body of BCI artworks that we are yet to discover. Finally, due to our curiosity to learn more about the artworks beyond the documentation that was available to us, we established contact with nineteen of the authors and asked them to share their experiences of working on the presented artworks, from challenges to technical details. We present their insights below.

3.3 Artists' Insights on Creating BCI Art: Survey Responses

During the process of conducting the presented research, we succeeded in establishing contact via emails/social media with nineteen out of thirty artists that we initially contacted. We asked these artists to share the nuances of their artworks via survey, mainly because some of the documentation available to us lacked details and precision. Our findings from the survey (and a few personal email exchanges with the artists) revealed valuable insights into their processes of creating BCI artworks and also the challenges and limitations they encountered while working in the field of brain-computer interface art. From these insights, we unfolded themes that relate to working with BCI devices and present them here through three categories: design of the devices, conceptual limitations, and the potentials recognized within the technological shortcomings of BCI.

Design of Devices The design of devices is critical when it comes to the reliability of the data readings. One artist reported that one of the BCI devices they used *"doesn't lay flat on people with a large brow and thus does not work for all head shapes and types"* [A7]. Along the same lines, [A3] reported that it was *"very difficult and time-consuming to attach electrodes and maintain contact limiting its use for multiple participant installations"* and continued to discuss the *"hygiene issues for multiple participants"*.

In real-time performances, the sample rate is crucial for maintaining feedback loops. However, one of the artists reported issues with Bluetooth transfer of data between the BCI device and a computer, causing a very slow sample rate of 0.8 Hz. This issue, according to the artist *"made neurofeedback very challenging, given the lag between samples"* [A1]. Due to software issues and failure in noise filtering [A3] *"stopped using this because data was not trustworthy"*. [A4] reported that overall the device they were using was more complex than needed; however, the biggest shortcoming was false positives *"especially since the sensors pick up much muscular activity in the face and scalp."* Finally, two artists reported on the attractiveness of the device as a critical factor in making decisions related to the overall aesthetics of the artwork: *"ABM is the scientifically highest quality device we experimented with, ability to monitor evoked potentials, etc.; however the form factor was impractical for audience throughout of installation experience. Also, unattractive which was a factor in our design."*[A17]

Conceptual Limitations The design of the devices, their hardware and software limitations, as well as human factors such as the presence of the audience, and the psycho-physical endurance of a performer all contribute to the articulation of the concept. These factors pose conceptual challenges and determine how the piece will evolve in space and over the exhibition's time-span.

One of the main conceptual challenges is how to make a long-lasting, engaging artwork if the technology is the core of it? [A4]'s observation is that *"because the BCI is the core of the whole concept the piece runs the risk of being a one-trick-pony. Still, I think both the contextualization that happens in the piece as well as the unique aesthetic experience offered to the viewer/user, makes it something more than a science-fair encounter"*. What distinguishes an artwork from "a science-fair encounter", as suggested by [A4], is the artist's intent, and their sensibility regarding aesthetics, interaction, as well as the context in which the artwork is presented. However, the presence of a BCI headset and its visibility, aesthetics, or perceived gadgetry influences how the audience will experience the artwork. [A9] disclosed that *"the theatric costuming or the scientific instrumentation adds to deflecting audience members' glances and obscuring the body through its unique gadgetry"* that can take the audience away from the other, less immediate values of the artwork.

Another device-related conceptual limitation lies in the sometimes unpredictable quality of the signal from BCI devices in natural, real-world settings (outside the lab). The majority of BCI artworks presented in this chapter rely on real-time data and are at risk of failing to achieve the prescribed outcome should there be a disconnect in the data transmission between the BCI device and the artwork's architecture. The artists face the question of whether all interaction and the outcome should be prescribed, planned, and programmed to account for the unpredictability of the signal. While we have no definite answer, [A19] presented an interesting perspective on the role of data, which they consider as co-Actors in the piece, emphasizing the beauty of its agency and ambiguity over prescribed outcomes. To that end, [A19] states that *"in a participatory process of sense-making, we invite participants to give meaning to the very abstract, sometimes mystifying E.E.G. data-visualization of their kisses. Instead of scientific interpretation and validation, people who kissed interpret the data-visualization based on their shared memories of kissing and on imagination. Often the kisses are remembered as intimate processes of 'co-creation' and the data are perceived as 'A portrait of our kiss'."*

Regarding human-factors, fluctuations in the performer's attention or mood directly translate into the final output of the piece and pose conceptual considerations about how those should be handled. If not accounted for through the design of the piece, this might be detrimental for the artwork. For example, artworks that require its participants to reach meditative states can be challenging if the performer is surrounded by the audience. As [A10] emphasizes *"when I had to perform in front of hundreds of audience members, it made me feel vulnerable by presenting myself, brainwave data translated in to sound"*. On the other side, there are pieces in which *"the BCI performers had to practice to stay focused in a concert (theatrical) environment. These pieces also investigate a state of being in which the performers ride a very thin line separating learning to consciously control their attention shifts and focus –(as represented in ERP P300 activity)—and being a part of a system larger than themselves. In other words, they had to make subtle decisions about when to try to be an initiator of action and when to be an active, imaginative listening processor in the larger system"* [A3]. Employing more than one BCI input allows more room for potential distraction and unwanted brain activity to be masked by the activity of others who are in the right state (more on multi-brain BCI input can be found in (Nijholt 2015).

Finally, some artists shared that public showing of their artworks often required an assistant that helps the audience with the headsets [A16]. Having to have an assistant can introduce conceptual considerations around their role, the meaning behind the assistance, and how the process of assisting is performed so that it becomes an organic part of the artwork.

Limitations seen as potential Even though the artists who responded to our survey prioritized discussing potentials over limitations, a few shared that shortcomings of the device or approach can be effectively employed as a potential. [A8] pointed out that they were trying to *"limit the effect of the BCI to the minimum due to the huge noise that the data has. The signal is translated into a laser pattern which beautifully shows the variability and noise. So again as in the last piece I use the limitation of this technology aesthetically"*. This account demonstrates that while not perfect, emerging technologies can be a fertile ground for exploration and meaning-making of EEG data and that interpretation does not always rely on precision of the device when it comes to creative processes.

3.4 Challenges and Potentials

In our search for BCI artworks, we did an extensive review of online sources, catalogs, books, journals and conference papers. In this process, we identified two main challenges to further the development and advancement of the BCI art field. The first main challenge we encountered is limited documentation of the artworks. In our process of collecting the information and analyzing the artworks, the documentation was crucial for understanding the specifics of particular works in order to analyze them through a lens of proposed categories in the taxonomy (Table 3.4). However, a large number of pieces we included here haven't been documented in great detail. Nevertheless, despite this challenge, we aimed to provide the reader with a bigger picture of BCI art based on the landscape of the pieces that we found, focusing on their shared features rather than on specificity of a particular concept.

The other challenge concerns work on BCI artworks itself and can be broken down into technical and experiential challenges.

3.4.1 Technical Challenges in BCI Art

Irreplicability—The challenges of this in-flux field are many. Lack of documentation can lead to not just misunderstood concepts and ideas but to failed attempts to replicate the project. Unlike in science, in which each step of inquiry is rigorously documented, in art-making that is not the case for the majority of the artworks we found. Lack of documentation makes the artworks irreplicable. While we can argue that unique artworks do not need or aim to be replicable, an overview of their technical nuances, approaches, and solutions adds value to the whole field of BCI art. Well-documented artworks help the field grow by breaking through the unknown into new possibilities for creative output.

The majority of the artworks we analyzed are documented in the form of a portfolio or a website presentation, with a brief description of the concept. Some artists use their websites to provide more conceptual and philosophical insights about the work, also outlining technical details about the artwork's setup, as was done in the documentation by Lancel/Maat www.lancelmaat.nl or David Rosenboom http://www.davidrosenboom.com. Finally, the most detailed technical descriptions can be found in published books and papers, often including a description of the artwork's hardware and software to a precise detail that would allow for replicability, as done by Rosenboom in multiple publications (Rosenboom 1976; Rosenboom and Number 1990).

Recently a positive shift came with the popularization and growth of a DIY culture that seems to reflect on the documentation practices within the BCI art field positively. Open-source EEG hardware and software enthusiasts, engineers, artists, and researchers that have been sharing knowledge, hacks and best practices via communities such as *Sourceforge*, *Open BCI* or *Brain Control Club*. These communities offer, besides documentation and project descriptions on their websites, workshops and hackathons. Similarly, some artists like Ursula Gastfall, Pascale Gustin and Gérard Paresys longitudinally documented their EEG artwork *In-Between* by posting the iterations of code, pure data patches, project documentation such as sketches and videos in the form of a web project diary (Gastfall et al. 2018). These examples of BCI artwork documentation offer a starting point that we wish more artists would embrace as a part of their practice.

Reliability—The other type of technical challenges relate to the shortcomings of the technology used in the artworks we discussed. Despite the advances of dry electrodes over wet electrodes, the latter is still in use mainly in research labs as they require wired connection and gel/paste application. However, addressing the drawbacks of hybrid electrodes (such as pressure and price) is crucial for the democratization of affordable and reliable measuring tools (Davis et al. 2013) that are easy to set up outside of labs. Even though there is no perfect solution to any of the imposed problems, the development of semi-dry and dry electrodes is directing the development of BCI devices that are more portable and affordable, compact and easy to use, making these devices appealing to the consumer market. Only then will BCI devices contribute to the shift of focus from medical applications (e.g., assistive technologies) towards various applications in art, gaming, and the entertainment industry.

While new generations of electrodes have improved reliability and signal quality, the field of EEG based BCI still has a few obstacles to overcome before it can be considered a completely reliable tool. For this to happen, some of the issues to be addressed are a change of EEG signals during BCI sessions, and noise and low output rate (Millán et al. 2010). The change of signals, or signal non-stationarity, during BCI sessions is discussed in papers such as (Schlögl 2000; Shenoy et al. 2006; Vidaurre et al. 2009). Some solutions to this problem involved the rejection of the signal change and maintaining levels of the stationary signal as proposed by (Kawanabe et al. 2009). The other approach—*adaptation*, is to choose EEG features that are stable over time (Galán et al. 2007) and feed the data of these features into the artwork. However, no instant solutions are available for any of these issues, and what works for one of the artworks might not work for others. These issues are even amplified if the participants are moving, which makes the use of BCI devices close to impossible. Decisions about which approach to take will depend on the nature of the artwork itself. However, while data artifacts usually present an obstacle for precise observation of brain functioning, these artifacts are a source of unpredictability that can add value to artistic explorations. We are looking forward to seeing these issues addressed with more variety of BCI devices and approaches employed in generating BCI-based performing art such as dance and theater.

Lastly, our analysis is bound to the artworks that employ EEG-based BCI devices to record brain activity. Only one artwork **E.E.G Kiss**, to our best knowledge, combined EEG BCI with an IMEC headset that utilizes fNIRS approach. While other non-invasive approaches in BCIs are more reliable, they are either robust or costly. Even though consumer-BCI have been considered somewhat reliable and potent for revealing humans' feelings and cognitive processes, these devices have severe limitations. Panoulas (Panoulas et al. 2010) mentioned that "all EEG-based BCI classes have to face the problem of separating the control signal from interfering noise signals that have two sources: non-EEG artifacts, such as recording noise, power line interference, eye movement, eye blinking, EMG and ECG; and EEG signals that are not used as control signals." An additional drawback lays in the fact that BCI devices cannot provide a complete picture of processes in the brain, as the brain produces electrochemical signals from which only electrical are recorded. Other challenges recognized by Panoulas are technically oriented and relate to the calibration that has to be done before every use, which is cumbersome and requires additional time. We expect that by solving these issues, low-cost BCI devices will be able to provide a reliable overview of cortical activity in real time, without long and demanding training time or complicated pre-use procedures. If these devices succeed in doing so, we speculate that BCI devices will have a broader range of applications as well as more applications in art.

Compatibility, transparency and open-sourcing—Finally, the last two limitations discussed here are the compatibility of BCI systems across platforms and the transparency of data. The majority of the applications for BCI devices are created for desktops exclusively. If those applications were included on smartphones/tablets as well (Millán et al. 2010) a broader range of applications that require greater mobility would be possible. Mobility is an especially important aspect to consider when creating performances and spatial installations, and artists' hacks are sometimes geared towards ensuring the compatibility of BCI setups across various platforms (desktop, tablets, mobile).

Regarding transparency, many of the BCI devices available on the market (Sect. 3.2.1) do not provide their EEG classifying software and raw data to its users. Therefore the artists are presented with two options. One option is to use the available devices and trust their algorithms in how they sort out raw data. The other option depends on the artists' knowledge of hardware/software hacks if they aim to use raw data and apply open-source algorithms. Our analysis yielded only five artists who used custom-made BCI. One of them, [A9] worked on the piece that involved a significant amount of time spent on the setup itself: *"about 50% of that time involved composition or circuit-building…and the other 50% was spent testing to see how many zillions of ways the software or circuit was likely to crash during performance"*. [A9]'s experience poses new questions: is the lack of skill to hack custom BCI setup, or the trust in stability and reliability of consumer BCI over custom-made devices, deterring artists from customizing and hacking BCIs in their work at a larger scale?

3.4.2 Experiential Challenges

The fascination and curiosity to understand the invisible processes of the brain often leads an audience to wish for more than is achievable with the current technology used in BCI artworks. Some of the common questions that regularly emerge from interaction with the audience are questions around *mind reading* and *privacy*. While it is certainly not possible to read one's mind (at least not to our best knowledge), lack of understanding of what BCI can and cannot do leads to confusion, fear, resentment, and often rejection of the artwork. One way to prevent this from happening is to inform the audience about what the artwork does, what it collects regarding physiological data, where it stores the data, and if recordings of brain activity are made, who will have access to them in the future.

However, this approach of disclosing all the details of the artwork to the audience poses the risk of undermining the audience's curiosity, exploration, and meaning-making in the moment of experiencing the artwork. What should be disclosed to the audience, to what level of detail, and should the disclosure be done prior, during, or after the experience? One of our interviewees shared that, compared to other biofeedback such as breathing and heartbeat which the audience can immediately relate to through feeling of their heartbeat or breath at the moment, *"brainwaves however are mystical, need to be explained in ways that are hard to avoid using confusing terminology, and almost always lend a glare of scientific endeavor"*[A9]. The lack of available information about how the artists listed in our taxonomy dealt with this challenge when they show their BCI artworks prevents us from looking at the best practices or even proposing some solutions. However, we recognize that this chapter would have benefited from such a discussion, and we leave it for future work.

Lastly, a challenge that emerges from a lack of understanding of the nuances of the artwork is: how to utilize real-time brain activity in the artwork in such a way that it does not appear superficial or fake? In other words, how should artists demonstrate that the EEG of brain activity obtained from the audience member/performer via the headset is utilized in the presentation of the final output of the artwork (Millán et al. 2010)? Unless the audience is familiar with the algorithms, mapping, and interaction design of the artwork, there might always be questions about the truthfulness of data lingering in the air. This is not necessarily a defect of the artwork, but rather a design opportunity that welcomes ambiguity in a meaning-making process that adds to the beauty of the unknown to be explored and discovered individually.

3.4.3 Potentials Within BCI Art

Despite the challenges we encountered, we recognized many potentials of BCI that can be explored further in BCI art. So far, many of the pieces we analyzed are focused on the meditational (self-reflective) aspect of one's experience. To our knowledge, it is mostly because BCI devices easily detect when the participants are meditating not just by recording brain activity from particular parts on the skull, but by recording the muscle activity that occurs while they keep their eyes closed. By acquiring more reliable devices that can give us more detailed insights into brain processes, it is to be expected that the main focus will shift from meditation towards many other conceptually different directions that include various states of altered consciousness.

Agency is yet another aspect that has much potential for further exploitation in BCI art. The current category of input control can be further expanded beyond passive, active and reactive input types. Including semi-active BCI could complement the previous input types by introducing the concept of *controlled unpredictability*. This approach to a participant's agency falls somewhere between boredom (predictability) and chaos (complete unpredictability) and is achievable with current BCI devices. This could be a crucial point for further exploration of BCI applications in art. This effect could even be amplified in the case of a few simultaneously employed BCI devices within the same piece, such as a collaboration between an artist and a few participants simultaneously generating the outcome. Thinking even further, by adding an artificial intelligence component, the possibilities for the evolution of BCI art are unlimited. However, while the possibilities are many, their materialization depends on the technological progress in the BCI field and adoption from creative minds.

Lastly, on that note, even Alvin Lucier recognized the potential of agency in real-time performances over pre-recorded sessions: *"I let the structure go, let the continuity of the Alpha pulses, as they flowed out of my head, determine the moment-by-moment form of the performance. Somebody suggested to record the Alpha waves and compose the piece, but then I decided to do it live, and that is a risk because it is not sure you can get them, the more you try, the less likely is to succeed. So the task of performing without intending to give the work an irony it would not have had on a tape"* (Lucier and Simon 1980).

3.5   Conclusion

We presented a structured overview of the expanding field of BCI art, with utilized approaches and BCI devices, and proposed a systematic way of categorizing
artworks based on their similarities in the presented taxonomy. The presented taxonomy encompasses sixty-one artworks; however this list is not exhaustive. Our goal
was to offer a list of artworks to serve as examples that illustrate nuances of the
categories in the presented taxonomy. Finally, our contribution is in the proposed
categories of the taxonomy and gathered insights from the artists. With this chapter
we aimed to provide an overview and analysis of the BCI art landscape from the
1960s until 2018, and we suggest that this work should be seen as an open invitation
to a discourse on not only present practices but what can be done differently in the
future.

3.6   Additional Materials

Illustrations of presented artworks, links to the artists and artworks, and other
resources can be found here: https://bci-art.tumblr.com/.

3.7   Taxonomy

The taxonomy consists of two tables. Table 3.3 introduces sixty-one artworks by their
titles, year, authors' names and provides references. Table 3.4 details each artwork
across taxonomy categories.

Table 3.3 BCI artworks

#	Title	Year	Author(s)	References
01	Music for Solo Performer	1965	Alvin Lucier	Straebel and Thoben (2014), Novello (2014b)
02	BrainWave Drawings	1973–2008	Nina Sobell	Sobell (2008)
03	On Being Invisible	1976–1977	David Rosenboom	Rosenboom and Number (1990)
04	Terrain 01	1993	Ulrike Gabriel	Whitelaw (2004)
05	Barriere	1993	Ulrike Gabriel	Whitelaw (2004)
06	On Being Invisible II	1994	David Rosenboom	Rosenboom and Number (1990)
07	The Mnemonic Body	1995–2001	Alan Dunning, Paul Woodrow	Dunning et al. (2001)
08	The Madhouse	1995–2001	Alan Dunning, Paul Woodrow	Dunning et al. (2001)
09	Pandaemonium	1995–2001	Alan Dunning, Paul Woodrow	Dunning et al. (2001)
10	Derive	1995–2001	Alan Dunning, Paul Woodrow	Dunning et al. (2001)
11	The Errant Eye	1995–2001	Alan Dunning, Paul Woodrow	Dunning et al. (2001)
12	The Shapes of Thought	1995–2001	Alan Dunning, Paul Woodrow	Dunning et al. (2001)
13	Body Degree Zero	1995–2001	Alan Dunning, Paul Woodrow	Dunning et al. (2001)
14	Terrain 02	1997	Ulrike Gabriel	Gabriel (1997)
15	BIOS	2002	Thomas Tirel, Sven Hahne, Jaanis Garancs, Norman Muller	Tirel et al. (2002)
16	Sitting.Breathing. Beating. [NOT] Thinking	2004	Adam Overton	Overton (2004)
17	UFO wave	2005	Mariko Mori	Mori (2005)
18	Naos	2008	Carlso Castellanos, Philippe Pasquier, Luther Thie, Kyu Che	Castellanos et al. (2008)
19	The Multimodal Brain Orchestra	2009	Sylvain Le Groux, Jonatas Manzolli, Paul F.M.J Verschure	Le Groux et al. (2010)
20	The subConch	2009	Mats J. Sivertsen	Sivertsen (2014)
21	Mind Pool	2010	Kiel Long, John Vines	Long and Vines (2013)
22	The Brain Noise Machine	2010	Greg Kress	Kress (2010)
23	Staalhemel	2010	Christoph De Boeck	Boeck (2010)
24	Hidden Rooms	2011	Marie-France Bojanowski	Bojanowski (2014)
25	Ascent	2011	Yehuda Duenyas	Duenyas (2012)
26	MoodMixer	2011	Grace Leslie, Tim Mullen	Leslie and Mullen (2011)
27	DECONcert series	2011	Steve Mann, James Fung, Ariel Garten	Mann et al. (2007)
28	BrainArt	2011	Daria Migotina, Carlos Isidoro, Agostinho Rosa	Migotina et al. (2011)
29	The Magic of Mutual Gaze	2011–2014	Suzzane Dikker, Marina Abramović, Matthias Oostrik, Jason Zevin	Dikker and Oostrik (2014)
30	Praystation	2012	Justin Love, Philippe Pasquier	Love and Pasquier (2011)
31	Brain Pulse Music	2012	Masaki Batoh	Batoh (2012)
32	The Escalation of Mind	2012	Dmitry Morozov	Morozov (2012)

(continued)

Table 3.3 (continued)

#	Title	Year	Author(s)	References
33	Dream Zone	2012	Karen Casey	Casey (2012)
34	Clasp Together (beta)	2012	Harry Whalley, Panos Mavros, Peter Furniss	Whalley et al. (2015)
35	Compatibility Racer	2012	Lauren Silbert, Jennifer Silbert, Suzzane Dikker, Mattias Oostrik, Oliver Hess	Silbert et al. (2012)
36	Cerebral Interaction and Painting	2013	Yiyuan Huang, Alain Lioret	Huang and Lioret (2013)
37	Alpha Lab	2013	George Khut, James P. Brown	Khut and Brown (2014)
38	Eunoia	2013	Lisa Park	Park (2013)
39	The Creation with Strobes	2013	Luciana Haill	Haill (2013)
40	Chromatographic Ballads	2013	Ursula Damm	Damm (2013)
41	The Mutual Wave machine	2013	Suzanne Dikker, Matthias Oostrik	Dikker and Oostrik (2013)
42	(un)Focused	2013	Alberto Novello	Novello (2013)
43	Eunoia II	2014	Lisa Park	Park (2014)
44	Activating Memory	2014	Eduardo Miranda	Miranda (2014)
45	Conductar	2014	Jeff Crouse, Gary Gunn, Aramique	Aramique (2014)
46	eeg–deer	2014	Dmitry Morozov	Morozov (2014)
47	Solaris	2014	Dmitry Morozov, Julia Borovaya, Eduard Rakhmanov	Morozov et al. (2014)
48	Turbo-Gusli	2014	Dmitry Morozov	Morozov (nd)
49	Fragmentation: a brain-controlled performance	2014	Alberto Novello	Novello (2014a)
50	Mind Art	2014	Jody Xiong	Xiong (2014)
51	State.Scape	2014	Mirjana Prpa, Svetozar Miucin, Bernhard Riecke	Prpa et al. (2014)
52	Vessels	2015	Grace Leslie	Leslie (2015b)
53	Eyes Awake	2015	Grace Leslie, Carolyn Chen	Leslie (2015a)
54	Behind Your Eyes, Between Your Ears	2016	George Khut	Khut (2015)
55	E.E.G KISS	2016	Karen Lancel, Hermen Maat	Lancel and Maat (2016)
56	Noor: a Brain Opera	2016	Ellen Pearlman	Pearlman (2017); Fedorova (2017)
57	You are the Ocean	2017	Özge Samanci, Gabriel Caniglia	Samanci and Caniglia (2018)
58	The Gender Generator	2017	Josh Urban Davis	Davis (2018)
59	NeuroSnap	2017	Ryan Lieblein, Camille Hunter, Sarah Garcia, Marvin Andujar, Chris S. Crawford, Juan E. Gilbert	Lieblein et al. (2017)
60	Harmonic Dissonance	2018	Matthias Oostrik, Suzanne Dikker	Oostrik and Dikker (2018)
61	The Moment	2018	Richard Ramchurn	Ramchurn (2018)

Table 3.4 Taxonomy of BCI art

	Input		Modality	Agency	Output	Presentation	Audience
	Device	EEG classification			Type	Format	Active/Passive
01	Custom: 2 + 1 reference	LTCW	EEG	Passive	Sound: Sonification	Performance	1/1+
02	Various: EEG Grass Valley, IBVA, Brainquiry	LTCW: Alpha, Beta, Theta, Delta	EEG	Passive	Visual: Image	Installation	2/1+
03	Custom[a]	LTCW + STTW: ERP	EEG	Passive + reactive	Sound: Sonification, ongoing musical form	Performance	1/1+
04	—	LTCW	EEG	Passive	P.O: Structure + Light	Installation	1/1+
05	—	LTCW	EEG	Passive	P.O: Structure + Light	Installation	2/1+
06	Custom[b]	STTW: P300	EEG	Reactive	Sound: ongoing musical form	Performance	2/1+
07	IBVA	LTCW: Alpha, Beta, Theta.	EEG	Passive	Audio-Visual	Installation	1/1+
08	IBVA	LTCW: Alpha, Beta, Theta.	EEG	Passive	Immersive: VR + Audio-Visual	Installation	1/1+
09	IBVA	LTCW: Alpha, Beta, Theta.	EEG	Passive	Immersive: VR	Installation	1/0
10	IBVA	LTCW: Alpha, Beta, Theta.	EEG/EKG	Passive	Audio-Visual	Installation	1/1+
11	IBVA	LTCW: Alpha, Beta, Theta.	EEG	Passive	Immersive: VR	Installation	1/0

(continued)

Table 3.4 (continued)

	Input		Agency	Modality	Output	Presentation	Audience
	Device	EEG classification			Type	Format	Active/Passive
12	Contact Precision sensor + 2 amplifiers Contact (n.d.)	STTW: ERP /VEP	Reactive	EEG/EKG/GSR /HR	Visual: Image	Installation	1/0
13	Contact Precision sensor + 2 amplifiers Contact (n.d.)	STTW: ERP /VEP	Reactive	EEG/EMG/GSR /HR	Audio–Visual	Performance	3/1+
14	–	LTCW	Passive	EEG	P.O: Structure + Light	Installation	2/1+
15	Custom: 16 electrodes, on back of the head	STTW: VEP	Passive	EEG	Audio–Visual	Installation	1/0
16	Custom: based on Ciarcia (1988) and Rosenboom (1976)	LTCW	Active	EEG/ECG	Sound: Sonification	Performance	1/1+
17	–	LTCW	Passive	EEG	Physical Object	Installation	1–3/0
18	Neurosky EEG	STTW: ERP/P300	Passive	EEG/EMG/GSR	Physical Object	Installation	1/1+
19	g.Tech	STTW: P300/SSVEP	Active	EEG	Audio–Visual	Performance	4/1+
20	Emotiv EPOC	LTCW	Passive	EEG	Physical Object + Audio + Light	Installation	1/1+
21	Emotiv EPOC	LTCW	Passive	EEG	Physical Object + Audio	Installation	1/1+
22	Neurosky EEG	LTCW	Passive	EEG	Physical Object + Audio	Installation	1/1+
23	IMEC	LTCW: Alpha	Passive	EEG	Physical Object + Sonification	Installation	1/1+

(continued)

Table 3.4 (continued)

	Input		Agency	Modality	Output	Presentation	Audience
	Device	EEG classification			Type	Format	Active/Passive
(24)	Custom	LTCW: Alpha, Beta	Active	EEG	Immersive: 180 panorama	Installation	1/—
(25)	NeuroSky	Hybrid: NeuroSky	Active	EEG	Physical Object	Performance	1/1+
(26)	NeuroSky MindSet	Hybrid: NeuroSky: Focus, Relaxation	Passive	EEG	Audio-Visual	Installation	2/1+
(27)	ThoughtTec	LTCW: Alpha	Passive	EEG	Sound: Sonification/Music	Installation	48/1+
(28)	—	LTCW	—c	EEG	Visual: Visualization	Screen-Based	1+/—
(29)	Emotiv EPOC	LTCW	Passive	EEG	Visual: generative Visuals	Installation/ Performance	2/1+
(30)	Neurosky Mindset	LTCW: Alpha, Beta	Active	EEG	Physical Object + Visuals	Installation	1/—
(31)	Custom	STTW: VEP	Reactive	EEG	Sound: Music	Performance	1/1+
(32)	Emotiv EPOC	LTCW	Passive	EEG	Audio-Visual	Performance	1/1+
(33)	—	LTCW	—d	EEG	Moving Image: Generative Video	Installation	1/1+
(34)	Emotiv EPOC	Hybrid: Affectiv Suite: Excitement, Frustration, Engagement, Meditation; facial expressions; head movement	Active	EEG/ ECG/ gyroscope	Sound: Music	Performance	1/1+
(35)	Emotiv EPOC	LTCW	Passive	EEG	Physical Object	Installation	2/1+

(continued)

Table 3.4 (continued)

	Input		Agency	Modality	Output	Presentation	Audience
	Device	EEG classification			Type	Format	Active/Passive
36	Emotiv EPOC	LTCW	Active	EEG	Visual: Image	Screen-Based	1/0
37	Myndplay	LTCW: Alpha	Passive	EEG	Sound: Sonification	Installation	1–3/0
38	Neurosky Mindwave	Hybrid: NeuroSky: Attention, Meditation	Passive	EEG	Physical Object	Performance	1/1+
39	Custom	LTCW	Passive	EEG	Physical Object	Performance	1/1+
40	Emotiv EPOC	LTCW	Passive	EEG	Moving Images: Live Video footage	Installation	1/1+
41	Emotiv EPOC	LTCW	Passive	EEG	Physical Object	Installation	2/1+
42	Emotiv EPOC	LTCW	Active	EEG	Audio-Visual	Performance: Butoh	1/1+
43	Emotiv EPOC	Hybrid: Affective Suite: Excitement, Engagement, Meditation, F-rustration	Active	EEG	Physical Object	Performance	1/–
44	g.Tech g.tech (n.d.)	STTW: VEP	Reactive	EEG	Sound: Music	Performance	4/1+
45	NeuroSky Mindwave Mobile	LTCW	Passive	EEG	Audio-Visual + Immersive: AR	Installation	1/1+
46	Modified Necomimi	LTCW	Passive	EEG	Audio-Visual	Performance	1/1+
47	Emotiv EPOC	LTCW	Passive	EEG	Physical Object	Installation	1/1+
48	Emotiv EPOC	LTCW	Passive	EEG	Physical Object	Installation	1/1+
49	Emotiv EPOC	LTCW	Active	EEG	Audio-Visual	Performance: Butoh	1/1+
50	–	LTCW	Active	EEG	Physical Object + Visual: Painting	Performance	1/1+

(continued)

Table 3.4 (continued)

Input Device	EEG classification	Agency	Modality	Output Type	Presentation Format	Audience Active/Passive
(51) Emotiv EPOC	Hybrid: Attention, Excitement, Meditation, Boredom	Passive	EEG	Audio-Visual	Installation	1/1+
(52) Muse	LTCW	Passive	EEG/ECG/EDA	Sound: Music	Performance	1/1+
(53) Muse	LTCW: Alpha	Passive	EEG	Sound: Music	Performance	1/1+
(54) Muse	LTCW: Alpha	Passive	EEG	Audio-Visual	Installation	1/1+
(55) IMEC (fNIRS), Muse	LTCW: Alpha, Beta, Theta	Passive	EEG	Audio-Visual	Installation/Performance	2/1+
(56) Emotiv EPOC	Hybrid: Affectiv Suite: Excitement, Interest, Meditation, Frustration	Passive	EEG	Sound: Opera + Audio-Visual	Performance	1/1+
(57) NeuroSky MindWave 2	Hybrid: eSense: Attention, Meditation	Passive	EEG	Audio-Visual	Installation	1/1+
(58) OpenBCI with OpenVibe	STTW: n250 ERP (modified P300)	Active	EEG	Visual: Visualization	Installation	1/1+
(59) Emotiv Insight	LTCW: Alpha, Beta	Passive	EEG	Visual: Camera Overlay	Screen-Based	1/—
(60) Emotiv EPOC	LTCW	Passive	EEG/HRV/GSR	Audio-Visual	Installation	5/1+
(61) NeuroSky	Hybrid: NeuroSky: Attention	Passive	EEG	Moving Image: Movie	Installation	1/1+

[a]2 electrodes + EEG amplifiers made by Princeton Applied Research, model PAR 133, Pre Amp
[b]2 electrodes at Cz for each of two performers
[c]Artworks 28 & 33 are pre-recorded
[d]Artworks 28 & 33 are pre-recorded

Acknowledgements   We thank all the reviewers for their insightful comments and suggestions that helped shape and improve this chapter through a "blind" peer-review process. A big thank you goes to all artists who responded to our survey and kept providing necessary information and images and without whom this chapter wouldn't be as comprehensive. We also thank: Bernhard Riecke whose insights in the early days of this chapter helped in framing the scope; Omid Alemi and Kivanç Tatar for the suggestions about the structure and chapter organization; Mohammad El-Ghanem M.D. for suggestions and revisions of technical parts of this chapter; Ellen Pearlman for last-minute reading and suggestions. Finally, we thank Ash Tanasiychuk for tireless proofreading over the span of a couple of years while this chapter was transitioning from "in-progress" to "completed".

References

A wireless low-power, high-quality EEG headset (n.d.) Kurzweil accelerating intelligence. http://www.kurzweilai.net/a-wireless-low-power-high-quality-eeg-headset. Accessed Sept 2018

Adrian ED, Matthews BHC (1934) The Berger rhytm: potential changes from the occipital lobes in man. Brain 57(4):355–385

Agency (nd) In Merriam-Webster's dictionary. https://www.merriam-webster.com/dictionary/agency. Accessed Oct 2018

Amft O, Lauffer M, Ossevoort S, Macaluso F, Lukowicz P, Troster G (2004) Design of the QBIC wearable computing platform. In: Proceedings 15th IEEE international conference on application-specific systems, architectures and processors, 2004. Chicago, pp. 398–410

Aramique (2014) Conductar. http://aramique.com/conductar-moogfest/. Accessed Oct 2018

B-Alert X series mobile EEG (nd) Advanced brain monitoring. https://www.advancedbrainmonitoring.com/xseries/. Accessed Sept 2018

Batoh M (2012) Brain pulse music. https://www.youtube.com/watch?v=XI4Mge8nLMw. Accessed Sept 2018

Boeck CD (2010) *Staalhemel.* http://www.staalhemel.com/. Accessed Sept 2018

Bojanowski M-F (2014) The Hidden room. http://artistsinlabs.ch/portfolio/marie-france-bojanowski/. Accessed Oct 2018

Braintone Art (2005–2019). http://www.braintoneart.com//. Accessed Sept 2018

Casey K (2012) Dream zone. http://www.realtimearts.net/article/119/11492. Accessed Sept 2018

Castellanos C, Pasquier P, Thie L, Che K (2008) Biometric tendency recognition and classification system: an artistic approach. Proceedings of the 3rd international conference on digital interactive media in entertainment and arts. ACM, New York, pp 166–173

Ciarcia S (1988) Computers on the brain, part 1. Byte 13(6):273–285

Contact Precision Instruments (nd). http://www.psychlab.com/EEG_8_amplifier.html. Accessed Oct 2018

Coyle SM, Ward TE, Markham CM (2007) Brain—computer interface using a simplified functional near-infrared spectroscopy system. J Neural Eng 4(3):219–226

Damm U (2013) Chromatographic ballads. http://ursuladamm.de/nco-neural-chromatographic-orchestra-2012/. Accessed Oct 2018

Davis G, McConnell C, Popovic D, Berka C, Korszen S (2013) Soft, embeddable, dry EEG sensors for real world applications. In: Schmorrow DD, Fidopiastis CM (eds) Foundations of augmented cognition. Springer, Berlin, Heidelberg, pp 269–278

Davis JU (2018) The gender generator: towards a machine-empathy inter-face for the evocation of gender dysphoria symptoms. Dartmouth College (master's thesis)

Dikker S, Oostrik M (2013) Mutual wave machine. http://todaysart.org/project/99/. Accessed Sept 2018

Dikker S, Oostrik M (2014) Measuring the magic of mutual gaze. In: Leonardo, Vol 47, pp 431–431

Duenyas Y (2011–2012). Ascent. https://www.youtube.com/watch?v=yGvsDD50cb8. Accessed Sept 2018

Dunning A, Woodrow P, Hollenberg M (1995–2001) The Einstein's Brain project. http://people.ucalgary.ca/~einbrain/new/main.html. Accessed Oct 2018

Ear EEG demo (2018) Neuro-Machine augmented intelligence lab, school of computing, KAIST. https://www.youtube.com/watch?v=08MuLufgpFM. Accessed Dec 2018

Ear EEG project (2018) Aarhus university. http://ear-eeg.org/. Accessed Dec 2018

Emotiv (nd) http://emotiv.com/. Accessed Oct 2018

Emotiv EPOC (nd) https://www.emotiv.com/epoc/ Accessed Sept 2018

Emotiv Insight (nd) https://www.emotiv.com/insight/. Accessed Oct 2018

Emotiv myEmotiv (nd) www.emotiv.com/myemotiv/. Accessed Oct 2018

Eyeborg project (2019) http://eyeborgproject.tv/. Accessed Jan 2019

Fedorova N (2017) The first neuroopera 'Noor': transparent brain and the end of humanistic ethics? Russ J Commun 9(3):310–314

Gabriel U (1997) Terrain 02. http://llllllll.de/terrain02e.html#b0. Accessed Oct 2018

Galán F, Ferrez PW, Oliva F, Guardia J, del R Millan J (2007) Feature extraction for multi-class bci using canonical variates analysis. In: IEEE international symposium on intelligent signal processing, 2007. WISP 2007, pp 1–6

Gastfall U, Gustin P, Paresys G (2018) Diary of EEG project–in between. http://gerard.paresys.free.fr/Projets/ProjetEEG.html. Accessed Dec 2018

George L, Lècuyer A (2010) An overview of research on "passive" braincomputer interfaces for implicit human-computer interaction. In: International conference on applied bionics and biomechanics icabb 2010-workshop w1 "brain-computer interfacing and virtual reality"

g.tech medical engineering (nd) http://www.gtec.at/. Accessed Oct 2018

Haill L (nd) *History of EEG in art*. https://lucianahaill.wordpress.com/history-of-eeg-in-art/. Accessed Oct 2018

Haill L (2013) *The creation with the strobes*. http://lucianahaill.wordpress.com/2014/05/08/the-creation-of-the-strobes/. Accessed Oct 2018

Harbisson N, Ribas M (2010) Cyborg foundation. https://www.cyborgfoundation.com/. Accessed Jan 2019

Holmes T (2016) Early "Live" moog modular artists: richard teitelbaum and the first moog modular synthesizer in Europe. http://moogfoundation.org/early-live-moog-modular-artists-richard-teitelbaum-first-moog-modular-synthesizer-europe. Accessed Oct 2018

Huang Y, Lioret A (2013) Cerebral interaction and painting. In: SIGGRAPH Asia 2013 art gallery. ACM, NY, USA, pp 21:1–21:7

Jurcak V, Tsuzuki D, Dan I (2007) 10/20, 10/10, and 10/5 systems revisited: their validity as relative head-surface-based positioning systems. Neuroimage 34(4):1600–1611

Kawanabe M, Vidaurre C, Scholler S, Muller K-R (2009) Robust common spatial filters with a maxmin approach. In: Annual international conference of the IEEE engineering in medicine and biology society, 2009. EMBC 2009, pp 2470–2473

Khut G (2015) In behind your eyes, between your ears. http://www.georgekhut.com/behind-your-eyes-between-your-ears/. Accessed Oct 2018

Khut G, Brown JP (2014) Alpha Lab. http://www.georgekhut.com/portfolio/alpha-lab/. Accessed Oct 2018

Kress G (2010) Brain noise machine. http://glkress.com/art-and-design/brain-art/brain-noise-machine/. Accessed Oct 2018

Kumar JS, Bhuvaneswari P (2012) Analysis of electroencephalography (EEG) signals and its categorization—a study. Procedia Eng 38:2525–2536

Lancel K, Maat H (2016) EEG kiss. http://www.lancelmaat.nl/work/e.e.g-kiss/. Accessed Sept 2018

Lebedev MA, Nicolelis MAL (2006) Brain-machine interfaces: past, present and future. Trends Neurosci 29(9):536–546

Le Groux S, Manzolli J, Verschure PF, Sanchez M, Luvizotto A, Mura A, Bernardet U (2010) Disembodied and collaborative musical interaction in the multimodal brain orchestra. In: proceedings of the conference on new interfaces for musical expression (NIME), pp 309–314

Leslie G (2015a) Eyes awake. http://www.graceleslie.com/Eyes-Awake. Accessed Jan 2018

Leslie G (2015b) Vessels. http://www.graceleslie.com/Vessels. Accessed Jan 2018

Leslie G, Mullen TR (2011) Moodmixer: EEG-based collaborative sonification. In: NIME pp 296–299

Lieblein R, Hunter C, Garcia S, Andujar M, Crawford CS, Gilbert JE (2017) NeuroSnap: expressing the user's affective state with facial filters. In: Schmorrow DD, Fidopiastis CM (eds) Augmented cognition. Enhancing cognition and behavior in complex human environments. Springer International Publishing, pp 345–353

Lin C-T, Liao L-D, Liu Y-H, Wang I-J, Lin B-S, Chang J-Y (2011) Novel dry polymer foam electrodes for long-term EEG measurement. IEEE Trans Biomed Eng 58(5):1200–1207

Long K, Vines J (2013) Mind pool: encouraging self-reflection through ambiguous bio-feedback. CHI '13 extended abstracts on human factors in computing systems. ACM, NY, USA, pp 2975–2978

Looney D, Kidmose P, Park C, Ungstrup M, Rank ML, Rosenkranz K, Mandic DP (2012) The in-the-ear recording concept: user-centered and wearable brain monitoring. IEEE Pulse 3(6):32–42

Love J, Pasquier P (2011) Aesthetic agents: a multiagent system for nonphotorealistic rendering with multiple images. In: Proceedings of the international symposium on electronic arts (ISEA), Istanbul, Turkey, pp 47–54

Lucier A, Simon D (1980) Chambers. Scores by Alvin Lucier, interviews with the composer by Douglas Simon. Weslayan University Press

Mann S, Fung J, Garten A (2007) DECONcert: bathing in the light, sound, and waters of the musical brainbaths. In: ICMC, p 8

Migotina D, Isidoro C, Rosa A (2011) Brain art: abstract visualization of sleeping brain. In: Proceedings of GA 2011—14th generative art conference

Mikkelsen KB, Kappel SL, Mandic DP, Kidmose P (2015) EEG recorded from the ear: characterizing the ear-EEG method. Front Neurosci 9:438

Millán JdR, Rupp R, Mueller-Putz G, Murray-Smith R, Giugliemma C, Tangermann M, Mattia D (2010) Combining brain computer interfaces and assistive technologies: state-of-the-art and challenges. Front Neurosci 4:161

MindPlay MyndBand (nd) http://store.myndplay.com/products.php?prod=28/. Accessed Oct 2018

Miranda E (2014) Activating memory. https://vimeo.com/88151780. Accessed Oct 2018

Miranda E, Castet J (eds) (2014) Guide to brain-computer music interfacing. Springer

Mori M (2005) UFO wave. http://www.ibva.co.uk/Templates/mariko.htm. Accessed Oct 2018

Morley A, Hill L, Kaditis AG (2013) 10–20 system EEG placement. http://www.ers-education.org/lrmedia/2016/pdf/298830.pdf. Accessed Oct 2018

Morozov D (nd) Turbo-gusli. http://vtol.cc/filter/works/turbo-gusli/. Accessed Oct 2018

Morozov D (2012) The escalation of mind. http://vtol.cc/filter/works/ The-Escalation-Of-Mind. Accessed Oct 2018

Morozov D (2014) eeg_deer. http://vtol.cc/filter/works/eegdeer/. Accessed Oct 2018

Morozov D, Borovaya J, Rakhmanov E (2014) Solaris. http://vtol.cc/filter/works/solaris/. Accessed Oct 2018

Moss D (2003) Handbook of mind-body medicine for primary care. Sage

Muse by Interaxon (nd) http://www.choosemuse.com/. Accessed Oct 2018

Muse Documentation (nd) http://developer.choosemuse.com/hardware-firmware/headband-configuration-presets. Accessed Oct 2018

Nakamura J, Csikszentmihalyi M (2014) The concept of ow. In: Flow and the foundations of positive psychology. Springer, pp 239–263

Nam CS, Choi I, Wadeson A, Whang M (2018, January) Brain-computer interface: an emerging interaction technology. In: Nam CS, Nijholt A, Lotte F (eds) Brain-computer interfaces handbook: technological and theoretical advances. CRC Press, pp 11–52

Naseer N, Hong K-S (2015) fNIRS-based brain-computer interfaces: a review. Frontiers in Human Neurosci, 9

Neuroelectrics ENOBIO (nd) http://www.neuroelectrics.com/products/enobio/. Accessed Oct 2018

NeuroSky (nd) Mindset EEG headset. http://support.neurosky.com/kb/mindset. Accessed Sept 2018

NeuroSky Algorithms (nd) http://neurosky.com/biosensors/eeg-sensor/algorithms/ see: NeuroSky algorithms. Accessed Oct 2018

NeuroSky MindWave (nd-a) http://neurosky.com/biosensors/eeg-sensor/biosensors/. Accessed Oct 2018

NeuroSky MindWave (nd-b) http://store.neurosky.com/products/mindwave-1. Accessed Oct 2018

Nicolas-Alonso LF, Gomez-Gil J (2012) Brain computer interfaces, a review. Sensors 12(2)

Nijholt A (2015) Competing and collaborating brains: multi-brain computer interfacing. In: Brain-computer interfaces. Springer, pp 313–335

NIRSIT (2018) Soterix Medical. https://soterixmedical.com/research/nirsit. Accessed Dec 2018

Novello A (2013) (un) focused. https://vimeo.com/161614592. Accessed Sept 2018

Novello A (2014a) Fragmentation: a brain-controlled performance. http://jestern.com/. Accessed Oct 2018

Novello A (2014b) From invisible to visible. LAP LAMBERT Academic Publishing

Oostrik V, Dikker S (2018) Harmonic disonance. http://plplpl.pl/hd/. Accessed Sept 2018

OpenBCI (nd) http://forums.ni.com/t5/Community-Documents/LabVIEW-OpenBCI-Toolkit/ta-p/3495333. Accessed Oct 2018

Overton A (2004) Sitting. Breathing. Beating. [Not] Thinking. http://archive.org/details/Sitting.Breathing.Beating.NotThinking/. Accessed Oct 2018

Panoulas KJ, Hadjileontiadis LJ, Panas SM (2010) Brain-computer interface (BCI): types, processing perspectives and applications. In: Tsihrintzis GA, Jain LC (eds) Multimedia services in intelligent environments. Springer, Berlin, Heidelberg, pp 299–321

Park L (2013) Eunoia. http://thelisapark.com/. Accessed Sept 2018

Park L (2014) Eunoia II. http://www.thelisapark.com/. Accessed Oct 2018

Pearlman E (2014) The world's first skull transmitted painting. https://artdis.tumblr.com/post/96868284536/the-worlds-first-skull-transmitted-painting. Accessed Jan 2019

Pearlman E (2015) I, cyborg. PAJ: A J Perform Art 37(2):84–90

Pearlman E (2017) Brain opera: exploring surveillance in 360° immersive theatre. PAJ: A J Perform Art 39(2):79–85

Prpa M, Riecke B, Miucin S (2014) State. scape: a brain as an experience generator. In: Brains in electronic arts [Session 32] ISEA 2015 proceedings. Disruption—ISEA 2015

Psychophysiological Recording (2000) 2edn. Oxford, England, New York

Quick-20 (nd) Cognionics https://www.cognionics.net/quick-20. Accessed Jan 2019

Quick-30 (nd) Cognionics https://www.cognionics.net/quick-30. Accessed Jan 2019

Ramchurn R (2018) The moment. https://www.firstpost.com/news%20/this-ai-based-movie-changes-music-scenes-and-animations-based-on-what-you-are-thinking-4499281.html. Accessed Sept 2018

Rosenboom D (1976) Biofeedback and the arts, results of early experiments. Aesthetic Research Centre of Canada

Rosenboom D (1990) Extended musical interface with the human nervous system. Leonardo monograph, vol. 1. The MIT Press, Cambridge, MA

Samanci Ö, Caniglia G (2018) You are the ocean. In: ACM SIGGRAPH 2018 art gallery. pp 442–442

Schlögl A (2000) The electroencephalogram and the adaptive autoregressive model: theory and applications. Shaker Germany

Shenoy P, Krauledat M, Blankertz B, Rao RP, Müller K-R (2006) Towards adaptive classification for bci. J Neural Eng 3(1)

Shimamura AP, Palmer SE (2012) Aesthetic science: connecting minds, brains, and experience. Oxford University Press, USA

Silbert L, Silbert J, Dikker S, Oostrik M, Hess O (2012) Compatibility racer. http://compatibilityracer.blogspot.com/. Accessed Sept 2018

Silva FLD, Niedermeyer E (2012) Electroencephalography: basic principles, clinical applications, and related fields (Fifth edn; FLDSM PhD, Ed). Lippincott Williams, Wilkins

Sivertsen M (2014) The SubConch. http://www.subconch.net/. Accessed Oct 2018

Sobell N (1973–2008) Brainwave drawings. http://colophon.com/ninasobell/parkbenchdocs/portfolio/3/frame.html. Accessed Dec 2018

Steriade M, Gloor P, Llinas R, Lopes da Silva F, Mesulam M-M (1990) Basic mechanisms of cerebral rhythmic activities. Electroencephalogr Clin Neurophysiol 76(6):481–508

Straebel V, Thoben, W (2014) Alvin Lucier's music for solo performer: experimental music beyond sonification. Organised Sound 19 (Special Issue 01):17–29

Szafir DJ (2010) Non-invasive BCI through EEG (Unpublished doctoral dissertation). Boston College, College of Arts and Sciences

Thompson M, Thompson L (2003) The neurofeedback book: an introduction to basic concepts in applied psychophysiology. Association for applied psychophysiology and biofeedback, Wheat Ridge, CO

Tirel T, Hahne S, Garancs J, Muller N (2002) BIOS. http://bios.x-i.net/

Vidal JJ (1973) Toward direct brain-computer communication. Ann Rev Biophys Bioeng 2(1):157–180

Vidal JJ (1977) Real-time detection of brain events in EEG. In: Proceedings of the IEEE, vol 65(5), pp 633–641

Vidaurre C, Krämer N, Blankertz B, Schlögl A (2009) Time domain parameters as a feature for EEG-based brain-computer interfaces. Neural Netw 22(9):1313–1319

von Luhmann A, Muller K (2017) Why build an integrated EEG-NIRS? about the advantages of hybrid bio-acquisition hardware. In: Proceedings of the annual international conference of the IEEE engineering in medicine and biology society, vol 2017, pp 44–75

Wadeson A, Nijholt A, Nam CS (2015) Artistic brain-computer interfaces: state-of-the-art control mechanisms. Brain-Comput Interfaces 2(2–3):70–75

Whalley JH, Mavros P, Furniss P (2015) Clasp together: composing for mind and machine. Empir Musicol Rev 9(3–4):263–276

Whitelaw M (2004) Metacreation: art and artificial life. MIT Press, Cambridge, MA

Wolpaw J, Wolpaw EW (2012) Brain-computer interfaces: principles and practice. OUP USA

Wolpaw JR, Birbaumer N, Heetderks WJ, McFarland DJ, Peckham PH, Schalk G, Donchin E, Quatrano LA, Robinson CJ, Vaughan TM (2000) Brain-computer interface technology: a review of the first international meeting. In: IEEE transactions on rehabilitation engineering, vol 8(2). A Publication of the IEEE Engineering in Medicine and Biology Society, pp 164–173

Xiong J (2014) Mind art. http://thecreatorsproject.vice.com/blog/this-art-project-lets-anyone-paint-with-brainwaves. Accessed Sept 2018

Zander TO, Kothe C (2011) Towards passive brain-computer interfaces: applying brain-computer interface technology to human-machine systems in general. J Neural Eng 8(2)

Zander TO, Kothe C, Jatzev S, Gaertner M (2010) Enhancing humancomputer interaction with input from active and passive brain-computer interfaces. In: Brain-computer interfaces. Springer, pp 181–199

Chapter 4
More Than One—Artistic Explorations with Multi-agent BCIs

David Rosenboom and Tim Mullen

Abstract In this chapter, the historical context and relevant scientific, artistic, and cultural milieus from which the idea of brain-computer interfaces involving multiple participants emerged is discussed. Additional contextualization includes descriptions of the intellectual climate from which ideas about brain biofeedback led to pioneering applications in music and its allied arts. The chapter then proceeds with more in-depth explanations of what are termed *contingent* and *non-contingent* feedback schemes, along with descriptions of early artistic applications and how those might be differentiated. Effects ensuing from the qualitative nature of the feedback signals in brainwave music are also briefly discussed. Following this, substantial space is devoted to describing selected examples of relatively recent musical and artistic pieces that employ multi-agent BCI. These are described with more extensive technical details that illustrate how the ideas, some of which could only have been imagined in earlier times, are now made possible by advances in available technology and new methods for analyzing brain signals from both individuals and groups. These include: implementing biofeedback schemes in which feedback signals depend upon contingent conditions in electroencephalographic features measured among multiple participants, multivariate principal oscillation pattern detection, "hyper-brain" scanning, employing wearable technology, and other related methods. Complex brain-computer music systems are also described in detail. Key artistic concepts explored include the idea of active imaginative listening as performance and cooperative multi-agent artistic productions with BCIs. Some concluding commentary and ideas for future research are also offered.

Keywords Active imaginative listening · Artscience · BCI · BCMI · Bioart · Biofeedback · Brain-computer interface · Brain-computer music interface · Brainwave music · Contingent feedback · Cooperative brain-computer interface · EEG · Event related potentials · ERP · Hyper-brain · Hyperscanning · Listening as

D. Rosenboom (✉)
The Herb Alpert School of Music, California Institute of the Arts, Santa Clarita, USA
e-mail: david@calarts.edu; davidcharlesrosenboom@gmail.com

T. Mullen
Intheon Labs, San Diego, USA
e-mail: tim@intheon.io

performance · Live electronic music · Multi-agent brain-computer interface · Neuromusic · Principal oscillation pattern

4.1 Introduction—Historical and Philosophical Background

"Mister Science Meets Earth Mother"—an opening line from Rosenboom's 1971 presentation, called *Homuncular Homophony*, that was delivered to the Spring Joint Computer Conference in Atlantic City, the Audio Engineering Society Convention in Los Angeles, and the University of Illinois Festival of Contemporary Arts, broadcasted something about the spirit of how the inspiring emergence of biofeedback in the 1960s, with all its implications and intermingling with cybernetics, computer science, neuroscience, systems theory, artificial intelligence, evolution, complex adaptive systems, studies in cognition and consciousness, and epistemology, among other disciplines, offered a doorway into a space where science and art might meet in meaningful and substantial, deep theoretical territory (Rosenboom 1976, 1997). Today, we might refer to this as *artscience*. It was imagined then that new developments in neuro-technology might be building a potentially powerful bridge that could link what were thought of as the inner and outer spaces of individual experience, while simultaneously offering both reasonable measures of phenomenal objectivity and rich offerings for creative realization. From this point, and in the whirling historical-cultural context of that time, it was a natural and obvious step to also want to explore building such links among the experiences of more than one human individual. From this environment of inquiry, *multi-agent* biofeedback emerged early on as a natural and irresistible arena for investigation. The term BCI (Brain-Computer Interface) was first coined in 1973 by Jacques Vidal (Vidal 1973) to describe a direct link between observable neuroelectric signals in the brain and a computer system. Now, decades later, the term has become relatively widespread and even colloquially used. Consequently, we can now conveniently refer to this emergent phenomenon as multi-agent BCI (MABCI), and in music, multi-agent BCMI (Brain-Computer Music Interface).

Several critical concepts about systems organization penetrated this environment deeply. The nature of feedback in developing electronic music and video synthesis paradigms, for example, was—and still is—foundational. The qualities of resonance and resonant emergence, also driven by feedback and observed in a wide range of natural phenomena, including the physical, cosmological, psychological, historical, biological, sociological, and cultural arenas, to name a few, have remained unbroken threads. Investigations into the behavior of systems (Foerster 1981), and more recently, self-organization and non-linear dynamics in the brain and human functioning (Kelso 1995), new understandings about the emergence of order (Holland 1995), and the adjacent possible in models of evolution (Kaufman 2000), are examples of continuous sources of inspiration from science crossing over into music and related arts. Multi-agent BCI in the arts was seen as a manifestation of interconnectivity, a broadening of self-reference to encompass multiple selves.

The meeting of science and art in this arena should not be confused as being one or the other; it is a joint space in which concepts can be exchanged and perhaps influence each other, ideas tried out with the freedom that artistic practice brings, and techniques tested for their relevance to rigorous practice and their potential to illuminate new theoretical models. Rosenboom has written extensively about an approach to composition he terms *propositional music*. Propositional music involves building proposed models of worlds, universes, evolution, brains, consciousness or whole domains of thought and life, and then proceeding to make dynamical musical embodiments of these models, inviting us to experience them in spontaneously emerging sonic forms (Rosenboom 2000c, 2018). Artistic license allows us to build these propositional models without requiring that they must completely correspond or explain some idea of reality. As Stephen Hawking is reported to have said, "I don't demand that a theory correspond to reality because I don't know what it is … All I'm concerned with is that the theory should predict the results of measurements." (Holt 2018). There are many challenges to predictive model building when linking complex self-organizing systems via mappings in multi-modal stimulus domains. Propositional music, and by extension, proposition art making, may help open *artscience* conversations where some of our deepest theoretical questions lie: in our languages of description, how we describe what we experience, deduce, induce, propose, and believe that we know. A quote from *Biofeedback and the Arts* asks, "What is the place of nonverbal communication in the scientific method?" (Rosenboom 1972).

4.2 Qualities of Sounds in Non-contingent and Contingent Feedback Paradigms

In musical biofeedback paradigms, the qualities of the feedback signal and the nature of the auditory environment became important subjects of investigation. It was soon observed that the degree of success in achieving some control over the feedback signal—increasing the ability to influence the presence and coherence of alpha brainwave bursts, for example—was influenced by the nature of the sounds (Rosenboom 1976, 1997, 2003). Though this relationship might seem to be obvious—sounds conducive to the mental states being associated with particular frequencies of brainwaves extracted from the EEG—deeper investigation in both laboratory and performance situations revealed that the relationship is, indeed, not a simple one. It was found that aspects of attention, the dynamics of focused attention, the musical backgrounds of subjects and their facility with active imaginative listening strategies, all had profound effects on how subjects were able to interact with sonic environments and achieve a measure of success in a biofeedback control setup, be they simple or complex sonic worlds. The effects of musical and artistic backgrounds on affective judgments of aesthetic qualities had already been investigated in what was known as experimental aesthetics (Berlyne 1971). Subjects—or in the case of brainwave music, brainwave performers—could also develop considerable facility in feedback setups with extended practice. This seemed directly parallel to the way musicians gain performance facility with instruments through long, intensive, extended practice. Furthermore, those with experience in meditation, particularly as found in Zen practice, often brought skills that enhanced their comfort with complex sound environments and their facility in manipulating brain states within them, again, particularly with practice (Rosenboom 1976, 1997). The matrix of possible relationships among sound worlds and the cultural backgrounds of participants in biofeedback paradigms is also a complex one. A vast and continuously interesting territory for neuro-musical investigation in BCMI and the qualities of sounds in musical forms remains to be traversed. Elsewhere, Rosenboom has offered an agenda with questions and suggestions for future research in this territory (Rosenboom 2014).

With multi-agent BCMI setups, the qualities of the sound environments seemed at first to be particularly important. Early investigations involved practicing biofeedback exercises in groups. Early examples include various projects by composer Richard Teitelbaum—described in his article, *In tune: some early experiments in biofeedback music (1966–74)*, and in Rosenboom's early 1970s, carefully structured and immersive *Three day biofeedback learning experience for Brown University* (both are contained in Rosenboom 1976). These were mostly *non-contingent*, group biofeedback music setups. That is, the electronic musical feedback did not depend upon features of the performers' EEGs being detected simultaneously. These group experiences soon lead to what was termed *contingent*, multi-agent biofeedback setups (Fehmi and Rosenboom 1971). In these situations, various methods of observing EEG features that were synchronous, or simultaneously detected, among two or more participants were developed and used to generate the auditory feedback signals. Sounds that were initially conducive to a group achieving simultaneous, synchronous brain states—simultaneous enhancement of alpha brainwave production, for example—were particularly important for the group to practice effectively. Again, though, more recent work has shown that multiple brainwave performers in multi-agent BCMI setups can achieve positive results in complex sound environments, especially if they have strong musical backgrounds and are active, imaginative, creative listeners. Setups like this will be described in detail later with technical descriptions of recent brainwave music works.

4.3   Historical Roots for the Development of Multi-agent BCMI and BCI in the Arts

Multi-Agent BCMI has long roots. Around 1969–1970, Rosenboom programmed an interactive game of *Alpha Checkers*, in which a computer screen would display a checkerboard for two players only when they produced EEG alpha wave bursts sufficient to cross an amplitude threshold at the same time. The players could only play the game when the checkerboard was visible. The task proved impossible to carry out, because when either player was looking at the checkerboard and trying to make a move in the game, the presence of alpha waves would decrease. This was an example of contingent, multi-agent BCI. Access to the necessary technology at the time was limited; so, it wasn't possible to see whether with sufficient practice the players could master the system and play the game, while continuing to produce simultaneous and sustained alpha wave bursts. Interestingly, similar questions were explored nearly 30 years later in the context of a popular multi-agent BCI game and research project called *BrainBall*. Created in 1999 at the RISE Interactive Institute AB, the game has two players compete to increase their level of relaxation—thereby increasing alpha and theta EEG activity—which in turn controls movement of a magnetically coupled ball towards an opponent's "goal" area. The game could also be played collaboratively, wherein players must alternately increase and decrease relaxation levels to move the ball towards the center of the table. Playing the game competitively reportedly resulted in reduced stress, as measured by galvanic skin response, and the players' attitudes towards the game were reported as generally quite positive with user tests suggesting that players were able to successfully "competitively relax" (Ilstedt Hjelm and Browall 2000). In addition to *Alpha Checkers*, throughout the late 1960s and early 1970s other explorations with contingent alpha biofeedback setups involving only sound-based feedback were carried out. These proved more successful. Some of these took place in an EEG lab at the State University of New York at Stony Brook in collaboration with psychologist Lester Fehmi (Fehmi and Rosenboom 1971).

Rosenboom's more substantial work in the multi-agent BCI arena began around 1970, with an environmental, demonstration-participation-performance event, called *Ecology of the Skin*, which was held at Automation House in New York City. In this exhibition, up to ten participants could wear EEG electrodes connected to portable EEG preamplifiers, filters, and amplitude envelope followers that were connected to an electronic music generating system. In addition, some EKG monitors were available, and stations for electrical stimulation of visual phosphenes were installed around the exhibition space. Most of these employed non-contingent feedback setups. Subsequent iterations and spinoffs from the original *Ecology of the Skin*, however, did begin to employ contingent biofeedback setups (Fig. 4.1).

Soon after *Ecology of the Skin*, Rosenboom and collaborators built a facility called the Laboratory for Experimental Aesthetics at York University in Toronto. Here, students and faculty, notably Richard Teitelbaum, Barbara Mayfield, C. Mark Nunn, and others, developed systems for exploring both contingent and non-contingent, multi-agent BCMI on a regular basis. Various artists developed installation pieces, such as Jacqueline Humbert's *Brainwave Etch-A-Sketch*, in which low-frequency envelope followers tracked the amplitudes of alpha brainwaves from two participants, one of which moved a dot on a storage oscilloscope along its x-axis and the other along its y-axis, to create a shared drawing. This was an example of a *non-contingent* feedback system, as the presence of absence of feedback did not depend on the contemporaneous detection of a specific EEG feature in both participants. Conversely, *contingent* feedback was employed in another Humbert installation, *Alpha Garden*, wherein

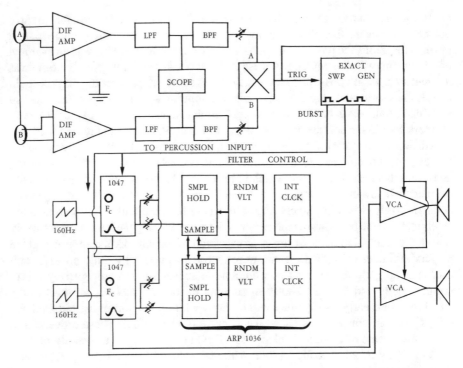

Fig. 4.1 Diagram of an early example of a contingent multi-agent feedback system (reproduced from Rosenboom 1972). Synchronous alpha bursts from two participants triggered slow rising sweeps of a harmonic series with slight, automatically induced sequence changes through initiation of voltages that determined the starting pitches of resonant filters and randomly introduced bell-like accentuations of various harmonic tones produced by shocking the resonant filters with narrow pulses at the attack initiated by each synchronous alpha burst and throughout the sequence

simultaneous alpha bursts from two participants would turn on pulses of water from a lawn sprinkler that irrigated a piece of artificial turf.

In 1972, Rosenboom expanded the scope of contingent biofeedback art with his *Vancouver Piece*. A darkened, sound-isolated room was built inside the Vancouver Art gallery during an exhibition of sound sculpture pieces. Inside the room were subtle types of visual and auditory displays and equipment to detect the EEGs of two participants at a time. In one of the room's most intriguing setups, two participants could sit on either side of a two-way mirror with red and green lighting arranged so as to subtly illuminate each participant's face when they produced alpha brainwave bursts that exceeded a preset threshold. Each participant would see their own face reflected in the two-way mirror when they produced sufficient alpha; but when the two produced simultaneous increases in alpha, their faces would appear to switch positions, so that each player would see their own face seemingly positioned on the other player's shoulders. The intended effect was to open the participants' con-

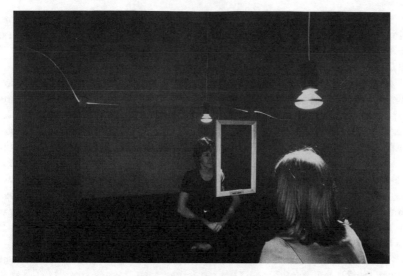

Fig. 4.2 Two museum attendees participating in Rosenboom's *Vancouver Piece* at the Vancouver Art Gallery in 1972

sciousness of self to enable them to explore ideas about shared identity. The result was strongly engaging. It was another example of contingent feedback (Fig. 4.2).

Around 1969–1970, Rosenboom organized a multi-agent biofeedback ensemble, called the *New York Biofeedback Quartet*. The idea was to gather a group that could practice biofeedback music together regularly. As personal circumstances played themselves out, this ensemble was short lived. However, it lead to several substantial biomusic compositions that have become regarded as early classics.

In 1972, Rosenboom created two works called *Portable Gold and Philosophers Stones (Music With Trills)* and *Portable Gold and Philosophers' Stones (Music From Brains in Fours)* (Rosenboom 1976). The score for the first piece describes a "dual-contingent" feedback system in which musical results depend on simultaneous theta or alpha brainwave activity. The electronic music equipment includes a device for generating sub-harmonically related tone complexes, the spectra of which are scanned with resonant band pass filters that are being tuned by the brainwaves. The resulting tones are mixed into an immersive electronic music texture that is broadened further via an accumulation tape-delay system. This piece was performed, though never recorded.

The second work, *Portable Gold and Philosophers Stones (Music From Brains in Fours)*, expanded the multi-agent BCMI paradigm in significant ways. This time, an ensemble of four biomusic performers is specified in the score. EEG signals from the four performers are routed and processed through a coarse-grained Fourier analysis device, in order to track several EEG frequency bands, and a correlation function computer that measures the coherence times of signal bursts in selected EEG frequency bands. At the time this piece was created, both these analysis functions were accomplished with analog equipment. A fifth performer operates the electronic equipment and routes the outputs of the analyzers to an electronic music generating system, which again includes a set of sub-harmonically related tone complexes. This time the tone complexes are fed into bank of resonant band pass filters, known as a *holophone*. The holophone idea was inspired by Longuet-Higgins's description of a scheme for non-local storage in the time domain, analogous to non-local storage in the spatial domain with holograms (Longuet-Higgins 1969). The performers know that as the coherence times of their selected EEG frequency band bursts increases—as measured by the correlation function computing circuits—, the range of their control over the holophone is also increased. Furthermore, by changing various time constants in the holophone circuitry, the detail of control they can affect is also increased. Thus, initially, slowly moving effects—gradually evolving, drone-like sounds, for instance—may become broader with more fast moving detail—wider pitch excursions with trill-like sounds, for example—as the corresponding performer's EEG band bursts become longer and smoother (Fig. 4.3).

Over years, *Portable Gold and Philosophers' Stones (Music From Brains in Fours)* has been performed many times. One particular performance from 1972 has been released and re-released on vinyl records, CDs, and digital distribution (Rosenboom 1975, 2000a, b, c, 2006, 2019a, b).

In the mid-1970s, Rosenboom's work shifted towards investigating what can be done with auditory event related potentials (AERPs) extracted from the EEGs of participants in a biofeedback paradigm. Of particular interest was how AERPs might

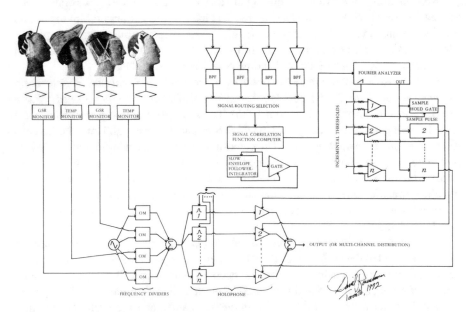

Fig. 4.3 Signal flow diagram from the score for Rosenboom's *Portable Gold and Philosophers' Stones (Music From Brains in Fours)*

provide information about attention shifts that could be related to various kinds of changes in an evolving sonic form. Through many experiments, this led to another major biomusical work called *On Being Invisible*, in which AERPs in a biofeedback scheme were used to guide the emergence of a self-organizing musical form. Full technical description of this project and the nature of AERPs is beyond the scope of this article. However, *On Being Invisible* has been documented extensively elsewhere, performed, and recorded (Rosenboom 1977, 1997, 2000a, b, c, 2019a, b). Mostly, this work was presented as a solo performance, not a multi-agent work, and the technical descriptions will not be presented here. However, it is worth mentioning that a follow-up piece, called *On Being Invisible II (Hypatia Speaks to Jefferson in a Dream)*, which used two performers in a multi-agent AERP feedback paradigm, was created and realized in 1994 (Rosenboom1997). Building on the first iteration of *On Being Invisible*, the artistic concept was partly to try to see if the idea could be extended so as to create a self-organizing opera, one in which the pathways through the opera's non-linear narrative would be guided by the AERPs detected from the two performers together. The performers would react primarily to auditory events and be shielded from visual stimulation; however, AERP events with strong P300 components—(a peak in the AERP, occurring approximately 300 ms after the onset of a highly differentiated stimulus event, that is commonly associated with aspects of attention)—would also be used to essentially edit sampled voices delivering bits of text and stored visual sequences for the audience. The results would be different in every performance. A recorded example is available (Rosenboom 2000b) and program notes are available online (Rosenboom 1994).

4.4   More Detailed Descriptions of Selected Recent Works Produced with Multi-agent BCI and Multi-agent BCMI Paradigms

4.4.1   *Ringing Minds*

Ringing Minds is a collaborative work created by David Rosenboom, Tim Mullen, and Alexander Khalil. A first version was produced and performed in 2014, and a detailed technical description was published in (Mullen et al. 2015). *Ringing Minds* is a complex multi-dimensional, multimedia, multi-agent BCI project in the arts which explores new possibilities in contingent and non-contingent feedback, concepts of "audience-as-performer," complexity and structural forms in music and the brain, and resonance within and between listeners and performers. *Ringing Minds* uses real-time "hyperscanning" techniques to model event related potentials (ERPs) and resonant properties of neural activity simultaneously measured from a group of individuals engaged in active imaginative listening during a live musical performance.

The EEG signal processing builds on multivariate principal oscillation pattern (POP or *eigenmode*) analysis methods for identifying resonant properties of a time-varying dynamical system. Each POP characterizes the response to a specific input of an independent, stochastically forced, damped harmonic oscillator or relaxator. Another way to think about the dynamics of a POP is as equivalent to an idealized string "plucked" with a specific force plus additive random excitation. POP analysis methods had previously been applied to multi-electrode electrophysiological data to identify characteristics of spatiotemporal oscillatory modes in single individuals (Mullen et al. 2012). For *Ringing Minds*, each of four participants' single-electrode EEG time series (sampled at the 10–20 Cz location) were instead treated as if generated by a common dynamical process—a "hyper-brain" sampled by four sensors. Within a sliding window, the multi-brain EEG time-series were decomposed into a set of forty POPs, spanning the EEG frequency spectrum. In this manner, each POP may be regarded as an extended neuronal process (e.g. a coherent network) spanning the four brains, oscillating at some frequency and/or exponentially decaying in response to an excitatory input (e.g. a musical event), or reflecting a resonant/synchronous state of this "hyper-brain". Each POP was characterized by seven dynamical parameters, including frequency, initial amplitude (excitation), and decay (damping) time, which were mapped onto a software-based electronic music instrument, the central core of which is a very large array of complex resonators. These respond to the POP data in a way that generates a vast, spatialized sound field of ringing components, analogous to ways neural circuits might also "resonate" and sustain modes of behavior within and between individuals. POP-to-resonator mappings were chosen to produce an aesthetic interpretation of the precise meaning of oscillator/relaxator for POPs. Periodically, the shapes and temporal positions of important peaks in ERPs, averaged across the four brains within a 1 s sliding window, were applied to modulate the resonant auditory field, sounding as if a stone had been tossed onto the surface of a sonic lake (Fig. 4.4).

A second version of *Ringing Minds* was produced and performed at the Whitney Museum of American Art in 2015 during a fifty-year retrospective of Rosenboom's work. In this version, the work was expanded with the collaboration of visual designers, Matt Wachter and Glenn Snyder, to include elaborate video projection displays showing components that paralleled the EEG analysis and music generation systems.

The concept for the visual display began with the idea that the POP resonances detected from the four brainwave performers were analogous to stones being dropped onto the surface of a still lake; and the nature of the ripples that spread out from the location of the stones impact on the water was analogous to the properties of each POP. This also paralleled how in the computer music instrument, the POPs were mapped onto a large array of complex digital sound resonators.

For each POP, a splash of color was displayed on a screen, forming a visual backdrop in the performance space. The spatial positions of the color splashes were determined by the dominant frequencies and spatial distributions of each POP. The vertical position was determined by a corresponding POP frequency, and the horizontal position was determined by the spatial distribution of the energy contributing to the POP across the four brainwave performers (Fig. 4.5).

Another new feature in the Whitney Museum performance of *Ringing Minds* was the addition of a contingent feedback component based on detection of contemporaneous ERPs amongst the participants. ERPs were extracted by averaging EEG signals, sampled at the standard 10–20 Cz electrode location, across the members of the N-Brain group, rather than the traditional method of averaging over succes-

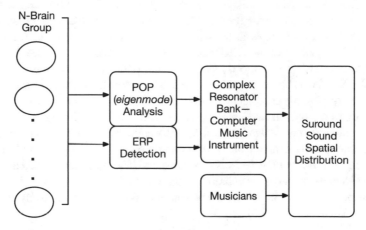

Fig. 4.4 Simplified diagram of the Ringing Minds system. Hyper-scanning analysis techniques are used with an N-Brain Group. Single-trial average ERPs (Event-Related Potentials) are captured via spatial averaging across the group, rather than the traditional approach of averaging across multiple trials (i.e. repeated events) within a single individual

Fig. 4.5 Image from the *Ringing Minds* performance at the Whitney Museum of American Art in New York in 2015 showing color splashes initiated by individual POP (eigenmode) detections from the N-Brain Group. Photo by Paula Court

sive occurrences of a stimulus in time. A simple template matching procedure was then used to detect the presence of a contemporaneous ERP response across the N-Brain group to unexpected auditory events. The group-averaged EEG time-series were convolved in a sliding window with an ERP template reflecting the average of ERPs from 48 individuals elicited in response to rare *deviant* (unexpected) tones interspersed within otherwise predictable sequences of *standard* tones which differed from the deviant tones in their duration. When the correlation between the N-Brain ERP and the template ERP exceeded a pre-determined threshold the shape of the detected ERP was mapped onto a musical pitch space and sent via MIDI to a Yamaha Disklavier™ grand piano. The ERPs were, thus, played automatically on a piano, providing contingent feedback to listeners, as well as to musicians, and adding to the overall musical experience. This occurred during sections of the music in which two musicians (Khalil and Rosenboom) played a lithoharp (a kind of xylophone made of carved stone bars) and an electronically processed violin, in interaction with the brainwave music performers. The musicians could attempt to create and violate musical expectation in the listeners and thereby elicit collective neural responses, which in turn would be sonified. For the brain artists, listening was the performative act (Fig. 4.6).

Fig. 4.6 Four listening brainwave performers participating in *Ringing Minds*. The shapes of their evoked responses (ERPs) are being played on the Yamaha Disklavier™ piano on the right. Photo by Paula Court

4.4.2 Portable Gold and Philosophers' Stones (Deviant Resonances)

Following his 2015 Whitney retrospective, Rosenboom composed a third piece in the *Portable Gold...* series, *Portable Gold and Philosophers' Stones (Deviant Resonances)*. It was premiered in The House performance space at Plymouth University, UK, as part of a BCMI Workshop in association with the 2015 Computer Music and Multimedia Research Conference. It has been performed many times since then and has recently been recorded for release (Rosenboom 2019a).

Portable Gold and Philosophers' Stones (Deviant Resonances) is structured for two active imaginative listening brainwave performers and a computer-electronics performer. It employs both non-contingent and dual contingent biofeedback paradigms. The brainwave performers' task is to remain still and listen actively with eyes closed, allow their attention to be drawn to any features of the sound texture, to actively direct their attention to specific sonic features they may choose, and to continuously notice when they observe that these listening actions may be related to how features of the sound texture evolve. If possible, they may also choose to direct features of the sound texture with their shifting attention; and, in any case, they are to practice immersing themselves in the sound texture and attempt to increase the degree to which they can actively interact with it. This is called *listening as performance*.

A complex software instrument was created for this piece using the Reaktor Core digital signal processing and synthesis platform. The instrument is designed to receive raw EEG signals from two performers via the OSC (Open Sound Control) data format. Each of the decoded EEG signals is parsed into three individual frequency bands, the upper and lower band limits of which may be freely set and adjusted. Each of the filter outputs is made available individually and is also fed into an amplitude envelope follower algorithm with adjustable time constants for rise and fall. The frequencies of the filter outputs are also tracked and provided as outputs. In addition, they are mapped onto a selected range of MIDI pitch values with controllable scaling and offsets. These signals may be used to *play* other modules in the overall, complex instrument. The envelope outputs are also made individually available for use by other modules. Other modules in the overall instrument will be described below, along with a tour through the compositional structure.

The composition begins with a sound texture created with sub-harmonically related tonal complexes made with pulse waves feeding banks of resonant band pass filters. Just as in 1972 with *Portable Gold and Philosophers' Stones (Music from Brains in Fours)*, this bank of resonant filters is referred to as a holophone, this time with refinements. The holophone can isolate and recall multiple tones from the several overlapping harmonic series in the pulse-wave chords that are fed into it. The pitches of the pulse wave chords can be determined in advance and may be changed during performance. (So far, performers have mostly chosen to keep them pretty stable, though this is not required by the composition.) The envelopes of selected EEG bands are patched into the holophone in a manner that gives them control over movement among the tones it produces. The performers are informed ahead of time about which musical voices from the holophone are responding to each of their EEG signals and where those sounds will tend to be located in a simulated 3D sound diffusion space. They know which sounds are responding to which performer and where in space they are likely to be heard. The ensemble may decide ahead of time, or in the moment of improvisation, exactly how to choose among the optional control paths from EEG envelopes to holophone movement. The electronics performer must carefully monitor and adjust the EEG signals received via OSC to make sure they fall within acceptable ranges, adjust the time constants of the envelope followers, and monitor signal flow into the holophone. This section unfolds slowly, usually starting simply and growing more complex as the performers settle into the nature of the exercise.

At a certain point, the electronics performer may choose to activate a second and eventually a third layer of sound elements enabled by the overall instrument. This is analogous to activating second and third sections of an electronic orchestra, all of which are responding to signals from the brainwave performers, while also being guided by the electronics performer. Before describing how these sections evolve in a musical performance, some further technical description is needed. Refer also to the Fig. 4.7.

The amplitude envelope values from each performer's filtered EEG are patched into a module called EEG2MIDI, the algorithm of which includes a delta function that responds when the rate of change in the amplitude envelope exceeds a delta threshold set by the electronics performer. When the signal exceeds the delta threshold, its value is mapped onto an adjustable range of MIDI pitch values and a scalable range of MIDI

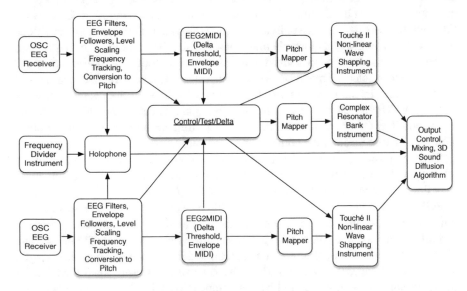

Fig. 4.7 Signal flow among primary components of the *Portable Gold and Philosophers' Stones (Deviant Resonances)* system

velocity values. These MIDI pitch values are sent to a pitch-mapping module and used as index values into a lookup table of pitches. This table contains data defining four musical scales that have been pre-composed to contain particular pitch interval sequences: (1) a scale made with pentagonal numbers and interval sequences in numbers of semitones (4 5 2 …), (2) interval sequences that do not repeat at the octave (1 1 1 1 2 2 1 1 1 2 …), (3) interval sequences based on hexagonal numbers (1 2 …), and (4) superimposed sub-harmonic and harmonic sequences on the pitch C.

Three such pitch mappers are used in the system. The outputs of the first two are patched to two very particular synthesis instrument modules, one corresponding to each performer. These instrument modules are known as Touché II. The original Touché digital keyboard instrument that was developed by Donald Buchla in collaboration with David Rosenboom in 1978–1979 inspired their design. Both instruments employ nonlinear wave shaping algorithms as their primary sound synthesis method. Linked to this, and a key to the instruments' success, is the ability to program many, very complex control envelopes of arbitrary length, with decision logic that can be applied to every breakpoint in an envelope design. These envelopes can be used to modulate every parameter in the instrument's synthesis algorithms. The original Touché was a hybrid, digital-analog hardware instrument. The Touché II, developed in 2007 by Rosenboom in collaboration with Martijn Zwartjes, is an entirely software-based instrument. Describing its full technical details lies beyond the scope of this article. Suffice it to say that it is an extraordinarily powerful, live electronic music instrument that can traverse a tremendously rich sonic terrain.

The third pitch mapper is connected to another interesting instrument made of a bank of complex digital resonators, with both deterministic and stochastic control functions. These resonators are activated by a set of exciter functions with variable slope controls and ways to inject indeterminacy with various kinds of noise into their behaviors. One can think of these digital resonators as complex bells that can be rung by complex exciter functions. The design of these resonator banks began with the composition of *Ringing Minds*. Those used in *Portable Gold and Philosophers' Stones (Deviant Resonances)* are derived from those used first in *Ringing Minds*, though they are slightly simplified and reduced in number for practical reasons.

To recapitulate, the electronics performer can activate and deactivate what may be thought of as three layers or sections in an electronic orchestra. The first is the one described above with sub-harmonic pulse wave complexes and the holophone's resonant band pass filter bank. The second is the complex digital resonator bank with its exciter algorithms. The third consists of the two Touché II synthesis instruments, each of which is also preprogrammed with an array of preset algorithms that can be called up instantly with MIDI program change signals.

A module in the system called Control/Test/Delta is key in managing a performance. It enables the routing of control information and triggers for sounds around the instruments in the electronic orchestra in several ways. First, direct EEG pitch tracking from either brainwave performer can be selected and routed to the Touché II instruments and/or the complex resonator bank. Second, EEG amplitude envelope delta threshold crossings from either performer can be selected and routed to the

Touché II instruments. When this is done, each delta threshold crossing will initiate a program change in its corresponding Touché II instrument. Delta threshold crossings can also be merged so that signals from both performers affect both Touché II instruments. Finally, and very importantly, a function can be selected which only routes program change signals to the Touché II instruments when delta threshold crossings from both performers occur at precisely the same time. This introduces the contingent feedback paradigm into a performance.

A word about these delta threshold crossings and their possible relation to attention shifts is pertinent here. It is commonly thought that when a subject is in the process of producing increasing amounts of coherent brainwave signals, such as highly coherent alpha waves, the interruption of these signals can often accompany a significant shift of attention in that subject. The instrument developed for this piece includes an ability to react to the rapid onset of alpha or other coherent brainwave bursts, and also to their quick interruption. In addition, it can respond when changes of this type occur simultaneously in the two performers. In the highly controlled conditions of a laboratory, the validity of these assumptions can be tested. In the environment of a brainwave music performance, they are considered to be very interesting phenomena to explore when intensively engaged in a biofeedback music environment.

It is also important to underscore that it may not be productive to engage with a brainwave music instrument of this complexity casually. It is, rather, an instrument to be mastered with extended practice, just as one might practice any musical instrument with high levels of discipline to achieve mastery. In addition, experience shows that performers experience more rewarding results when they are truly able to engage in active imaginative creative listening. Often such individuals bring prior experience in sound arts and/or music to bear, and perhaps, also techniques for meditation. With this in mind, it is important to differentiate among multi-agent BCI or BCMI designs that require extensive practice to achieve the desired results and those that do not. It is perfectly possible and legitimate to design experiences that are not based on practice, which can generate enriching experiences for participants who do not bring particular kinds of experience to bear—for example in installation-based or audience participation works. Those that do require extensive practice are made with different intentions. Both offer fertile territory to explore.

To date, typical performances of *Portable Gold and Philosophers Stones (Deviant Resonances)* have proceeded in a kind of arch form, beginning simply with the holophone layer, then adding the other sections, one at a time, until a peak of musical complexity is reached. Then, one by one, the layers are often reduced in intensity and eventually deactivated, until an ending section is reached with textures reminiscent of the opening. Optionally, the electronics performer might play an auxiliary instrument to interact with the brainwave performers, particularly in the central section when the contingent delta threshold detections are active. For this purpose the composer has often used an electric violin to trigger analog circuits designed to exhibit somewhat unstable, chaotic behaviors (Fig. 4.8).

Fig. 4.8 Image from a performance of *Portable Gold and Philosophers' Stones (Deviant Resonances)* at Fleet Science Center in San Diego presented as part of San Diego Art Institute's AMT (Art, Music, Technology) Festival in 2017. Rosenboom is seen behind active imaginative listening brainwave performers Susanne Thorpe (l) and Bonnie Jones (r), founders of TECHNE, an education organization emphasizing gender equity and social justice in arts and technology. Photo by Tom Erbe

4.4.3   The Experiment from Hopscotch

In the fall of 2015, an extraordinary opera, called *Hopscotch–a mobile opera for 24 cars*, was produced by The Industry, an opera production company in Los Angeles (The Industry 2015). In this extraordinary project, directed by Yuval Sharon, a story was presented in a non-linear fashion as audience members were driven around Los Angeles in 24 limousines along three different routes. Various scenes in the opera were performed inside each limousine and at specific, iconic locations in the Los Angeles cityscape. Audience members would experience the scenes of the opera in different orders, depending upon which route they were on and at what location they began their journey. Rosenboom was one of five principle composers commissioned to create music for various scenes in the opera. *The Experiment* was one of those scenes.

The Experiment was performed inside one of the *Hopscotch* limousines. In this scene, as audience members entered their limo, they heard spoken and sung explanations, accompanied by electronic backgrounds, of what they were about to experience, while individual brainwave monitors were affixed to their heads. As the scene unfolds, one of the opera's principle characters, Jamison, pursues an obsession with understanding the nature of consciousness by singing eleven questions to the audience that progress in nature from seemingly innocent inquiries to somewhat more confrontational probing. Concurrent patterns among the brain signals of the audience members are then detected with signal analysis techniques and used to gauge their collective responses to each question. The results are translated into an immersive mix of soprano voices that sing three possible answers for each question with different musical qualities, representing: (1) an agitated state, (2) shifting attention or alertness, and (3) being focused on one's inner self and disinterested. These were presumed to come from the inner group psyche of the audience. In the end, instead of finding the answers he seeks, Jamison snaps. After an extraordinarily successful performance run of *Hopscotch* in 2015, Rosenboom created a concert version of *The Experiment*, a recording of which is available (Rosenboom 2019a), and a score for which is published online (Rosenboom 2015). Writer Erin Young wrote the texts. A relatively detailed technical description follows.

Raw EEG signals from four audience members are recorded with Muse™ brainwave monitoring headbands and transmitted to a computer via Bluetooth. The signals are received via the Muse–I/O program and sent to software written with the Reaktor Core digital signal processing and synthesis platform using the OSC (Open Sound Control) protocol. The raw signals from each audience member are parsed into three commonly used, brainwave frequency bands: theta (5–8 Hz), alpha (9–13 Hz), and broadband beta (14–30 Hz). Amplitude envelopes of the resulting twelve bands (three for each of four audience members) are detected with variable time constant, low frequency envelope followers. The four envelopes corresponding to each EEG filter band are then averaged to produce collective audience envelopes for theta, alpha, and beta EEG frequencies.

Prior to the performance, eleven sets of vocal parts, corresponding to each of the character Jamison's eleven questions were recorded by a soprano. Within each set, three kinds of answer texts were also set for the solo soprano voice and recorded. All the recordings were stored in a computer. In a performance, after Jamison sings a question, he pauses. During the pause, the averaged audience EEG frequency band envelopes are used to control the playback amplitudes of the prerecorded answers to that question, sung by the soprano. The dramatic operatic result is that the brainwaves control an audio mix of the three types of answers for each question, as specified in the narrative: theta controls the singing related to an imagined condition of being focused on one's inner self and disinterested, alpha to shifting attention or alertness, and broad-band beta to an agitated state. A performance then proceeds through the sequence of eleven sung questions and eleven mixes of multiple soprano voices (all recorded by a single vocalist), the qualities of which are modulated by the collective brainwaves of the audience members (Fig. 4.9).

What has been described thus far is an example of non-contingent, multi-agent feedback. However, a contingent, multi-agent feedback mode is also included in *The Experiment*. The alpha frequency amplitude envelopes are also connected to a delta function detector, much like that described above for *Portable Gold and Philosophers' Stones (Deviant Resonances)*. If the absolute value of the rate of change of the envelope signal crosses a settable threshold, a trigger signal is generated. A *hold* time can also be adjusted to determine how much time must pass after detections

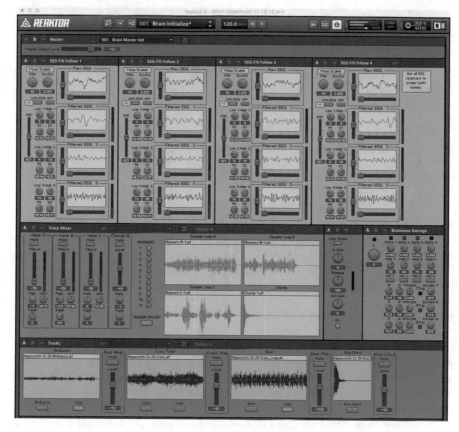

Fig. 4.9 User interface showing software control panels for *The Experiment*

before subsequent detections can be made. The system will respond both to the onset of rapid alpha bursts and to the sudden interruption of alpha bursts, depending on the hold time selected. Finally, the system tests to see if triggers from all audience members occur at the same time. If they do, the triggers from all four are simultaneous, then a special electronic chord is sounded over the mix of the soprano voices. This chord is meant to signal the possibility of a simultaneous shift in the group mind of the audience, whatever that might ultimately mean in their experiences.

4.4.4   Concurrent Complexity

Uncertainties in stimulus detection, uncertainties in response measurements, and uncertainties in the generation of feedback responses exist at some level in all interactive systems, in both research setups and interactive art works. Deviant resonances

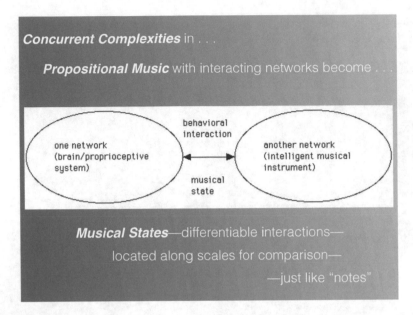

Fig. 4.10 Concurrent complexities among networks as musical states

emerge in all systems of differentiation among presumably identified component parts: in the instruments of technology, natural organisms, energy fields, time and space, to name a few. Often, the space in which deviant resonances emerge is where the greatest interest lies, in both the making of art and in the refining of theoretical models. Often new understandings of complexity emerge from deviant resonances. As many master musicians have often noted, the real interest in music lies not in the notes, but in the spaces between the notes.

At this stage in the development of multi-agent BCI in the arts, it may be productive to approach biofeedback and related pursuits as interactions among complex systems. In propositional music, it can be useful to describe musical *states* as particular behavioral interactions among networks, for example among a brain/proprioceptive system and an artificially intelligent musical instrument. These musical states become differentiable interactions located along scales for comparison, just as if they were musical *notes*. One may compose with these *states* as *notes* (Fig. 4.10).

One interesting subject high on the agenda for future development in the authors' work is investigating ways to correlate the complexity of a stimulus environment with the complexity of EEG signals and apply the results in feedback paradigms. Already, preliminary evidence indicates that correspondences among the dimensional complexity of brain activity and the complexity of music stimuli may be observable (Birbaumer et al. 1996). Previously, Rosenboom has experimented with complexity measures applied to musical parameters in some compositions (Rosenboom 1992, 1996, 2000a). In order to make progress in this realm, more work is needed on how to refine the meaning of complexity and ways to measure it. This is a common goal in many fields now, where analysis of complexity is of interest. Also important to this is continuing work on refining paradigms for sonification of biological data in feedback paradigms, especially where the phenomena of interest are emergent (Choi 2018; Novello 2012; Rosenboom 1997, 2003, 2014).

4.4.5 MoodMixer

The *MoodMixer* project was created in 2011 by Grace Leslie and Tim Mullen and explores new possibilities for multi-agent BMCIs that respond to, and influence, the mental state of multiple participants. There are three distinct versions of the system. The first of these was presented at the 2011 New Interfaces for Musical Expression conference in Oslo, Norway, with subsequent realizations presented at various venues between 2012 and 2014. The project is described in detail in (Leslie and Mullen 2011) and (Mullen et al. 2015).

MoodMixer employs a non-contingent audiovisual BCI system which reflects, and expands on, elements of much earlier works described in this chapter, particularly Humbert's 1974 *Brainwave Etch-A-Sketch*. Two normalized cognitive (e.g. focused attention, relaxation) or affective (e.g. arousal, valence) state indices are simultaneously and continuously calculated from each participant's EEG. These define a set of coordinates within a two-dimensional mental state space. The locations of all participants in the state space determine the evolution of a music composition, either through a dynamic spatial quadrophonic mix (*MoodMixer 1.0* and *3.0*) or an algorithmic composition procedure reminiscent of John Adam's *Phrygian Gates* piano piece (*MoodMixer 2.0*). In each of its three instantiations, a visual display also provided real-time feedback on the participants' individual and/or combined states. *MoodMixer* explores concepts of both collaborative and competitive, as well as active and passive, approaches to real-time EEG-based music generation within a multi-user design that promotes social interaction in the experience of the installation (Fig. 4.11).

4.4.6 Assembly Cognogenesis

Assembly Cognogenesis is a multi-agent BCI work created by Sheldon Brown and his lab at the Arthur C. Clarke Center for Human Imagination, in collaboration with Tim Mullen. *Assembly Cognogenesis* is a shared virtual reality environment in which two users use neural and gestural interfaces to collaborate within an artificial life world and cultivate the symbiotic relationship between imagination, engagement, and the evolving environmental system. A first version of the installation debuted in 2015 as part of the Mozart and the Mind festival in San Diego with a subsequent version shown at the Filmatic festival in 2016. A short demonstration video can be found in (Brown 2016).

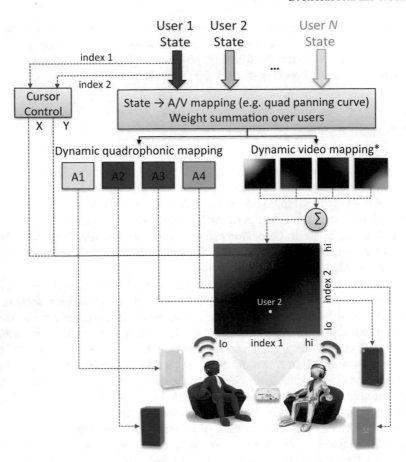

Fig. 4.11 Architectural diagram of the MoodMixer installation and its typical dual-user quadrophonic instantiation. A two-dimensional neural state-space is explored simultaneously by two or more users. A1–A4 represent four dynamically mixed audio tracks each composed to reflect an extremum of the state space. For MoodMixer 1.0 and 2.0, the users' positions in the state space are visually represented by a moving dot superimposed on a weighted sum of four colored spatial gradients. *For MoodMixer 3.0, this video mapping was replaced with dynamic blending of video footage from *Four Stream Mind* by Grace Leslie and Maxwell Citron

Assembly Cognogenesis is based on the *Assembly* emergent behavior platform created by Sheldon Brown and his lab (Brown 2015). In *Assembly*, collections of entities evolve over time in relationship to their environment and each other, with guidance provided by one or more viewers. In *Assembly Cognogenesis*, both contingent and non-contingent BCI elements were added to the system. Two participants are situated within a common environment, but with each person occupying a vastly different spatial scale. EEG power spectral measures associated with attention or engagement are calculated for each participant. Hand movements and gestures are tracked, enabling gestural interaction within the virtual environment. In one instantiation, the environment consists of a unicellular amoeba-like organism inhabited by hundreds or thousands of 'molecular' entities, which interact and evolve according to a set of rules akin to the chemistry and physics of a real world environment. One participant resides outside the organism, at the "macro" spatial scale, and can manipulate the shape of the amoeba using their hands. An energy field surrounding the organism increases in intensity proportional to the participant's engagement or attention measure. When this measure is sufficiently large, the participant can direct energy to specific locations within the organism. The second participant resides inside the organism, at the "micro" spatial scale, and can interact directly with the entities. When this participant's neural measure is sufficiently large, they can use gestural interactions to channel energy available at their location into one or more entities. This causes the selected entities to reproduce and evolve at an increased rate, propagating these entities' traits and enabling new variations to emerge. New possibilities within the environment thereby only emerge when a sufficient degree of attention or engagement is contemporaneously attained by both participants. In one variation, participants could also cooperate to evolve and guide the entities to where they could pass through the organism's membrane and transit outside the organism, at which point the environment would reset and participants would reverse roles.

In *Assembly Cognogenesis*, the participants must learn to cooperate across two vastly different levels of description, by observing, and responding to the effect of each other's actions on their shared environment. In doing so, they must maintain a common mental representation and goal structure, while also maintaining a common neural state. Contingent feedback facilitates maintenance of this common state, enabling life to evolve and new creative possibilities to emerge within the virtual world.

Assembly Cognogenesis incorporates a number of cutting-edge technologies including virtual reality, hand and gesture tracking, unobtrusive wearable EEG sensing, and (for MATM and Filmatic exhibits) real-time distributed computation of neural measures in the cloud using NeuroScale™. These afford possibilities for immersion, mobility, and scalability that would have been impossible twenty years ago. Nonetheless, one must appreciate the thematic parallels between this work and the earliest historical multi-agent BCI systems described in this chapter—for instance Rosenboom's *Alpha Checkers*. Between these and many other works spanning nearly five decades, contingent feedback plays a similar and central role in establishing and coordinating interaction and cooperativity between multiple agents. From these contemporaneous interactions, interesting new emergent behaviors and perspectives may arise as individuals within the group learn to function as a cohesive unit and indeed become more than one (Fig. 4.12).

4.5   Conclusions

The highly interdisciplinary terrain in which multi-agent BCI in the arts and multi-agent BCMI reside is populated now by a growing and wide-ranging field of practitioners, who are exploring very interesting phenomena, making stimulating works

Fig. 4.12 Image from a public exhibition of *Assembly Cognogenesis* in Sheldon Brown's lab at UC San Diego's Arthur C. Clarke Center for Human Imagination as part of the 2016 Filmatic festival. The participant on the left is occupying the macro-scale position outside the organism, while the participant on the right is occupying the micro-scale position inside the organism. The two views of the environment can be seen on the respective displays. Photo by Tim Mullen

of art, and illuminating a landscape of *artscience* investigations. It has not been the purpose of this chapter to provide a survey of this work. Rather, by concentrating on examining the environment within which some of these ideas emerged, particularly those focused on the multi-agent concept, and describing a few examples of recent realizations, the authors hope to contribute to the growing literature guiding the evolution of this field.

The multi-agent artistic BCI and BCMI systems we have described involved relatively few agents. However, the availability of cost-effective, wearable technology for measuring brain and body activity (Liao et al. 2012), as well as the emerging use of mobile computing, and scalable cloud and fog computing for BCI (Zao 2014a, b; Intheon 2018) create new possibilities for large-scale multi-agent artistic BCI systems. At the *Regen3* event in 2003, alpha-band EEG activity from 48 participants was simultaneously measured and used to control musical parameters of a jazz performance (Mann 2007). The *My Virtual Dream* installation at the 2013 Scotiabank Nuit Blanche arts festival in Toronto (Kovacevic et al. 2015) situated groups of 20 participants at a time (a total of 523 active participants over the duration of the 12-h event) within an 18 m geodesic dome accompanied by 360° projections of dream-like artistic visuals and soundscapes driven by the collective brain activity of all participants. In a 2017 event organized by Terra Mater Factual Studios, EEG activity from several hundred individuals in two movie theaters in Los Angeles and New York City were simultaneously measured, decomposed into spectral components and correlated in near real-time across the group using cloud computing. These measures were in turn used to drive a visual feedback display telecast to the audience members. These are but a few examples of emerging possibilities for multi-agent BCI systems as biosensing and computing technologies continue to advance.

Clearly, the idea of multi-agent BCI in the arts stimulates the imagination and suggests potentially rich paradigms for both disciplined research and imaginative exploration. Indeed, perhaps this is a place where science and art can meet in significant theoretical territory and in new avenues for materializing ideas. More generally, one can view artistic and scientific domains as complementary systems for investigating and understanding the nature of reality. Art provides a propositional "sandbox" in which one can freely explore what is *possible*—a realm in which new concepts and systems can be flexibly created and prototyped, without necessarily demanding a rigorous explanatory foundation, or even physical realization. Science can expand on such concepts to develop and rigorously test hypotheses, produce empirical evidence regarding what is *probable*, and ultimately enable the realization of some of these concepts within the physical world. Art in turn, can leverage scientific knowledge and discovery as a basis for further ideation and exploration, creating a virtuous cycle.

Thus, art and science support each other as co-evolving forms of practice and discipline. In most arenas of human exploration, balances shift among how propositional and empirical modes of speculation and verification are emphasized, a normal response to how any given era sharpens its focus on goals, aspirations, and practical needs. Decades ago, as imaginations were fueled by new waves of discovery in the fields discussed in this chapter, a spirit of futuristic optimism emerged that might be captured by this quote from *Biofeedback and the Arts*, "Through the use of computers as appendages of man's brain and methods of learning with biofeedback, rates of information processing will be achieved that approach the speed of light, ergo, conception will be bound less necessarily with action, elicited or observed, and life will eventually be embodied in information-energy networks creating non-physical art; spiritual art will be revived as established networks connect us firmly." (Rosenboom 1972).

In subsequent decades, scientific research has provided us an increasingly clearer understanding of the means to extend human cognition and communication beyond the central nervous system using neurotechnology. Advances in electromagnetic sensing and stimulation, optical physics, nanotechnology, and biocompatible materials are yielding new possibilities for measuring and modulating brain activity at far greater spatiotemporal resolution than previously possible. We have continued to increase our understanding of both the practical utility, as well as the limitations, of various forms of bio/neuro feedback and closed-loop neuromodulation. Although we have yet a great deal to learn about brain structure and function, and the neuroscience and neurotechnology fields are still relatively embryonic, artistic applications of neurotechnology provide a means for us to envision, explore, and discuss possible roles and implications for such technology within present, near future, and far future societies.

References

Berlyne DE (1971) Aesthetics and psychobiology. Appleton-Century-Crofts, New York

Birbaumer NW et al (1996) Perception of music and dimensional complexity of brain activity. Int J Bifurc Chaos 6:267–278

Brown S (2015) Assembly. Arthur C. Clarke Center for Human Imagination. http://www.imagination.ucsd.edu/assembly/

Brown S (2016) Cognogenesis. YouTube, 8 May 2016, youtu.be/MFno-ww8XV0

Choi I (2018) Interactive sonification exploring emergent behavior applying models for biological information and listening. Frontiers in neuroscience, vol 12, Neuroprosthetics. https://www.frontiersin.org/article/10.3389/fnins.2018.00197

Fehmi L, Rosenboom D (1971) Group contingent feedback. Talk given at the Spring 1971 Convention of the Association for Humanistic Psychology, Washington, D.C.

Holland JH (1995) Hidden order–how adaptation builds complexity. Addison-Wesley Publishing Company, Reading, MA

Holt J (2018) When Einstein walked with Gödel–excursions to the edge of thought. Farrar, Straus and Giroux, New York

Ilstedt Hjelm S, Browall C (2000) Brainball–using brain activity for cool competition. In: Proceedings of NordiCHI 2000, pp 177–188. http://www.mindball.pl/pdf/brainballChi.pdf

Intheon (2018) NeuroScale, www.neuroscale.io

Kaufman S (2000) Investigations. Oxford University Press, New York

Kelso JAS (1995) Dynamic patterns—the self-organization of brain and behavior. The MIT Press, Cambridge, MA

Kovacevic N, Ritter P, Tays W, Moreno S, McIntosh AR (2015) 'My Virtual Dream': collective neurofeedback in an immersive art environment. PLoS ONE 10(7):e0130129

Leslie G, Mullen T (2011) MoodMixer: EEG-based collaborative sonification. Proceedings of the international conference on new interfaces for musical expression. Department of Musicology, University of Oslo, Norwegian Academy of Music, pp 296–299

Liao L-D, Lin C-T, McDowell K, Wickenden AE, Gramann K, Jung T-P, Ko L-W, Chang J-Y (2012) Biosensor technologies for augmented brain–computer interfaces in the next decades. In: Proceedings of the IEEE 100, special centennial issue, no. (May 2012), pp 1553–1566

Longuet-Higgins HC (1969) The non-local storage and associative retrieval of spatio-temporal patterns. In: Leibovic KN (ed) Information processing in the nervous system. Springer, New York

Mann S, Fung J, Garten A (2007) DECONcert: bathing in the light, sounds, and waters of the musical brainbaths. In: Proceedings of the 2007 international computer music conference (ICMC2007), vol 2, August 27–31, 2007. Copenhagen, Denmark, pp 204–211

Mullen T, Worrell G, Makeig S (2012) Multivariate principal oscillation pattern analysis of ICA sources during seizure. In: Proceedings of the 34th annual international conference of the IEEE, EMBS San Diego, CA

Mullen T et al (2015) MindMusic: playful and social installations at the interface between music and the brain. In: Nijholt A (ed) More playful user interfaces. Interfaces that invite social and physical interaction. Gaming media and social effects. Springer, Singapore, pp 221–229. ISBN 978-981-287-545-7, ISBN 978-981-287-546-4 (eBook)

Novello A (2012) From invisible to visible, the EEG as a tool for music creation and control. Thesis. https://www.researchgate.net/publication/329072229

Rosenboom D (1972) Homuncular homophony. In: Rosenboom D (ed) Biofeedback and the arts, results of early experiments. Aesthetic Research Centre of Canada Publications, Vancouver

Rosenboom D (1975) Brainwave music. Vinyl record. Aesthetic Research Centre of Canada, Maple, Ontario

Rosenboom D (ed) (1976) Biofeedback and the arts, results of early experiments. Aesthetic Research Centre of Canada Publications, Vancouver

Rosenboom D (1977) On being invisible. Vinyl record. Music Gallery Editions, Toronto

Rosenboom D (1992) Complex adaptive systems in music. (Abstract from invited presentation.) J Acoust Soc Am 92(4), (2):2403

Rosenboom D (1994) Music from–On Being Invisible II (Hypatia Speaks to Jefferson in a Dream). https://davidrosenboom.com/compositions-19912000

Rosenboom D (1996) B.C.–A.D. and two lines, two ways of making music while exploring instability, in tribute to Salvatore Martirano. Perspect New Music 34(1):210–226

Rosenboom D (1997) Extended musical interface with the human nervous system: assessment and prospectus. Revised electronic monograph: https://davidrosenboom.com/writings/. (Original (1990), San Francisco: Leonardo, Monograph, 1)

Rosenboom D (2000a) Invisible gold, classics of live electronic music involving extended musical interface with the human nervous system. Audio CD. Pogus Productions 21002-2. Chester, New York

Rosenboom D (2000b) Music from–On Being Invisible II (Hypatia Speaks to Jefferson in a Dream), on transmigration music. Audio CD. Centaur Records, Inc., Consortium to Distribute Computer Music, vol. 30, CRC 2940, Baton Rouge, LA

Rosenboom D (2000c) Propositional music: on emergent properties in morphogenesis and the evolution of music, essays, propositions, commentaries, imponderable forms and compositional methods. In: Zorn J (ed) Arcana, musicians on music. Granary Books/Hips Road, New York, pp 203–232

Rosenboom D (2003) Propositional music from extended musical interface with the human nervous system. In: Avazini G et al (eds) The neurosciences and music, annals of the New York academy of sciences, vol 999. New York Academy of Sciences, New York, pp 263–271

Rosenboom D (2006) Brainwave music 2006. Audio CD. EM Records EM1054CD, Osaka, Japan

Rosenboom D (2014) Active imaginative listening – a neuromusical critique. Frontiers in neuroscience, auditory cognitive neuroscience, vol 8, The Musical Brain. http://journal.frontiersin.org/article/10.3389/fnins.2014.00251

Rosenboom D (2015) The experiment–from Hopscotch. Score available online: https://davidrosenboom.com/compositions

Rosenboom D (2018) Propositional music of many nows. In: Bogdanovic D, Bouvier X (eds) Tradition and synthesis–multiple modernities for composer-performers. Doberman-Yppan, Lévis, Québec, Canada, pp 121–142

Rosenboom D (2019a) Deviant resonances—live electronic music with instruments, voices & brains. Audio CD and digital distribution. Ravello Records RR8009, North Hampton, NH

Rosenboom D (2019b) Brainwave music. Double vinyl records. Black Truffle Records. http://www.blacktrufflerecords.com

The Industry (2015) Hopscotch a mobile opera for 24 cars. http://hopscotchopera.com. Los Angeles

Vidal JJ (1973) Toward direct brain-computer communication. Annu Rev Biophys Bioeng 2:157–180

von Foerster H (1981) Observing systems. Intersystems Publications, Seaside, CA

Zao JK, Gan T-T, You C-K, Mendez SJR, Chung C-E, Wang Y-T, Mullen T, Jung T-P (2014a) Augmented brain computer interaction based on fog computing and linked data. In: Proceedings—2014 international conference on intelligent environments, IE 2014

Zao JK, Gan T-T, You C-K, Chung C-E, Wang Y-T, Méndez SJR, Mullen T et al (2014b) Pervasive brain monitoring and data sharing based on multi-tier distributed computing and linked data technology. Front Hum Neurosci 8(June):370. January 2014

Chapter 5
Evaluating BCI for Musical Expression: Historical Approaches, Challenges and Benefits

Duncan A. H. Williams⊚

Abstract A recurring challenge in the use of BCI (and more generally HCI) for musical expression is in the design and conduct of appropriate evaluation strategies when considering BCI systems for music composition or performance. Assessing the value of computationally assisted creativity is challenging in most artistic domains, and the assessment of computer assisted (or entirely computer generated) music is no different. BCI provides two unique possibilities over traditional evaluation strategies: firstly, the possibility of devising evaluations which do not require conscious input from the listener (and therefore do not detract from the immersive experience of performing, creating, or listening to music), and secondly in devising neurofeedback loops to actively maneuver the creator or listener through an expressive musical experience. Music offers some unusual challenges in comparison to other artistic interfaces: for example, often it is made in ensemble, and there is evidence to suggest neurophysiological differences are evident in ensemble measurement when compared to solo performance activities, for example see (Babiloni et al. in cortex 47:1082–1090, 2011). Moreover, a central purpose of music is often to incite movement (swaying, nodding head, dancing)—both in performer and audience—and as such this also offers up challenges for BCI/HCI design. This chapter considers historical approaches as well as making proposals for borrowing solutions from the world of auditory display (also referred to as sonification) and psychoacoustic evaluation techniques, to propose a hybrid paradigm for the evaluation of expression in BCI music applications.

Keywords Music · Sound · Sonification · Multi-criteria decision aid

D. A. H. Williams (⊠)
University of York, York, UK
e-mail: Duncan.williams@york.ac.uk

© Springer Nature Switzerland AG 2019
A. Nijholt (ed.), *Brain Art*,
https://doi.org/10.1007/978-3-030-14323-7_5

5.1 Introduction

Music has been described as a language for emotional expression (Lin and Cheng 2012) and is comprised of both communication, and interaction. Music allows communication from the composer or performer to an audience of listener(s), and symbiotically between performer and audience, and performer(s). Music as a form of artistic expression is ubiquitously popular, perhaps because listeners need no special musical training to enjoy or understand musical expression (Bailes and Dean 2009; Bigand and Poulin-Charronnat 2006). BCI and other biophysiological sensor techniques have gradually been adopted by the research community involved in the design of new musical instruments, music information retrieval, and computationally-assisted musical creativity (for example, algorithmic composition systems, automated accompaniment systems and the like). One term gaining traction amongst the community for this field is Brain-Computer Music Interfacing, or BCMI, (Miranda and Castet 2014), though this does not tend to include the full range of possible musical experiences, and indeed multimodal sensors are more commonplace (motion tracking, galvanic skin response, heart rate measurement) than BCI alone. Computational creativity is an emerging field, and, like BCMI, does not have established methodologies for robust evaluation. Simply put, a BCI generated composition may be designed by engineers or composers, but then be unobjectively rendered, or perhaps explored with 'Turing style' testing to establish convincing algorithms. The potential use of BCI to offer meaningful and responsive control signals for music generation has yet to be fully realized, though BCI has been used by some to adapt the design of generative music techniques that respond to brain signals. For example, to offer platforms for music making to improve the lives of people with physical disabilities, as well as in the more common design of applications for artistic purposes.

In these contexts, BCI offers some unique possibilities over traditional music making, particularly in the design of expressive systems with emotionally-congruous mappings between brain derived control signal, and musical feature selection or performance. In the long term, this may be useful for commercial applications, functional music selection, and to provide tools for individuals with particular communicative problems to create aids for communicating emotional state (e.g., people with Asperger's syndrome).

This chapter will provide a brief review of systems for BCI and music, before considering the challenges involved in the design of such systems and the need for specific and context dependent evaluation methodologies. Therefore for our purposes we will assume that You the reader will already have experience with the vast majority of the particular BCI methods described here. Musical applications tend to borrow from and build upon existing robust strategies, rather than improve upon or develop new hardware or software methods for measurement. For example the P300 or 'oddball' paradigm, the use of steady-state visually evoked stimuli (SSVEP), asymmetry measurement and filter based techniques from electroencephalogram (EEG) measurement, as well as hybrid systems incorporating the other biophysiological measurements above, and traditional psychometric evaluation techniques (self assessment in both qualitative and quantitative domains). Here, there are advantages in terms of unconscious response, the potential for neurofeedback, and designing a sense of agency over the performance, and music specific challenges including the tendency of music to induce a motile response (dancing, head nodding), or a common tendency to be designed for ensemble performance (such as live music concerts).

Pleasingly, a number of the challenges presented by traditional evaluation of computationally assisted creativity in music might actually be solved by the use of BCI. Typically evaluation is rare, or might be simplistic (did the audience enjoy a performance, did the music 'sound good', did a recording sell well, or similar questions which are highly variable and subjective). Therefore we conclude with some suggestions for future evaluation strategies which borrow from the world of auditory display (often simply called sonification). There are several examples of music created by means of sonifying EEG (or other biophysiological) data, either in real time or through more complex systems.

Some suggestions for further work are also volunteered, including development of collaborative platforms for music performance by means of BCMI. The field, though small at first glance, is steadily growing, and this chapter focuses on a discrete group of research in the context of the field—inclusive but by no means exhaustive—a great variety of existing work is taking place at the time of writing. Music remains an exciting and challenging application, particularly at this time, for the BCI community.

5.2 Historical Review and Possibilities for BCI in Music Making

BCI for music making is not common amongst music technologists, instrument designers, and the like, in comparison to the large research communities actively engaged in new musical instrument or music information retrieval problems. Nevertheless the community investigating the use of BCI for music has slowly gained traction over the past two decades. Typical systems might analyse a real-time input, subject it to a range of signal processing (perhaps filtering or more complicate statistical reductions) and use the resulting signal to choose or create from scratch a musical stimulus. The potential for such systems includes provision of aesthetic communication tools through music for users who are not musically confident or trained in performance to a level where they might engage in traditional music-making (Clair and Memmott 2008; Fagen 1982; Hanser 1985). Engaging with music making activities has been shown to be therapeutic in the treatment of both physical and mental impairments (Aldridge 2005; Hanser 1985).

Early pioneers made use of EEG to create contemporary music performances in concert settings, such as Alvin Lucier's 1965 *Music for Solo Performer* (Lucier 1976) which used a single electrode to distribute amplified alpha waves to a number of percussion instruments, which are then essentially stimulated 'hands free' by the performer, who mediates their mental state to give some degree of control over the performance itself. Richard Teitelbaum explored the use of an amplified EEG signal as a control source for analog sound synthesis in an improvised performance in the 1967 piece *Spacecraft* (Teitelbaum 1976). Ideas regarding the use of adaptive biofeedback in music were explored by Eaton (1971), who combined visual and auditory stimuli.

David Rosenboom was inspired by this work and continued its explorations in his *Brainwave Music* (1974), an interesting example as it was designed to incorporate the use of biofeedback in the performance process (Rosenboom 1990; Teitelbaum 1976).

Biomuse (Knapp and Lusted 1990) mapped the acquisition of low-level neuroelectric and myoelectric signals via statistical feature extraction to the real-time generation of music notation (musical structures in MIDI format). *Biomuse* also used other physiological readings (muscle tension and eye tracking). Whilst such signals are tangential to BCI, there is a growing field of work using non-nervous physiological signals, such as heart rate, galvanic skin response, and so on, in the design of systems for creative music technology (Daly et al. 2015; Nirjon et al. 2012; Pérez and Knapp 2008). One of the earliest examples of similar work combining signals for musical performance can be seen in Richard Teitelbaum's *In Tune* (1967), which used two EEG inputs alongside heartbeat and breathing sensors to give the performers control of a variety of analog synthesis functions.

BCI offers the possibility of directly translating brain activity (for example, motor or visual cortex activity, or more abstractly, emotional state for expression) to inform performance in music making. For example, particular frequencies of brain activity could be correlated with fixed musical parameters, so that the performer is required to mediate their own brainwave frequencies to achieve the intended musical output from the system (e.g., actively attempting to mediate brainwave amplitudes and frequencies as collected by the EEG). For the purposes of this chapter, we will consider this *parameter mapping* (forming control signal links between established BCI metrics and musical parameters).

These parameters might be musical control signals; temporal (start or end a playback) dynamic (adjust volume) or spectral (frequency equalization) for example. An overview of specific mapping techniques for digital instrument design is given by Goudeseune (2002). More recently, an overview of different types of musical parameter mapping from complex biomedical data and possible evaluation strategies is given in Williams (2016), but design and evaluation of such mappings for maximal musical expression remains a significant area for further work at the time of writing. Various combinations of mapping strategies exist, including one-to-one, one-to-many, and many-to-many combinations (Hunt and Kirk 2000), and indeed the linear mapping of alpha waves to particular acoustic instruments in *Music for Solo Performer* is significantly different to the more complex mappings employed latterly, including ensemble performances in examples by the *Biomuse Trio* (Knapp et al. 2009; Lyon et al. 2014) (see, e.g., their 2011 piece *Music for Sleeping and Waking Minds*). Whilst on-the-fly mapping is theoretically possible, musical parameter mapping is predetermined at the stage of system design and generally considered a part of the compositional process. It is in the mapping stage that systems for music composition generally derive their variety. Both the format of the output and the parameter selection, and ratios between control and parameter are considered valid, with many different types of mappings explored by those working with BCI for music (Brouwer and van Erp 2010; Chew and Caspary 2011; Daly et al. 2014c). Further opportunity for musical expression and variety can be given at the performance stage. BCI measurement has been combined with real-time sound synthesis in musical performance contexts (Hinterberger and Baier 2005). The use of BCI informed musical stimulus selection to mediate or entrain the listeners' brain activity (i.e., neurofeedback) is also a fertile area for research activity (Daly et al. 2014a, 2016).

Recently, machine learning techniques are being used to inform hybrid adaptive processing of control signals for music generation and performance (AlZoubi et al. 2008, 2009). Neurofeedback is particularly suitable for the specification of combined composition and performance music systems.

It is perhaps not surprising that of the biophysiological measurement techniques which are often adapted to music making, EEG is prevalent, due to the cost and accessibility of the relevant hardware. Amongst EEG based systems, both event related potentials (ERP) and spontaneous input are common. The P300 ERP (or "oddball paradigm") has been deployed in a system to allow active control over note selection for real-time synthesis (Grierson 2008; Grierson and Kiefer 2011)—techniques which are not dissimilar to the commonplace ERP typing or spelling systems, but used for the selection of musical notes rather than text input. Similarly, stimulus-responsive input measures such as the SSVEP (Middendorf et al. 2000), have been adapted to real-time control of musical parameters such as volume, or even limited selection of pre-composed score. However such systems are markedly different to approaches *sonifying* or *musifying* brainwave data (Baier et al. 2007a, b; Hinterberger and Baier 2005), wherein EEG (or other BCI data) is directly transmitted by auditory means (Toharia et al. 2014). Indeed, many existing EEG mappings for sonification are now in use (Väljamäe et al. 2013). The link comes again in the mapping between musical parameters (ruleset or other compositional decision making processes) so that the BCI input is constrained in some musically meaningful manner to create a performance with compositional intent, and aesthetic expression for the listener. One system for musifying EEG data mapped the rate of alpha wave activity to the cadence of the rhythm structure in a series of musical segments, while mapping the variance in the EEG to musical chord selections on a bar-by-bar basis, and the note position of a melody to the amplitude of the EEG waves per analysis window (Wu et al. 2010). Rhythm is an interesting musical property with specific brain cortex associations (Baier et al. 2007a) and, as such, has also been utilized in EEG analysis of musical rhythm, for example, in the evoked gamma band (20–60 Hz) by rhythmic tone sequences (Snyder and Large 2005). This type of mapping has also been explored in reverse, where the rhythmic properties of the resulting material are directly controlled by BCI input (Daly et al. 2014c). Making music in ensemble has a rich history (Le Groux and Verschure 2009; Manzolli and Verschure 2005), and has begun to be explored in BCI informed music making, for example a system which provides the ability for two users to collaborate in musical performance by mapping BCI measures of affect to the control of amplitude of two separate musical features (Leslie and Mullen 2012). Mullen et al. also survey systems which are designed to give agency to multiple performers, which they describe as *social* installations (Mullen et al. 2015). This work is closely aligned with the spirit of musical performance as communicative and interactive.

5.2.1 Possibilities

In all of the systems described above there remains a separation between the use of BCI as a cognitive control (active control) or the deliberate mapping of composition or performative generative music techniques in a passive (unconscious) manner. Recent research has suggested a number of unconscious cognitive performance benefits for the listener when music which is particularly evocative is played (Franco et al. 2014).

An example of affective state mapping to unconscious musical feature selection can be seen in the world of musical information retrieval (Lin and Cheng 2012). The potential to create systems for functional music (selection, performance, or even creation) in an unconscious manner (i.e., without the need for active management by the user) is enormous and perhaps the largest likely avenue for BCI music creation in terms of broad user base. Levels of emotional engagement, as measured via BCI, have been adapted to musical control by Ramirez and Vamvakousis (2012). They analysed EEG recordings elicited from listeners who were played a database of music which they considered to be emotionally charged, across a two-dimensional affect space (a commonly used space in psychometric evaluation, the arousal-valence, or circumplex model, of affect), defining affective (emotional) states from EEG (Chanel et al. 2006, 2007). For the original source of the circumplex model the interested reader is referred to (Russell 1980). The overarching tendency is to spend time creating complicated mappings but not exploring how successful these were in communicating artistic or aesthetic intent—which, perhaps ironically, is one of the most promising areas of BCI in the arts (as explored elsewhere in this book) as a tool for evaluating aesthetic experience, —in other words, an emotional response *to music* (Lin et al. 2010).

5.2.2 Overcoming the Self-report Confound

In music psychology, a great deal of attention has been paid to determining listeners emotional responses to certain types of music. This has significant implications for the use of BCI in evaluation of music. For example, "sad" music—or music which listeners report to communicate sadness—has been shown to be enjoyable (Vuoskoski and Eerola 2012; Vuoskoski et al. 2012) and subsequently, to have similar neural correlates when measured by EEG (Daly et al. 2014b). This research hinges on the distinction between perceived and induced emotions, wherein *perceived* emotions pertain to the understood meaning the listener perceives in the music (the compositional intent), and *induced* emotions are actually conveyed or felt, by the listener

(Juslin and Laukka 2004). Thus, a listener may report a piece of music as sounding "sad" but in fact enjoy listening to it.

With recent advances in affective response measurement, for example in determining neurophysiological correlates of affect (Mühl et al. 2015), it appears that the distinction between perceived and induced emotion is a challenge which BCI may help to address in this musical context. In a visual context or multimodal context several systems harness this potential in a variety of BCI for arts systems—see (Gürkök and Nijholt 2013) for a summary of systems including audification, musification, instrument control and emotional expression through BCI art. While visual examples can help differentiate aesthetic responses, music offers perhaps one of the strongest ways to explore this affective phenomenon. The temporal nature of music also lends itself well to the illustration of the changing pattern and transient nature of emotions, and many neurophysiological responses in general. The paradox between enjoyment, perception, and emotional induction has been well explored in musicological research (Hunter et al. 2010; Huron 2011; Manuel 2005) and would be a logical area for further exploration given the startling advances in BCI technology for estimation of affective state; such applications are uniquely afforded by BCI—for example if used to generate music that gradually improves the mood of the patient in an autonomic process without the need for a therapist (Daly et al. 2014b, 2016).

5.2.3 An Example System: MINDMIX, a Hybrid BCI Interface for Music Production

There are many reasons why audio engineers prefer tactile control of mixing processes (Merchel et al. 2010), which partially explains the significant interest, and progress being made in the field of haptic augmentation in audio and musical instrument design (Picinali and Katz 2010; Merchel et al. 2012).

MINDMIX is a hybrid system (combining active and passive control) using EEG metrics in a many-many mapping of to parameters on an audio mixer by generation of synchronous MIDI Machine Control messages. In this case, end-users might have little or no experience of music mixing, and a such careful mapping to ensure agency and congruence between neurophysiological metric and music parameter is vital. The general methodology for design and application might be equally suited to a wide variety of artistic applications.

In this case the application is ultimately to facilitate control of music production apparatus. Previous attempts to use BCI to control audio mixing parameters have been designed to use alpha and beta activity to control the amplitude of two separate faders (Miranda 2010). In the case of music mixing, there are many application-specific goals that need to be considered. In, for example, a music therapy context, one advantage of a BCI system is that it might be used by a person with no a priori experience or musical training, in order to engage in music production in context. However, in order to do this the BCI must be capable of performing music which is well correlated with the signal being analysed as a control signal (e.g., BCI parameters mapped according to constraints of melody, harmony, rhythm, or genre) yet also allows the user enough degrees of freedom to feel that they are truly the agent of their performance. The challenge, then, is in devising and evaluating mappings which are most suited to task-specific control—in this case, audio engineering processes, more specifically, mixing processes. *MINDMIX* control mappings were selected according to this philosophy. For example, once a particular channel has been selected, left or right motor imagery can be actively engaged to adjust the panorama of an audio source to move a sound image between left and right loudspeakers in a 2-channel stereo configuration. This is a many-many mapping wherein the channel is first selected by means of SSVEP, then the pan control selected by ERP, before the pan value is adjusted according to Mu L/R balance.

The range of tactile functions the *MINDMIX* prototype aims to augment are as follows: Transport control (play, stop, fwd, rev), fader select and level (individual channels, buss, and FX return), potentiometer select and adjust (pan, parametric EQ), and channel switching (solo, mute, insert, EQ in/out). Each of these parameters has been mapped to a sequence of actively controllable metrics, combining motor imagery (left and right), SSVEP, and ERP.

The *MINDMIX* prototype focusses solely on mixing (including remixing, and post-production tasks), rather than on source capture or recording. Combinations of mappings (i.e., many-many mapping) allows for a channel to be selected using SSVEP, followed by a potentiometer (e.g., pan, or semi-parametric EQ frequency/gain) to be selected according to ERP, before the value of the potentiometer itself is set according to imagined motor imagery (i.e., left, or right). SSVEP allows users to make a selection by focusing their gaze on a visual stimulus oscillating at a given rate. As well as initial parameter selection, SSVEP also allows for second level of control by mapping the duration of the gaze with non-linear features, for example amplitude, allowing for a degree of continuous control i.e., after selecting a specific channel the duration of a user's gaze can be used to adjust the fader for the selected channel accordingly. A similar effect could be achieved using eye-tracking in a hybrid system, using duration of gaze as a secondary mapping for amplitude. The parameters which are most useful for broad user participation in terms of transport across the digital audio workstation are play, stop, select, and various level parameters. It is important to consider the most meaningful signal type for each parameter in the mapping; some of these control signals have analogous actions in a mixer, for example, motor cortex with transport controls (stop, go, fast forward, rewind), and some have analogous parameters in music (SSVEP to non-linear adjustment of amplitude via faders).

Beyond encouraging inclusivity and participation through facilitating access to audio engineering processes via linear mapping strategies, the potential to harness unconscious processes (passive control) suggests that augmented audio engineering, for example, individually adaptive, responsive, or context-dependent remixing, may be a possibility. Such technology could be married together with the significant advances in music information retrieval (MIR), non-linear music creation (Berndt 2009), and context-adaptive music selection in the future. For example, creating systems for unsupervised music selection based on individual preferences and brain activity. Of most concern to the prototype described here is the appropriateness of the mapping and the relevance and usefulness of the user interaction with the application. Established methodologies for the evaluation of these types of systems are few. In the traditional audio engineering domain, this would be comparable to evaluating decisions such as whether, for example, a rotary potentiometer or a fader was most appropriate for control of a discrete audio parameter. The remaining sections of this chapter will consider appropriate methods for evaluating musical expression and the design of congruent musical parameter mapping with BCI derived control metrics.

5.3 Musical Expression: Challenges

Evaluation of creative computing generally is challenging, and in the case of music, highly context dependent (experience, history, timing, memory, and a whole host of other multimodal factors are involved in experiential evaluation of music). However, a common thread can be drawn between system design across creative computing applications, including music use cases. In order for the performer to feel engaged with the system there must be a sense of agency, which in the case of BCI for music is imbued by the aforementioned parameter mappings. Put plainly, we want the user to feel like a performer, to have some sense of active control over the ensuing musical interaction. The mapping between neurophysiological cues and audio parameter must be intuitive for a neophyte audience (i.e., one without prior training or the physical skills developed by professional audio engineers when working with tactile interfaces).

In the case of performance, the dream of many musicians, particularly musicians who also engage in composition activity, is to be able to bypass the physical intermediary in the process; that of notation or transcribing ideas for performance. Highly talented musicians are able to do this to some extent when creating and simultaneously performing (the process of musical improvisation). However, this requires a significant degree of musical training and becomes infinitely more complex when other musicians are also involved.

Those readers who have played musical instruments in isolation will likely find it axiomatic that in the process of *collaboration*, BCI for music might find a true niche as a viable and meaningful alternative to traditional paradigms. Again, the world of BCI for art has already made significant progress here with examples including work by De Smedt and Menschaert (2012), Casey and Smith (2013), Lee and Lee (2014), amongst others, which designers and practitioners creating music systems might look to for inspiration.

5.4 Evaluation Strategies from Auditory Display

A number of paradigms for the evaluation of BCI systems exist, however they often focus on technical or methodological details. There is a tendency in BCI work to prioritise technical implementation in research reporting, for example considering increased speed or accuracy of a system, rather than the application itself. For the purposes of work combining BCI with music, such evaluations are less relevant. In the design of such systems, it is important to consider the most meaningful signal type for each parameter in the mapping; some of the most common BCI control signals have analogous actions in a music performance, for example, motor cortex with physical actions (dancing, tempo, time signature, or starting and stopping an action), and some have analogous auditory parameters (dynamic control of instrument volume for example with amplitude of a frequency band in EEG). However, partly due to the infancy of the use of BCI for music making, the selection of these combinations is problematic and tends to become a question of 'taste'. The challenge, then, is in devising evaluation strategies for meaningful mappings which are most suited to task-specific control, in our case, aesthetic control of music parameters. Established methodologies for the evaluation of these types of systems are few, but borrowing from the world of sonification, multi-criteria decision aid analysis might be a particularly useful paradigm to explore the aesthetic success of a BCI music system, having previously been utilized in data-music mapping strategies.

5.5 Concluding Remarks

This chapter has attempted to give the reader a sense of the possible applications for music performance which the power of BCI might afford. The science fiction scenario is that a listener might 'imagine' a piece of music and through the use of BCI hear that piece realized. Readers of this book will be familiar with the reality will—rightly—be more sceptical, but nevertheless there appears to be a significant opportunity to explore the use of BCI to evaluate systems for creativity—especially complex mechanisms involving multimodal responses, such as music and the arts —in ways which traditional psychometric profiling might not otherwise offer. The possibilities for audience engagement with music, including emotional communication, physical motility, mood contagion, most importantly, *interaction*, are well placed as creative examples for BCI which the vast majority of the population might find interesting, even though such systems do not tend to contribute directly to the advancement of BCI technologically, as they are typically problems of engineering implementation rather than advancement. Evaluation strategies for BCI-to-music mappings, in general, are far from universally agreed upon and remain a significant area for further work. One approach would be to borrow from the world of auditory display the use of multi-criteria decision aid analysis technique to the evaluation of aesthetic success. In any case, a significant amount of further work remains in quantifying listener responses to music in terms of emotional or experiential communication, such as measurement of impact on induced emotional state versus perceived or self-reported emotional state, as traditional psychological approaches suggest that individual preferences and other environmental factors such as cultural expectations and musical training make emotional responses to musical stimuli highly variable (Scherer 2004).

An exciting area of BCI work which this chapter has not explored is the possibility of joint studies combining other neuroimaging techniques, for example fMRI and EEG. For music, such work will be particularly useful, given the spatial resolution with EEG, and the temporal restrictions with fMRI (which make feature correlation from dynamic stimuli such as music listening more challenging, as musical features can change radically over a comparatively short period of time, certainly smaller than the typical frame sizes afforded by fMRI studies). There are practical implications given the size and cost of such facilities but the potential for design of affectively adaptive systems in an artistic context, using such an apparatus, is hugely enticing. The possibility of developing affectively responsive BCI following rigorous evaluation of musical parameter mapping to neural correlates suggests that individual musical interactions might be facilitated by BCI technology in ways that had previously been thought impossible by music technologists, instrument designers, and music psychologists. We have presented a prototype here in *MINDMIX* — a mapping between active EEG control and a series of music production (mixing) tasks. Such technology could be married together with the significant advances in music information retrieval (MIR), non-linear music creation, and context-adaptive music selection in the future. For example, creating systems for unsupervised music selection based on individual preferences and brain activity.

References

Aldridge D (2005) Music therapy and neurological rehabilitation: performing health. Jessica Kingsley Publishers

AlZoubi O, Calvo RA, Stevens RH (2009) Classification of EEG for affect recognition: an adaptive approach. In: Australasian joint conference on artificial intelligence. Springer, pp 52–61

AlZoubi O, Koprinska I, Calvo RA (2008) Classification of brain-computer interface data. In: Proceedings of the 7th Australasian data mining conference, vol 87. Australian Computer Society, Inc, pp 123–131

Babiloni C, Vecchio F, Infarinato F, Buffo P, Marzano N, Spada D, Rossi S, Bruni I, Rossini PM, Perani D (2011) Simultaneous recording of electroencephalographic data in musicians playing in ensemble. cortex 47:1082–1090

Baier G, Hermann T, Stephani U (2007a) Event-based sonification of EEG rhythms in real time. Clin Neurophysiol 118:1377–1386

Baier G, Hermann T, Stephani U (2007b) Multi-channel sonification of human EEG. In: Proceedings of the 13th international conference on auditory display

Bailes F, Dean RT (2009) Listeners discern affective variation in computer-generated musical sounds. Perception 38:1386–1404. https://doi.org/10.1068/p6063

Berndt A (2009) Musical nonlinearity in interactive narrative environments. MPublishing, University of Michigan Library, Ann Arbor, MI

Bigand E, Poulin-Charronnat B (2006) Are we "experienced listeners"? A review of the musical capacities that do not depend on formal musical training. Cognition 100:100–130

Brouwer A-M, van Erp J (2010) A tactile P300 brain-computer interface. Front Neurosci. https://doi.org/10.3389/fnins.2010.00019

Casey K, Smith D (2013) Global mind field-a cybernetic perspective

Chanel G, Ansari-Asl K, Pun T (2007) Valence-arousal evaluation using physiological signals in an emotion recall paradigm. In: IEEE international conference on systems, man and cybernetics, 2007. ISIC, pp 2662–2667. https://doi.org/10.1109/icsmc.2007.4413638

Chanel G, Kronegg J, Grandjean D, Pun T (2006) Emotion assessment: arousal evaluation using EEG's and peripheral physiological signals. In: Multimedia content representation, classification and security, pp 530–537

Chew, YCD, Caspary E (2011) MusEEGk: a brain computer musical interface. In: Proceedings of the 2011 annual conference extended abstracts on human factors in computing systems. ACM Press, New York, NY, pp 1417–1422. https://doi.org/10.1145/1979742.1979784

Clair AA, Memmott J (2008) Therapeutic uses of music with older adults. ERIC

Daly I, Hallowell J, Hwang F, Kirke A, Malik A, Roesch E, Weaver J, Williams D, Miranda E, Nasuto SJ (2014a) Changes in music tempo entrain movement related brain activity. In: 2014 36th annual international conference of the IEEE engineering in medicine and biology society. IEEE, pp 4595–4598

Daly I, Malik A, Hwang F, Roesch E, Weaver J, Kirke A, Williams D, Miranda E, Nasuto SJ (2014b) Neural correlates of emotional responses to music: an EEG study. Neurosci Lett 573:52–57

Daly I, Williams D, Hwang F, Kirke A, Malik A, Roesch E, Weaver J, Miranda E, Nasuto SJ (2014c) Brain-computer music interfacing for continuous control of musical tempo

Daly I, Malik A, Weaver J, Hwang F, Nasuto SJ, Williams D, Kirke A, Miranda E (2015) Towards human-computer music interaction: evaluation of an affectively-driven music generator via galvanic skin response measures. IEEE, pp 87–92. https://doi.org/10.1109/ceec.2015.7332705

Daly I, Williams D, Kirke A, Weaver J, Malik A, Hwang F, Miranda E, Nasuto SJ (2016) Affective brain–computer music interfacing. J Neural Eng 13:46022–46035

De Smedt T, Menschaert L (2012) VALENCE: affective visualisation using EEG. Digit Creat 23:272–277

Eaton ML (1971) Bio-music: biological feedback experimental music systems. Orcus

Fagen TS (1982) Music therapy in the treatment of anxiety and fear in terminal pediatric patients. Music Ther 2:13–23

Franco F, Swaine JS, Israni S, Zaborowska KA, Kaloko F, Kesavarajan I, Majek JA (2014) Affect-matching music improves cognitive performance in adults and young children for both positive and negative emotions. Psychol Music 42:869–887

Goudeseune C (2002) Interpolated mappings for musical instruments. Organ Sound 7:85–96

Grierson M (2008) Composing with brainwaves: minimal trial P300b recognition as an indication of subjective preference for the control of a musical instrument. In: Proceedings of international cryogenic materials conference (ICMC'08)

Grierson M, Kiefer C (2011) Better brain interfacing for the masses. ACM Press, p 1681. https://doi.org/10.1145/1979742.1979828

Gürkök H, Nijholt A (2013) Affective brain-computer interfaces for arts. In: 2013 Humaine association conference on affective computing and intelligent interaction (ACII). IEEE, pp 827–831

Hanser SB (1985) Music therapy and stress reduction research. J Music Ther 22:193–206

Hinterberger T, Baier G (2005) Poser: parametric orchestral sonification of eeg in real-time for the self-regulation of brain states. IEEE Trans Multimed 12:70

Hunt A, Kirk R (2000) Mapping strategies for musical performance. Trends Gestural Control Music 21:231–258

Hunter PG, Schellenberg EG, Schimmack U (2010) Feelings and perceptions of happiness and sadness induced by music: similarities, differences, and mixed emotions. Psychol Aesthet Creat Arts 4:47

Huron D (2011) Why is sad music pleasurable? A possible role for prolactin. Music Sci 15:146–158

555

555

55

Juslin PN, Laukka P (2004) Expression, perception, and induction of musical emotions: a review and a questionnaire study of everyday listening. J New Music Res 33:217–238

Knapp RB, Jaimovich J, Coghlan N (2009) Measurement of motion and emotion during musical performance

Knapp RB, Lusted HS (1990) A bioelectric controller for computer music applications. Comput Music J 14:42–47

Le Groux S, Verschure P (2009) Neuromuse: training your brain through musical interaction. In: Proceedings of the international conference on auditory display, Copenhagen, Denmark

Lee HY, Lee WH (2014) A study on interactive media art to apply emotion recognition. Int J Multimed Ubiquitous Eng 9:12

Leslie G, Mullen T (2012) MoodMixer: EEG-based collaborative sonification. In: Proceedings of the international conference on new interfaces for musical expression, pp 296–299. http://www.nime.org/proceedings/2011/nime2011_296.pdf. Accessed 19 Nov

Lin C-Y, Cheng S (2012) Multi-theme analysis of music emotion similarity for jukebox application. In: 2012 International conference on audio, language and image processing (ICALIP). IEEE, pp 241–246

Lin Y-P, Wang C-H, Jung T-P, Wu T-L, Jeng S-K, Duann J-R, Chen J-H (2010) EEG-based emotion recognition in music listening. IEEE Trans Biomed Eng 57:1798–1806. https://doi.org/10.1109/TBME.2010.2048568

Lucier A (1976) Statement on: music for solo performer. Biofeedback and the arts, results of early experiments. Aesthetic Research Center of Canada Publications, Vancouver, pp 60–61

Lyon E, Knapp RB, Ouzounian G (2014) Compositional and performance mapping in computer chamber music: a case study. Comput Music J 38:64–75

Manuel P (2005) Docs sad music make one sad? An ethnographic perspective. Contemp Aesthet 3

Manzolli J, Verschure PFMJ (2005) Roboser: a real-world composition system. Comput Music J 29:55–74

Merchel S, Altinsoy E, Stamm M (2010) Tactile music instrument recognition for audio mixers. In: Audio engineering society convention 128

Merchel S, Altinsoy ME, Stamm M (2012) Touch the sound: audio-driven tactile feedback for audio mixing applications. J Audio Eng Soc 60:47–53

Middendorf M, McMillan G, Calhoun G, Jones KS et al (2000) Brain-computer interfaces based on the steady-state visual-evoked response. IEEE Trans Rehabil Eng 8:211–214

Miranda ER (2010) Plymouth brain-computer music interfacing project: from EEG audio mixers to composition informed by cognitive neuroscience. Int J Arts Technol 3:154–176

Miranda ER, Castet J (eds) (2014) Guide to brain-computer music interfacing. Springer, London

Mühl C, Heylen D, Nijholt A (2015) Affective brain-computer interfaces: neuroscientific approaches to affect detection. Oxford handbook of affective computing. Oxford University Press, Oxford, pp 217–232

Mullen T, Khalil A, Ward T, Iversen J, Leslie G, Warp R, Whitman M et al (2015) MindMusic: playful and social installations at the interface between music and the brain. In: More playful user interfaces. Springer, pp 197–229

Nirjon S, Dickerson RF, Li Q, Asare P, Stankovic JA, Hong D, Zhang B, Jiang X, Shen G, Zhao F (2012) Musicalheart: a hearty way of listening to music. In: Proceedings of the 10th ACM conference on embedded network sensor systems. ACM, pp 43–56

Pérez MAO, Knapp RB (2008) BioTools: a biosignal toolbox for composers and performers. In: Computer music modeling and retrieval. Sense of sounds. Springer, pp 441–452

Picinali L, Katz BF (2010) Spectral discrimination thresholds comparing audio and haptics. In: Proceedings of haptic and auditory interaction design workshop, Copenhagen, pp 1–2

Ramirez R, Vamvakousis Z (2012) Detecting emotion from EEG signals using the emotive epoc device. In: Zanzotto FM, Tsumoto S, Taatgen N, Yao Y (ed) Brain informatics. Lecture notes in computer science, vol 7670. Springer, Berlin, Heidelberg, pp 175–184

Rosenboom D (1990) The performing brain. Comput Music J 14:48–66

Russell JA (1980) A circumplex model of affect. J Pers Soc Psychol 39:1161

Scherer KR (2004) Which emotions can be induced by music? What are the underlying mechanisms? And how can we measure them? J New Music Res 33:239–251

Snyder JS, Large EW (2005) Gamma-band activity reflects the metric structure of rhythmic tone sequences. Cogn Brain Res 24:117–126. https://doi.org/10.1016/j.cogbrainres.2004.12.014

Teitelbaum R (1976) In tune: some early experiments in biofeedback music (1966–1974). In: Biofeedback and the arts, results of early experiments. Aesthetic Research Center of Canada Publications, Vancouver

Toharia P, Morales J, Juan O, Fernaud I, Rodríguez A, DeFelipe J (2014) Musical representation of dendritic spine distribution: a new exploratory tool. Neuroinformatics: 1–13. https://doi.org/10.1007/s12021-013-9195-0

Väljamäe A, Steffert T, Holland S, Marimon X, Benitez R, Mealla S, Oliveira A, Jordà S (2013) A review of real-time EEG sonification research

Vuoskoski JK, Eerola T (2012) Can sad music really make you sad? Indirect measures of affective states induced by music and autobiographical memories. Psychol Aesthet Creat Arts 6:204

Vuoskoski JK, Thompson WF, McIlwain D, Eerola T (2012) Who enjoys listening to sad music and why? Music Percept 29:311–317

Williams D (2016) Utility versus creativity in biomedical musification. J Creat Music Syst 1

Wu D, Li C, Yin Y, Zhou C, Yao D (2010) Music composition from the brain signal: representing the mental state by music. Comput Intell Neurosci 2010:14

Part II
Exploring Our 'Self' with Brain Art

Chapter 6
Using Synchrony-Based Neurofeedback in Search of Human Connectedness

Suzanne Dikker, Sean Montgomery and Suzan Tunca

Abstract In this chapter, we explore whether brain-computer interface (BCI) applications can embody the elusiveness of human connectedness. Concretely, we discuss a series of art/neuroscience works that track and visualize the extent to which brainwaves and physiological responses become synchronized between people and their environment. From a neuroscientific point of view, we ask whether such synchrony is 'meaningful', i.e., do these data streams (brainwaves, heart rate, movement) tell us something about how connected we feel to each other ("when we feel in sync with someone, are our brainwaves literally on the same wavelength?"). From an artistic, experiential point of view, we ask whether these works can raise critical questions about our often unsatisfactory quest to connect to ourselves and each other: via face-to-face interactions, scientific inquiry, tech-based communication tools, big data; about the exclusionary nature of groups, both in the real and virtual world. Finally, do the works stands on their own—independent of such research questions—as immersive, interactive aesthetic experiences, allowing visitors to gauge and explore their own interactions in a visceral, intuitive way?

S. Dikker (✉)
Utrecht University, Utrecht, The Netherlands
e-mail: sdikker@gmail.com

New York University, New York City, USA

S. Montgomery
Produce Consume Robot, New York City, USA

S. Tunca
ICK Amsterdam, Amsterdam, The Netherlands

CODARTS University of the Arts, Rotterdam, The Netherlands

PhDArts Leiden University, Leiden, The Netherlands

S. Dikker
DIKKER + OOSTRIK, Amsterdam, The Netherlands

S. Montgomery
Produce Consume Robot, New York City, USA

© Springer Nature Switzerland AG 2019
A. Nijholt (ed.), *Brain Art*,
https://doi.org/10.1007/978-3-030-14323-7_6

161

Keywords Art-science interface · Hyperscanning · BCI · Crowdsourcing
neuroscience · Synchrony · Brain rhythms · EEG · Biofeedback art

6.1 Introduction

What does it mean to lose yourself in someone else? How is it possible that the mere
physical presence of another human can make us believe we can conquer the world,
or conversely, make us feel lonely and incapable? We know, both scientifically and
intuitively, that relationships are crucial for our physical and mental wellbeing. But
they are also sources of frustration in their fluid, messy mix of internal inconsis-
tencies: love and hate, inclusion and exclusion, fascination and comfort, challenge
and familiarity. What are some possible consequences, either negative or positive,
of these processes? Does human interaction mediated by technological interfaces
improve our feeling of connectedness? Or perhaps we project our own intentions
onto these interfaces, creating merely the illusion of a connection? Balancing at the
intersection of art and science, our work aims to explore the nature of human connectedness, a topic that lies at the very core of artistic, scientific, and technological
inquiry.

In this chapter, we discuss a series of art/science Brain-Computer Interface (BCI)
works that aim to achieve true, substantive *interdisciplinary synergy* at the interface
of perception, cognitive neuroscience, social psychology, education, engineering,
and art. Moreover, and crucially, our goal is for the concepts, results, and analytic
methodology to transform ideas in each domain, as summarized in Fig. 6.1.

For example, using art installations to collect neuroscience data provides a host
of unique novel opportunities to the scientific community: through art installations
neuroscience data can be collected from large groups of visitors (*crowdsourcing
neuroscience*) outside of the laboratory (*real-world neuroscience*) while participants engage in direct, naturalistic face-to-face interaction (*hyperscanning*: recording brain activity from multiple people at the same time). This enables scientists to
investigate the neurobiological basis of dynamic, naturalistic social interactions.

This art/science cross-pollination is designed to be bidirectional. We incorporate a BCI component into the work not only to test the effects of the BCI on the
experience of human connectedness (*synchrony neurofeedback*) but we also hope
to enrich the audience experience of the artwork and allow audience members to
shape their own (*immersive and interactive*) art experience with their brainwaves.
In addition, the works provide a rich opportunity for *neuroscience/STEM outreach
and education* to people who may not otherwise directly experience the scientific
process. By combining a personally moving aesthetic experience with active neuroscience research, viewers become participants and hopefully gain an opportunity to
think about the mind and the brain and how to begin to understand the role synchrony
plays in our daily lives. Finally, we will argue that the cross-pollination between art,
neuroscience and education can lead to *technological innovation*. For example, software tools from the interactive arts 'accidentally' provided solutions for scientific

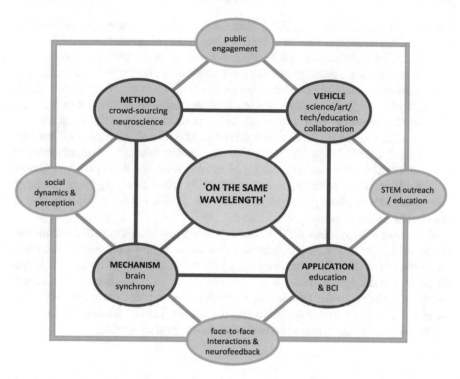

Fig. 6.1 Interdisciplinary synergy. In an effort to understand the functional role of brain-to-brain synchronization during human interaction, we bring together expertise from science, technology, education and art in a crowd-sourcing neuroscience approach to advance scientific insights into group dynamics in real-world environments, such as educational settings. We aim for substantive interdisciplinary synergy, with ideas at the interface of perception, cognitive neuroscience, social psychology, education, engineering, and art. Moreover, and crucially, our goal is for the concepts, results, and novel analytic methodology to transform and enrich ideas in each domain

hyperscanning research that the neuroscience community was struggling with. Each of these advantages will be discussed in detail in the Discussion.

With the projects discussed below, we hope to illustrate these interdisciplinary benefits, and show that the art-science interface provides a unique space to explore ways to capture, with 'objective' analytical tools, subjective experiences during human interactions. All the projects use brain-computer-interface technology to visualize/sonify the extent to which participants' biometrics 'synchronize' with the audiovisual environment, each other, or both. In the Discussion we will decompose and refine the term 'synchrony' from an art-science interface perspective, but for now let's consider 'synchrony' as a process that involves the convergence of two (or more) entities. In the case of social connectedness, the entities are two or more human agents, but, as will become clear below, synchrony is also achieved between humans and their environment, within our bodies, and even within our brains (e.g., cross-frequency coupling; Canolty et al. 2006; Lakatos et al. 2005). As is the case for most projects discussed in this volume, our works use electroencephalography (EEG), which allows us to measure brain activity at a millisecond timescale. Measuring brain activity (*brainwaves*) at the temporal resolution at which the brain operates permits us to look in a detailed way at the circumstances under which our brainwaves synchronize with people and stimuli in our environment. Thus, from a neuroscientific analysis perspective, synchrony involves the temporal alignment or (mutual) adaptation of brain rhythms. From a human experience perspective, temporal coupling as a prerequisite for synchrony need not be immediately obvious, or at least not experienced as such: being 'in sync' or 'on the same wavelength' is usually taken to imply interactive alignment at the level of mental representations (see e.g., Garrod and Pickering 2009; Pickering and Garrod 2013), often involving more 'abstract' constructs such as sharing viewpoints (e.g., Van Berkum et al. 2009) that may or may not be linked to convergence at the temporal level.

Below, we will first discuss the different processes of 'synchrony' that are relevant to our work. Then, in the Projects section, we describe how our works use art/BCI environments to measure synchrony and to probe the conditions under which synchrony might emerge (and may be stimulated) from both an artistic and neuroscientific perspective. In the Discussion, we will revisit the 'collateral benefits' of an interdisciplinary structure outlined in Fig. 6.1 in more detail from the perspective of our artistic and research programs discussed in the Projects section.

6.1.1 Synchrony Within the Brain

The brain is an oscillator. More specifically, each of the brain's 100 billion neurons is an individual oscillator. These neuronal oscillators communicate by synchronizing their individual patterns into coherent rhythms that range in scale from small coalition "chat groups" to large-scale synchronous activity involving nearly all regions of the brain (Buzsaki 2006). Neuronal oscillations are believed to play a role in various perceptual and cognitive tasks, including attention (Lakatos et al. 2008), navigation (Buzsaki 2006), memory (Gruber et al. 2008), motor planning (Donoghue et al. 1998), and spoken-language comprehension (Bastiaansen and Hagoort 2006; Schroeder et al. 2008; Meyer et al. 2016; Peelle and Davis 2012). The modulation of oscillatory activity during perceptual tasks is typically seen in distinct, "privileged" frequency bands, including delta (<3 Hz), theta (4–8 Hz), alpha (8–14 Hz) beta (15–25 Hz) and gamma (>30 Hz), all well captured by EEG recordings.

6.1.2 Synchrony During Perception: Brain-Stimulus Coupling

Speech and other dynamically changing auditory signals (as well as visual stimuli, including sign language) contain critical information required for successful decoding that is carried at multiple temporal scales (e.g. intonation-level information at the scale of 500–2000 ms, syllabic information that is closely correlated to the acoustic envelope of speech in the ~150–300 ms range, and rapidly changing featural information occurring around ~20–80 ms). These different aspects of signals (frequency spectrum composition combined with slow and fast temporal modulation) must be analyzed for successful recognition. (E.g., Giraud and Poeppel 2012, Nature Neuroscience.) In the laboratory it has been demonstrated that the endogenous rhythms of the brain can be **entrained** by visual (Gomez-Ramirez et al. 2011; Mathewson et al. 2012; de Graaf et al. 2013; Spaak et al. 2014) and auditory stimuli (Jirakittayakorn et al. 2017; Beauchene et al. 2017). Furthermore, exogenous stimuli that alter the EEG rhythms of the brain have been shown to affect perception, reaction times, attention, memory and mood (Williams 2001; Mathewson et al. 2012; Spaak et al. 2014; Jirakittayakorn ct al. 2017). This research suggests that brainwaves track the temporal dynamics of the world around us to permit successful decoding of the information it contains. Although there remains ongoing work to determine when rhythmic exogenous stimuli alter EEG oscillations by entraining (i.e. highjacking) endogenous rhythms through phase synchronization versus just passively reflecting the stimulus frequency via sensory inputs to the cortex (Keitel et al. 2014), the impact of exogenous stimuli on both physiology and behavior speaks to the important role that synchrony with our surroundings likely plays in our everyday lives.

Crucially, sharing the same audio-visual input does not guarantee that we 'see' the world the same way: many (top-down) factors are known to affect perceptual processes. These can stem from individual differences or from contextual or task demands. For example, musicians show tighter neural entrainment to musical stimuli at low beat rates than participants who are not musically trained (Doelling and Poeppel 2015) and Assaneo and colleagues recently demonstrated that about half of the population spontaneously synchronizes speech production with the rhythm of heard speech, a behavioral finding that is correlated with structural differences in neural pathways between the two groups (Assaneo et al. 2019). From a contextual/task perspective, selective attention is shown to be predictor of brain-stimulus synchrony. For example, as outlined above, when we listen to a story or conversation, our brainwaves track the envelope of the speech signal. Now take a situation where you're at a party with many ongoing conversations: most of us are perfectly able to selectively listen to one speaker and ignore all the others in the room, even if they're physically closer to us than our conversational partner. This phenomenon, called the 'cocktail party effect', has been very nicely demonstrated at the neural level in the lab. For example, Golumbic and colleagues (Golumbic et al. 2013) played two stories, one told by a male speaker and one by a female speaker, to their study participants. Crucially, the two speech streams were played simultaneously, seemingly causing a cacophony. Participants were then asked to listen to only the male or the female speaker. Listeners' brainwaves followed the speech signal of the person they were attending to, showing that attention is a strong predictor of entrainment, and that our brains prioritizes certain auditory input over others. In the project Hive Mind described below, a biofeedback cocktail party effect is created. The brains rhythms of two participants generate two distinct visual entrainment patterns and the audience members' brainwaves may selectively entrain with one of the patterns or be a wash in the cacophony of biofeedback conversation. It has been shown that such 'frequency tagging' can be used to decode where in a visual scene a participant is paying attention (e.g., Wu et al. 2011; Zang et al. 2010).

Contextual factors also affect brain responses to our environment in that they allow us generate predictions about upcoming audio-visual input (e.g., Bar 2007). For example, a sentence frame like "grass is …" strongly predicts for "green", but this is not true for e.g., "my favorite color is …" (e.g., Federmeier 2007; Dikker et al. 2009, 2010; Van Berkum et al. 2005; Van Berkum 2013, and many others). However, while there exists consensus that anticipation plays a role in processing, the extent to which the entrainment patterns observed in our brainwaves prove that they are either "predictors" or "followers" is a topic of contention (e.g., Haegens and Golumbic 2017).

In sum, intrinsic brain rhythms, expectations, and external influences can alter the synchrony of neuronal firing in the brain and create selective perceptual filters that affect the way we perceive the world around us (e.g. Enns and Lleras 2008; Desimone and Duncan 1995).

6.1.3 Synchrony During Social Interactions: Joint Action and Brain-to-Brain Coupling

As already illustrated with the cocktail party effect and linguistic predictions, the mechanisms described in the previous paragraph also apply to situations that involve human interaction. Successful social interactions require tight spatio-temporal coordination between their participants at motor, perceptual and cognitive levels. Using a variety of techniques, around a decade ago several labs began to study human social dynamic situations (ensembles performing music, multiple people performing actions together, or carrying on a conversation; e.g. Babiloni et al. 2006; Zamm et al. 2018; Tognoli et al. 2007; Dumas et al. 2010; Yun 2013; see e.g., Hari et al. 2013; Babiloni and Astolfi 2014 for review). For example, there exists a growing body of work measuring pairs of participants interacting while being recorded using fMRI, EEG and/or eye-tracking. In some of our own work (see Projects) we use hyperscanning in EEG while pairs or groups of participants interact directly. We then quantify the extent to which their brainwaves become 'synchronized' in real time. The moment-to-moment interbrain coupling is then translated into an audio-visual environment in real time.

Similar to stimulus-brain coupling, the interactive 'alignment' that supports joint action (doing something together) relies on our ability to anticipate both our own (linguistic and behavioral) actions and those of others (Pickering and Garrod 2013; Dikker et al. 2014; Sänger et al. 2011; Sebanz et al. 2006; Konvalinka et al. 2010; Ramnari and Miall 2003): As already mentioned above, our brains are naturally built to function as 'prediction machines' (Bar 2007). Given the importance of collaboration and social structures to our survival (e.g., Wilson 2012), it is unsurprising that our brains are particularly tuned to test expectations against external cues derived from others during social interaction (Hari and Kujala 2009; Keysers and Gazzola 2009). Such cues can be subtle (e.g. rhythmic movement or gaze direction) or obvious (e.g. explicit statements about beliefs/convictions). Crucially, this information feeds back into our 'internal' processing, as such allowing us to adjust our perceptual models of others and optimize interpersonal synchronization. This process may lead to shared viewpoints, but also to behavioral synchronization at lower levels such as walking pace, breath, gestures and facial expressions, and smooth turn-taking behavior during conversation (e.g. Sun et al. 2011; Richardson et al. 2008; Stivers et al. 2009; Paxton and Dale 2012). Even a slight mismatch in timing of audiovisual input, like a temporal lag in a long-distance phone conversation can lead to a suboptimal social experience (e.g., Powers et al. 2011).

A number of studies have shown that the mere presence of another person can alter brain activity (Verbeke et al. 2014). Social presence further triggers individuals to prioritize stimuli that are thought to be relevant to the social context (Shteynberg 2010; Böckler et al. 2012) as well as inducing adaptive behavior during joint action. Crucially, such interactive alignment can be either spontaneous or planned (Richardson et al. 2007; Knoblich et al. 2011; Marsh et al. 2006), and has been shown to persist when people merely believe that they are performing a joint action (e.g., Atmaca et al. 2011).

Interactive alignment between humans at both the behavioral and neural level may further be modulated by both contextual factors and individual differences (again, just like stimulus-brain coupling), in line with the intuition that some people are better at 'fitting in' than others. For example, Basnakova et al. (2013) suggest that listeners who have a stronger overall empathic disposition are more likely to take speaker intentions into account during language comprehension (also Van den Brink et al. 2012) and these assumed intentions can alter the predictive process (for example, you may know that your friend's favorite color is blue). Further, individual differences in prosocial versus pro-self orientation have been shown to affect spontaneous interpersonal coordination (Lumsden et al. 2012), and children with Autism Spectrum Disorder do not engage in spontaneous rhythmic movement synchronization with others (Marsh et al. 2013). In our own work, we found that interpersonal synchrony at the neural level is also predicted by social traits in typically developing children (Dikker et al. 2017).

Hyperscanning EEG studies comparing neural oscillations between dyads have shown that brain-to-brain synchrony—quantified in various ways—is correlated with a range of factors, including social closeness (Dikker et al. in revision; Kinreich et al. 2017), pain perception/touch (Goldstein et al. 2018), eye contact (Dikker et al. in revision) and cooperation versus competition (Babiloni et al. 2007); (see also Sänger, Lindenberger, and Müller 2011; Hari et al. 2013; Hasson et al. 2012; Babiloni and Astolfi 2014 for reviews, and Hasson and Frith 2016 for a theoretical account linking coupled dynamics to social interactions). For example, a number of scholars have reported alpha/mu coherence during interpersonal coordination tasks (~10 Hz; e.g., Dumas et al. 2010; Tognoli et al. 2007). Alpha/mu has been associated with attention (e.g. Anderson and Ding 2011), motor control (Pfurtscheller and Lopes da Silva 1999; Pfurtscheller and Neuper 1994) as well as motor simulation (Neuper and Pfurtscheller 1999). Our own research also contributes evidence to the relationship between interbrain synchrony and (a) individual differences, (b) interpersonal factors and (c) contextual factors (Dikker et al. 2017; Bevilacqua et al. 2018; Dikker et al. in revision).

6.1.4 Synchrony and Synchronicity

While most of our works define *synchrony* along the lines of the social (neuro)science perspective outlined above, other Arts/BCI installations, like Mariko Mori's UFO (2001) work with a different concept of *synchronicity*, as defined by Jung (see Goede 2017 for a very thoughtful analysis comparing Mariko Mori's UFO and the Mutual Wave Machine, discussed in detail below). Jung (2012) defined synchronicity as "meaningful coincidence". In his essay "On Synchronicity" (originally titled "Über Synchronizität" and originally presented as a lecture at an 1951 Eranos conference in Ascona, Switzerland) he writes that "Although meaningful coincidences are infinitely varied in their phenomenology, as acausal events they nevertheless from an element that is part of the scientific picture of the world. Causality is the way we explain the link between two successive events. Synchronicity designates the parallelism of time and meaning between psychic and psychophysical events, which scientific knowledge so far has been unable to reduce to a common principle. The term explains nothing, it simply formulates the occurrence of meaningful coincidences which, in themselves, are chance happenings, but are so improbable that we must assume them to be based on some kind of principle, or some property of the empirical world. No reciprocal causal connection can be shown to obtain between parallel events, which is just what gives them their chance character. The only recognizable and demonstrable link between them is a common meaning, or equivalence."

As discussed in more detail below, Harmonic Dissonance is a collaboration with International Choreographic Arts Centre ICK Amsterdam, who have been thinking of notions of synchronicity for decades. In the choreographic work of Emio Greco and Pieter C. Scholten (ICK Amsterdam) the term 'synchronicity' indicates a specific quality of shared intentionality actualized via movement. It is a central notion within their choreographic signature where an intuitive dancing body is cultivated that communicates via a dramaturgy derived from what Greco and Scholten term a "language of the flesh". This language articulates itself intuitively, sparked by a polarization between instinctive movement and movement controlled by the brain—*Fra Cervello e Movimento*—between brain and movement. In this choreographic paradigm, creativity is taken to be incarnate in the body and movement is taken to be self-sufficient (not in need of any externally imposed narrative or conceptual structure to generate meaning) and able to create time and space.

"Synchronicity" in the work of Greco and Scholten is considered as an ever to be actualized "dual utopia" ("not yet") as well as a concrete and palpable state and dimension of danced experience that expands the animated articulation of bodies in danced motion towards another state of being above and beyond mere movement in unison. A key working principle that functions as a condition for the actualization of "synchronicity" is the notion of being "one body". To achieve the quality of motion as "one body" for example in a duet, the cultivation of a double dual awareness is required of both one's own movements reflected by conscious awareness of the movements as well as of the movements and their reflections in the conscious awareness of the other dancer. The strength and the aimed at quality of the movement is generated by the shared striving towards "synchronicity", aiming the intentions towards the same direction and amplifying one's sense of self with the sense of self of the other. Something like a complete unity is strived for but not actually occurring. The individual characteristics of each dancer become even more perceivable through the individual and shared striving towards synchronicity. In C.G Jung's definition of "synchronicity" meaning is the central issue that suggest an "acausal connecting principle correlating mental and physical events" (Atmanspacher and Fuchs 2014, p. 5). In the artistic work of Greco and Scholten, "synchronicity" refers to a specific state and dimension of being in movement related to moving in "synchrony" but qualitatively above and beyond merely moving at the same time doing the same movements. In their choreographic practice, aiming at becoming "one body" with other bodies appears to imply or to be conditioned by an extreme expansion of affective and empathic capacity while maintaining a coherent and contained sense of self.

6.1.5 Synchrony Art/Neuroscience BCIs

In sum, our projects draw on the large body of research demonstrating that our brains are intrinsically rhythmic (Buzsaki 2006), predictive (e.g. Arnal and Giraud 2012; Friston 2003; Bar 2007), and adaptive, probing the interplay between audio-visual integration, prediction, and social adaptation during BCI experiences. Some of the projects described below additionally assign a critical role to 'interactive alignment' between individuals in explaining communicative success during (non-)verbal joint action.

Before diving into the detailed discussions of our projects, it is worth pointing out that the work discussed here is by no means exhaustive when it comes to art/BCI work involving multiple agents: for a comprehensive review, see Nijholt 2015 and Chap. 4 in this book by Rosenboom and Mullen. In fact, this line of work was initiated at least four decades ago: In this volume, Flora Lysen discusses pioneering work in the field of multi-agent BCIs such as David Rosenboom's *Ecology of the Skin* (1970; a "group encounter brain biofeedback performance system" for ten participants), and Nina Sobell's *Interactive Brainwave Drawing: EEG Telemetry Environment* (1975; involving the translation of "shared" EEG data between two participants into line-drawings). Also discussed in this volume is *EEG Kiss*, where EEG is measured simultaneously from two participants with the goal to highlight notions of intimacy, exposure, vulnerability, and surveillance. In another example, *Ringing Minds* (2014) creates a music performance based feedback loop with the audience (see Mullen et al. 2015 for a review of multi-agent BCIs focusing on sound and music). Similar to these works, we do aim for our work to stand on its own as interactive art, allowing visitors to gauge and explore their own interactions in a visceral, intuitive way. And we also wish to raise critical questions: about the technological communication interfaces in our everyday lives; about our often unsatisfactory quest to understand ourselves and each other, either through scientific inquiry or through direct face-to-face communication; about the exclusionary nature of groups. However, in a slight shift of emphasis from some other hyperscanning BCI work, the biometric data that we collect does not exclusively serve an artistic, societal, educational, or self-reflective goal, but mutually serves a (neuro)scientific goal. The data that is collected with our installations are used to test neuroscientific hypotheses. Any neuroscience findings, in turn, inform future iterations of the installations. The research questions are informed by both neuroscientific and psychological constructs, but also by humanities and artistic questions (see the discussion on synchrony and synchronicity above and in the Discussion). As such, much of our work bears similarity to performance-led research in the wild (e.g., Benford et al. 2013). In sum, in what follows, we hope to show that through interdisciplinary synergy, both interactive art and neuroscientific inquiry can be enriched in numerous ways.

6.2 Projects

6.2.1 Produce Consume Robot Projects

Sean Montgomery's work under the moniker *Produce Consume Robot* endeavors to bring deep synergy into the relationship between art and science. The work attempts to make an immersive aesthetic experience based on scientific phenomena and through the process of engaging viewers in a museum or gallery, generate novel datasets and ideas that can be used to test further scientific hypotheses. Working in the domain of neuroscience and affective computing while utilizing BCI technology gives the viewer the opportunity to ask questions about themselves, their own perception and memory, subjectivity and objectivity, and at the same time be an active human participant in a study in which physiological and behavioral data is gathered to address new scientific inquiries. The works described here examine the dynamic interplay between the endogenous rhythms of the body and the external environment to reflect on how we engage with the world around us and how that's likely to change as we head into the 21st century.

6.2.1.1 Telephone Rewired (2012)

Produce Consume Robot and LoVid (Kyle Lapidus and Tali Hinkis)

Telephone Rewired (Fig. 6.2) is an art installation and active scientific experiment based on a phenomenon in the neuroscience literature demonstrating that exogenous pulses of light and sound can entrain the endogenous rhythms of the brain and alter perception, reaction times, and memory (Williams 2001; Williams et al. 2006; Gomez-Ramirez et al. 2011; Mathewson et al. 2012; de Graaf et al. 2013; Spaak et al. 2014; Jirakittayakorn et al. 2017; Beauchene et al. 2017). To generate these conditions inside the installation, the viewer enters an immersive space in which the lighting and sound are pulsing at one of five brain relevant frequencies (delta ~2 Hz, theta ~6 Hz, alpha ~10 Hz, low beta ~15 Hz, high beta ~25 Hz). Even though the lights simply pulse on and off, viewers often report seeing colors, patterns, and shapes and having a feeling of being in an altered state of mind as their brain shifts into a different mode of operation. While the viewer is cycled through different brain-states, some sequences create a subjective sense of calm and clearing of the mind, while others might deliver a heightened sense of focus. During the experience the viewer may consider the implications for increasing the capacity of human learning and synchronizing multiple people for greater collaboration and empathy, and generally ask what possibilities exist for a future in which you can toggle the switches of human cognition.

Telephone Rewired is also an active scientific experiment in which EEG data, reaction times, memory scores, and subjective reports have been collected from over 1700 participants at the Science Gallery at Trinity College Dublin and additional EEG-only datasets at Harvestworks in New York City and the Daejeon Museum of Art in Daejeon South Korea. The experiment was designed to record continuous EEG as the installation entrained the participant's brain and to measure the participant's reaction time in response to words presented on-screen or as auditory stimuli. Figure 6.2 shows a participant in the Science Gallery at Trinity College performing the psychometric reaction time test while his brainwaves were being entrained by the strobing of lights and sound in the room. To facilitate the operation of the experiment with minimal intervention by gallery staff, we collected EEG data from the easy to use Zeo sleep headband. Although the installation necessarily had warnings about strobe lights and photo-sensitive epilepsy, by fully automating the reaction time collection and utilizing a unique code to cross reference with participants' subsequent memory test and subjective reports, the process was streamlined to allow as many viewers as possible to experience the work and participate in the study.

As a scientific experiment, Telephone Rewired successfully captured data from an enormous number of participants to address questions about exogenous synchronization of the brain in "real-life" situations and analysis of the data is ongoing to

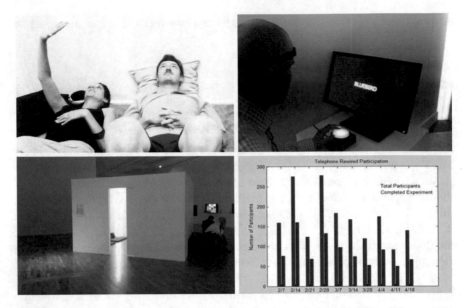

Fig. 6.2 Telephone Rewired. Telephone Rewired at Harvestworks, New York City, USA (top left), Science Gallery at Trinity College, Dublin, Ireland (top right) and in the Daejeon Museum of Art, Daejeon, South Korea (bottom left). Total number of participants and number of participants that completed the experiment each week that it was on show at Science Gallery (bottom right). The Zeo EEG headband can be seen worn by participants in the top two images. All images by Sean Montgomery Web: http://produceconsumerobot.com/telephonerewired/

identify new patterns of activation. Even though the gallery setting had a relatively high drop-out rate as participants would walk out half-way through stimulus presentation, in under 3 months 866 participants completed the entire stimulus presentation and reaction time assessment, which is a very large participant pool for human neuroscience studies (Fox et al. 2016; Van Essen et al. 2013). A larger challenge we faced was getting all the data back from overseas locations when the installation was broken down and shipped back. We ended up losing most of our memory scores and written subjective reports, which is a critical consideration to address along with novel experimental paradigms in future showings of the work.

As an art installation, the neuroscience research-based entrainment utilized in Telephone Rewired proved to be a powerful aesthetic experience. One participant remarked that at some times his mind was racing and in other parts felt calm and relaxed and concluded by saying "similar to if you gave your muscles a vigorous massage… my own state of being felt massaged… I felt very thoroughly worked, but more capable for it." Interestingly, different participants would sometimes report very different subjective experiences. In response to the strobing white lights, some reported seeing colors, others only black and white patterns, some fully identifiable objects and others only abstract shapes. In this context where the objective stimulus is as bare as a strobing white light, it reveals the locus of the aesthetic experience existing fundamentally in inside the participant's brain/mind. Perhaps works like Telephone Rewired, where objective measurements including EEG are gathered in combination with subjective reports, may offer a unique window to dig deeper into the relationship between objectivity and subjectivity in the context of viewing art.

6.2.1.2 Hive Mind (2017)

Produce Consume Robot and LoVid (Kyle Lapidus and Tali Hinkis)

Hive Mind is a peek into the future of augmented cooperative cognition. Two performers engage the audience in an on-stage discussion in which no words are spoken. Instead, the brain rhythms of each performer directly generate pulses of light and sound that synchronize the brain oscillations of viewers and create an immersive environment that transports the audience to altered states of consciousness. Based on neuroscientific research showing that rhythmic stimuli can entrain neuronal oscillations to alter perception, reaction times, and memory formation, Hive Mind uses the performers' live EEG and data processing to directly convey the performers' brain states. As one performer's brainwaves become the stimuli that entrains the other performer's brain patterns, a public brainwave-driven conversation unfolds between the performers. Together the performers and audience go on a journey through different induced brain states and altered perceptions, ultimately considering broad implications for the future of human cognition and communication. Future performances with resources to measure the EEG of the audience as well as the performers may provide insight into the mechanisms of collective human synchronization via exogenously driven oscillations. As discussed above, this biofeedback cocktail party may provide a unique window to examine how the audience synchronizes their brainwaves with one or the other of the competing entrainment signals being delivered from the two performers (Fig. 6.3).

6.2.1.3 Emergence (2010)

Produce Consume Robot and Diego Rioja and Mustafa Bagdatli

Emergence is an interactive biofeedback art installation that invites the viewer to examine the relationship between biology and technology. Extending the scope of BCI beyond direct central nervous system measurement, Emergence (Fig. 6.4) measures the electrical pulses generated by the beating of the human heart as it accelerates and decelerates with changes in the sympathetic and parasympathetic nervous systems. When a viewer touches the installation, the electrical impulses generated by each beat of the viewer's heart propagate throughout the viewer's body and are detected and digitized by the installation. During this interaction, Emergence synchronizes its own electrical pulses with the viewer's heart to create a syncopated light and sound-scape that reflects its intimate experience with the viewer and also includes the installation's "memory" of previous viewers' heart rhythms. Through a single eye, Emergence retains a visual memory of its encounter with the viewer. With each beat of the viewer's heart, Emergence captures an image of the interaction and digitally pulses the memory through the surrounding infrastructure and uploads it to the internet where it can be seen on flickr or facebook. By drawing parallels between the human body and the infrastructure of the digital age, Emergence invites the viewer to think about what fundamentally differentiates the electrical impulses of the internet from those impulses constantly traveling throughout their own body.

After traveling to countries around the world, the piece has captured over 30,000 digital memories triggered by the beating of viewer's hearts. Embedded in these data is the life and emotion of each person in that moment, some with a steady drumbeat while others with racing heartbeats. It reminds us that we're all fundamentally connected through the electricity-driven pump in our chests and through the electrical impulses traveling throughout the internet.

6.2.2 DIKKER + OOSTRIK Projects

DIKKER + OOSTRIK take a slightly different angle when exploring the embodiment of synchrony and human connectedness, to investigate the factors that determine whether our communicative interactions, in pairs or in groups, are successful. To do this, we use an EEG hyperscanning approach, measuring the similarities and differences in real-time between the brain activity of people engaging in dynamic social interactions: When we feel 'on the same wavelength' during a communicative

Fig. 6.3 Hive Mind. Hive Mind at the RIXC Art Science Festival 2017. EEG read via OpenBCI hardware. Image by Sean Montgomery. Web: http://produceconsumerobot.com/hivemind/. Neural Issue #59, Winter 2018

exchange, are our brains actually 'in sync' in a more formal, quantifiable sense? Which factors affect such hypothesized alignment of brain activity? Does the ability to successfully tune in with other people's intentions, expectations, and experience increase the synchronization of brain activity? And how might art help inform these neuroscientific questions? Matthias Oostrik's independent work explores the impact digital technology has on our society and the influence it has on our relations to others. His work lies at the interface of computer art, installation art and film and makes use of experimental man-machine interaction. For example, plplpl.pl is an ongoing series of interactive video installations by Matthias Oostrik, in collaboration with Pandelis Diamantides and Diederik Schoorl (2016–ongoing; Video: https:// vimeo.com/95962192). The installment plplpl.pl::scrutiny is a tech-noir surveillance machine made from displays steel and wires. The machine uses technologies from the surveillance industry to interact with its visitors. It provokes them into weird and devious behaviour, while analysing, correlating and storing all of their actions. The installation has been shown, among others, at the V2 Institute for Unstable Media (NL), TodaysArt Festival (NL), Next Level Festival (DE) and the Dutch parliament. Suzanne Dikker's independent research focuses on the neural correlates of top-down modulations of sensory cortices as a function of linguistic prediction (e.g., Dikker

Fig. 6.4 Emergence. Emergence installation at Open House Gallery, New York City (left) and digital memories triggered by successive heartbeats and uploaded to Flickr (right). Left image by Sean Montgomery. Right images by Emergence courtesy of Sean Montgomery. Web: http:// produceconsumerobot.com/emergence/; Leonardo Electronic Almanac Vol 18 No 5 pp 6–9

et al. 2010), the role of language prediction in speaker-listener brain-to-brain coupling (e.g., Dikker et al. 2014) and, most recently, the role of brain-to-brain synchrony as a possible biomarker for dynamic social interactions and face-to-face communication in real-world settings (Matusz et al. 2018). For example, in a recent series of studies (Dikker et al. 2017; Bevilacqua et al. 2018), Dikker and colleagues collected simultaneous EEG data from two groups of high school students during their regular class activities (Dikker et al. 2017; also see video by Micah Shaeffer: https://vimeo.com/108921898). In collaborative work, we (DIKKER + OOSTRIK) combine our own approaches and expertise in a series of art/science BCI projects that have developed as illustrated in Fig. 6.5:

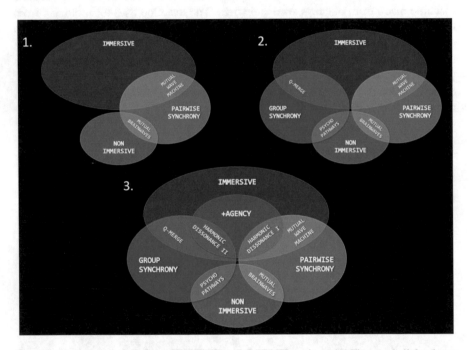

Fig. 6.5 Conceptual workflow of DIKKER + OOSTRIK projects. (1) First, we built hardware and software solutions that allowed us to record and compare brain activity from two people simultaneously (*pairwise synchrony*), with which we created 4 projects: three immersive installations (Measuring the Magic of Mutual Gaze, Compatibility Racer, and The Mutual Wave Machine) and one non-immersive neurofeedback game (Mutual Brainwaves Lab). (2) As a next step, we developed 'mirror projects' for Mutual Brainwaves Lab and the Mutual Wave Machine, expanding from pairs to groups (*group synchrony*). SocioPathways is a group synchrony non-immersive neurofeedback game and Q-merge will be a 5-person immersive group synchrony installation. (3) We then added yet another dimension to our projects in the Harmonic Dissonance series: by allowing participants to interact with the visualizations of their brainwave synchrony, we give them *agency*. Harmonic Dissonance: Synchron(icit)y is an immersive pairwise synchrony project, and Harmonic Dissonance: Act V is an immersive group synchrony project

1. First, we built hardware and software solutions that allowed us to record and compare brain activity from two people simultaneously (*pairwise synchrony*), with which we created 4 projects: three immersive installations (Measuring the Magic of Mutual Gaze, Compatibility Racer, and The Mutual Wave Machine) and one non-immersive neurofeedback game (Mutual Brainwaves Lab).
2. As a next step, we developed 'mirror projects' for Mutual Brainwaves Lab and the Mutual Wave Machine, expanding from pairs to groups (*group synchrony*). SocioPathways is a group synchrony non-immersive neurofeedback game and Q-merge will be a 5-person immersive group synchrony installation.
3. We then added yet another dimension to our projects in the Harmonic Dissonance series: by allowing participants to interact with the visualizations of their brain-wave synchrony, we give them *agency*. Harmonic Dissonance: Synchron(icit)y is an immersive pairwise synchrony project, and Harmonic Dissonance: Act V is an immersive group synchrony project.

6.2.2.1   Pairwise Synchrony

Measuring the Magic of Mutual Gaze (2011)

Suzanne Dikker, Matthias Oostrik, Marina Abramovic, and The Art & Science: Insights into Consciousness Group

Continuing Marina Abramović's interest in the transfer of energy between performer and public, performer and participant, Measuring the Magic of Mutual Gaze builds on her durational performance works Nightsea Crossing (1981–1987) and The Artist is Present (2010). During the latter, which was performed at The Museum of Modern Art (New York City; Fig. 6.6), Marina Abramović engaged in silent mutual gaze with museum visitors over a three-month period. Abramović reported feelings of pain, happiness and sadness, which resonated with those of the person sitting opposite her (Abramović 2010).

Measuring the Magic of Mutual Gaze (Fig. 6.6) restages The Artist is Present as an interactive art installation/neuroscience experiment, with the goal to investigate the relationship between human connectedness and brainwave synchronization. In a mix of scientific experiment and performance, participants sign consent forms and are then led to a stage by two assistants dressed in white lab coats, where they are fitted with EMOTIV EPOC wireless EEG headsets. Participants sit opposite each other and engage in eye-contact for 30 min. Meanwhile, two main features are extracted from the EEG signal in real time, and superimposed onto two rotating 3D brain models, displayed on screens hung directly behind the participants: (a) each participant's individual internal mental state and (b) synchronized EEG activity between the participants' respective brains (details below). The model brains pulsate at the frequency that is most prominent at any given moment: slowly pulsating brains are hypothesized to reflect more relaxed and meditative states of mind, while faster pulsations reflect agitation, excitement or concentration. Whenever brain synchrony

Fig. 6.6 Measuring the Magic of Mutual Gaze & The Artist Is Present. Top: Measuring the Magic of Mutual Gaze at The Garage Museum for Contemporary Art, Moscow in 2011. Photograph by Maxim Lubimov © Garage Center for Contemporary Culture. Video: www.youtube.com/watch? v=Ut9oPo8sLJw&t=73s. Bottom: Marina Abramović, The Artist is Present, Performance, 3 months, The Museum of Modern Art, New York, NY (2010), Photography by Marco Anelli. Courtesy of the Marina Abramović Archives

exceeds a certain threshold, this is shown through a lightning animation connecting the two model brains.

Using the EMOTIV EPOC device, EEG data was collected from ~150 visitors of the Garage Museum of Contemporary Art in 2011 and results from 50 participants can be summarized as follows. First, brain-to-brain synchrony (quantified as Pearson correlation coefficients over a 3-s sliding windows between amplitude modulations in delta, theta, alpha and beta [FFT] from each electrode in each headset [14 × 4 comparisons per window]) was significantly greater between participants engaging in mutual gaze than between participants who were not in the same room at the same time (i.e., sitting with someone else). Second, brain-to-brain synchrony was significantly correlated with pairs' average scores on the Personal Distress Scale (PDS) of the Interpersonal Reactivity Index (Davis 1996). These effects were most prominent in alpha (see Introduction), potentially indicating that people with high Personal Distress scores are worse at tuning into physical cues in their partners (e.g. coordinated eye-blinks, breath rate), in turn facilitating mutually adaptive alpha phase-synchronization. This result was later replicated in Dikker et al. (2017, in revision) for brain-to-brain synchrony in both groups and pairs of participants.

Measuring the Magic of Mutual Gaze was exhibited at the Garage Museum of Contemporary Art (Moscow, 2011) as part of Marina Abramovic' retrospective exhibit The Artist Is Present, and at SOFT CONTROL: Art, Science, and the Technologically Unconscious (Maribor, 2012).

The Mutual Wave Machine (2013–Ongoing)

DIKKER + OOSTRIK in collaboration with Peter Burr, Diederik Schoorl and Pandelis Diamantides

In Measuring the Magic of Mutual Gaze, the BCI component of the installation is geared toward the audience, not the participants. In the Mutual Wave Machine (Fig. 6.7; and in the rest of the Projects described here) synchrony neurofeedback is instead incorporated as an integral part of the experience. Enclosed by an intimate capsule and immersed in an audiovisual environment that responds and reflects their shared brain activity, two visitors can directly experience and manipulate their internal efforts to approach or distance themselves from each other. During the experience, greater brainwave synchronization is reflected in greater vividness and more coherent and recognizable audiovisual patterns, while lack of synchronization strays towards dark audio-visual chaos: a faint ringing in the ears and static in the retinas. Concretely, synchrony (quantified again as moving-window correlations in different frequency bands, see also Measuring the Magic of Mutual Gaze) is translated into expanding and retracting moire patterns. This aesthetic that is inspired by Peter Burr's video animation Alone With the Moon (2012) and is aimed at inducing the feeling of frustration in the absence of connectedness (darkness) and exhilaration (light) when synchrony is achieved: why is it that we can feel infinitely lonely in the presence of another human being when we fail to achieve a connection to this other human being? An additional layer of complexity is added via a real-time video feed: when there is more synchrony/light, a real-time video image of oneself emerges from the noisy moire patterns behind the person sitting opposite, alluding to questions of what it *means* to connect to someone: are we truly 'seeing' the other person, or is the process of synchrony achieved via mirroring our own thoughts and behavior; projecting them onto the other? The visualization of neural synchrony is supplemented by a direct audio translation of each participant's individual brain activity, creating an evolving composition of volume changes and harmonies that hover at the often uncomfortable perceptual boundary between rhythm and pitch. The audience can hear and observe the internal dance unfold through the semi-translucent shell of the Machine. Only a distant shadow remains of the participants' physical presence, their autonomy replaced by a new, shared identity.

Just like Measuring the Magic of Mutual Gaze, participants fill out pre-experiment and post-experiment questionnaires. These responses are then be correlated with brain synchrony during subsequent offline analysis. Over 4,000 people so far have participated in the Mutual Wave Machine across more than a dozen sites (see below). Results from 1,500 participants suggest that empathy, neurofeedback, and connectedness are strong predictors of brainwave synchrony. (In this case, the metrics used to quantify brain synchrony were imaginary coherence, Nolte et al. 2004; and projected power correlations, Hipp et al. 2012). For example, replicating findings from Measuring the Magic of Mutual Gaze and Dikker et al. (2017), pairs of participants with high Personal Distress exhibited less synchrony than those with low Personal Distress personalities (see also Goldstein et al. 2018 for results showing that empathy affects brain-to-brain synchrony). Similar effects were observed for social connectedness (using the Inclusion of the Other in the Self Scale; Aron et al. 1992), mood, and focus (Watson and Clark 1994). We also found that the 'connection strategies' used by participants influenced their brain synchrony: Pairs who used either eye contact or joint action as a connection strategy (mimicry, laughter, motion coordination) exhibited an increase in brainwave synchrony over time, in line with previous findings showing that joint action as well as eye contact can synchronize our brainwaves (see Introduction). Such an increase in synchrony was not observed for pairs who tried 'thinking about the same thing', or when they chose to focus on the audiovisual environment instead of each other. The latter is perhaps surprising in light of the brain-stimulus entrainment literature discussed in the introduction, although it is not clear if participants were also focusing on *the same* external audiovisuals.

Finally, and most relevant to the topic of this book, the neurofeedback environment affected brain synchrony in participant pairs. As discussed in more detail in Dikker et al. (in revision), those pairs who were explicitly told that the visuals were a direct reflection of their brainwave synchrony, also showed an increase in brain synchrony over time. Such differences were not observed when comparing 'real' versus 'sham' neurofeedback conditions (in the latter case the growing/shrinking of moire patterns was randomized), suggesting that the neurofeedback environment functions as a global motivational factor to maintain shared, focused attention on the task, rather than providing informative cues on a moment-to-moment basis that can be used to calibrate and improve upon the social interaction.

The Mutual Wave Machine has been exhibited at *TodaysArt Festival* (The Hague, 2013), *Eye Film Institute* (Amsterdam, 2013), *Lexus Hybrid Art* (Moscow, 2014), *Nemo Science Center* (Amsterdam, 2014), *Lowlands Festival* (The Netherlands, 2015), *Benaki Museum* (Athens Greece, 2016), *3LD Art & Technology Center* (New York City, 2016), *FORMS Festival* (Toronto, 2016), *Pioneer Works* (Brooklyn, 2017), *Pioneer Works* (Brooklyn, 2017), *OPUS 1, Merriweather* (Maryland, 2017) and *Espacio Fundación Telefónica* (Madrid, 2018), among others. EEG was recorded

Fig. 6.7 The Mutual Wave Machine. By Suzanne Dikker, Matthias Oostrik, Peter Burr, Diederik Schoorl and Pandelis Diamantides. Photographs by Sandra Kaas, Talia Hermann, Lexus Hybrid Art, Video: https://vimeo.com/96287858. Web: http://todaysart.org/project/99/

using the EMOTIV EPOC or the Muse headband (Madrid). Scientific findings are reported in Dikker et al. (in revision) (Fig. 6.7).

Mutual Brainwaves Lab (2014–Ongoing)

DIKKER + OOSTRIK

Mutual Brainwaves Lab (Fig. 6.8) is an interactive neurofeedback game that tracks and visualizes brain-to-brain synchronization as two heads merging in and out of each other. Whenever brainwave synchrony (again using the EMOTIV EPOC headset and quantifying synchrony as moving-window Pearson correlations, see above) meets a certain threshold, the heads perfectly overlap and participants' synchrony score goes up. As such, participants are challenged to get the 'highest' synchrony score during a 3 min period (a clock counts down the seconds on the display).

Just like the Mutual Wave Machine, hundreds of festival and museum visitors have participated in Mutual Brainwaves Lab, and similar forms were filled out by participant pairs. Unlike the Mutual Wave Machine, however, the data are not used for scientific analysis: Mutual Brainwaves lab is implemented as a neuroscience educational and outreach tool (venues include the American Museum of Natural History (New York City, 2013), World Science Festival (New York City, 2013, 2014); Basilica Hudson (Hudson, 2014), Pioneer Works (Brooklyn, 2016), among others).

NeuroTango used the Mutual Brainwaves Lab interface as a tool during an evening discussing connectedness in dance, and tango specifically (La Sala, Brooklyn NY, 2014). Two pairs of tango dancers competed against each other for the highest synchrony score during different 'conditions': dancing with familiar partners, switching partners, dancing to music, dancing without music, listening to music without dance, and perfect silence. While the data was not used for scientific purposes, the neurofeedback environment/performance/experiment served as a tangible illustration and conversation-starter for a discussion surrounding the neuroscientific constructs discussed in the Introduction and Discussion sections of this chapter: synchrony during joint action, entrainment, and emotional effects of connectedness during dance, etc. (see also Ballroom Brainwaves, *The Scientist Magazine*, 2014).

Compatibility Racer (2012–Ongoing)

Lauren Silbert, Jennifer Silbert, Oliver Hess, Suzanne Dikker, and Matthias Oostrik

Compatibility Racer (Fig. 6.9) is a competitive, interactive brain-robotics installation in which brainwave synchronization is translated into the speed of a cart: the more in sync participants' brains, the faster the cart moves along a track. The project was conceptualized by Lauren Silbert, following her fMRI research investigating the relationship between speaker-listener neural synchrony and communicative success (Stephens, Silbert et al. 2010), and developed in tandem with Measuring the Magic of Mutual Gaze.

Participants are outfitted with EMOTIV EEG headsets, pair up, and sit facing each other on the "bull." The "bull" moves according to an online correlation analysis between the brain activities of each person (same analyses as the other projects described in this section). Exploring the underlying brain mechanisms of interper-

Fig. 6.8 Mutual Brainwaves Lab. Top: View of Mutual Brainwaves Lab display with no synchrony (left) and 'full' synchrony (right). See www.youtube.com/watch?v=d64SeneJpgY for rendering. **Center**: Impressions from the American Museum of Natural History (2013). Photograph by Ellen Pearlman artdis.tumblr.com/post/45762224599/on-the-same-wavelength-the-brain-hits-the-museum. **Bottom**: NeuroTango (La Sala Brooklyn 2014). Screenshot from youtube.com/watch?v=d64SeneJpgY/www.the-scientist.com/daily-news/ballroom-brainwaves-37746

Fig. 6.9 Compatibility Racer. **Top/Center**: Compatibility Racer at Kulturpark Berlin. Photography Kate Moxham. compatibilityracer.blogspot.com. **Bottom left**: Marina Abramovic riding the Compatibility Racer in Rhinebeck, NY (2013) vimeo.com/71165002. **Bottom right**: Visualizations of 'results' for two sample pais

sonal communication through transportation, each "bull" moves as a direct result of increasingly shared brain activity. Conversely, movement slows or halts as a function of participants' lack of (brainwave) alignment. Thus, the participants' movement is literally fueled by successful communication and collaboration. Compatibility Racer was exhibited at Kulturpark Berlin (2011). Data was not used for scientific analysis.

6.2.2.2   Group Synchrony

In our scientific research in the classroom (Dikker et al. 2017; Bevilacqua et al. 2018), we implemented the simultaneous collection and visualization of (EMOTIV EPOC) EEG data in groups. In SocioPathways and Q-merge, we convert this extension from pairs to groups into neurofeedback experiences, essentially creating group-based versions of Mutual Brainwaves Lab and the Mutual Wave Machine.

The projects described below use either the EMOTIV EPOC+ or the Muse headset to collect EEG data (can be alternated within a project depending on the context). For Harmonic Dissonance: Synchron(icit)y EEG from audience members was additionally recorded using Brain Products LiveAmp 32 channel wireless amplifiers with RNet sponge-based electrode caps. We further switched to using Lab Streaming Layer (github.com/sccn/labstreaminglayer) to synchronize data recordings from multiple devices and integrate with our custom openFrameworks (openframeworks.cc) BCI environment.

SocioPathways

DIKKER + OOSTRIK

SocioPathways (Fig. 6.10) is a social network brain-computer interface game that visualizes brainwave synchrony between teams of 2–5 people. It is similar to Mutual Brainwaves Lab, but instead of visualizing synchrony between pairs of people, SocioPathways draws a real-time social network for a team of five people: individuals are represented as dots that move closer or further away from each other as a function of their brainwave synchrony. Team members with the strongest synchrony to the rest of the group are displayed as larger 'hubs.' Scores are kept at both the team level and the individual level. In addition to standing on its own as a neurofeedback experience to be used in similar contexts as Mutual Brainwaves Lab (both art & education/outreach), SocioPathways serves as the 'drawing board' for the larger-scale installations described below: It is the basis installation where we experiment with neural synchrony at the group level.

Q-merge (in Development)

DIKKER + OOSTRIK in collaboration with Diederik Schoorl and Peter Burr

Q-merge (Fig. 6.11) will be an immersive group synchrony installation that translates brain synchrony between 5 people into an audiovisual immersive experience. We will continue our collaborations with Peter Burr and Diederik Schoorl in expanding the Mutual Wave Machine into a larger-scale installation that can accommodate 5 people. We will also use this opportunity to incorporate our scientific findings into our analysis algorithms, ensuring that the experience we create accurately reflects our most current understanding of how different types of brainwaves predict interpersonal

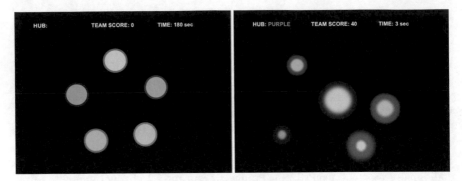

Fig. 6.10 Snapshots of SocioPathways display. Left: start position. **Right**: snapshot of an intermediate network configuration. Person-to-person synchrony is displayed as the physical distance between their bubbles; 'hubs' are indicated with the size of the bubbles; average 'hubness' over time is displayed through the darker haloes. Video: plplpl.pl/hd

Fig. 6.11 Q-merge render. Q-merge is an immersive art/neuroscience installation that translates group synchrony into patterns of light along a tangled (social) network of connections

experiences of social connectedness. The installation will consist of a network of LED light cables strung along a dome-like structure to form a tangled network reminiscent of organic patterns like veins, cerebral arteries, and lightning bolts. Each person sits at the 'root' of one light structure, the source from which light 'shoots' along the LED cables as a function of brainwave synchrony. Like with the Mutual Wave Machine, more light equals more synchrony. The installation is inspired by Mikey Siegel's HeartSync and Sean Montgomery's Emergence (see above).

6.2.2.3   Agency: The Harmonic Dissonance Series

In all of the interactive brainwave synchrony installations described above, participants are seated, their locations fixed. In the ***Harmonic Dissonance series*** we conjoin our brainwave synchrony technology with motion tracking. In the two works described below, participants are allowed to move freely through space while they are emerged in reflections of their brainwave synchrony. As such, the opportunity is created for participants to 'disobey' our installations and experiment, raising questions about agency in BCI (Brenninkmeijer 2013).

Harmonic Dissonance incorporates features of all the installations described above, with the use of FlowTools, an open source tool for interactive fluids developed by Matthias Oostrik in 2013 (github.com/moostrik/ofxFlowTools) Oostrik presented works made with the tool in e.g., the Eye FilmMuseum and Paradiso in Amsterdam. FlowTools is incorporated into the Harmonic Dissonance series as a tool to motivate people to interact freely with the installations.

Harmonic Dissonance: Group Installation (2018–Ongoing)

DIKKER + OOSTRIK with Arnoud Traa, Suzan Tunca, and Dana Bevilacqua

In Harmonic Dissonance (Fig. 6.12), groups of visitors are invited to explore human connectedness as it is mediated by their bodies, brains, and artificial algorithms. In interaction with the audience, Harmonic Dissonance investigates the friction—the harmony and dissonance—between our experience of connectedness and the algorithmic data systems that describe us and our relationships to others.

Harmonic Dissonance is shaped as an interactive, hexagonal playground; it is designed to be installed in various locations, like museums and festivals. In successive sessions, four people from the audience are invited to participate and engage with each other in a visceral, intuitive way. These participants are provided with sensors to measure their brain activity, heart rate, and skin conductance. On the basis of participants' biometric data and movements, the installation generates personalized audiovisual patterns. Combined these patterns reflect the level of biometric synchrony between the participants, as such serving as an impetus for social interaction: Movements, sounds, visuals, and the pulsating rhythm of continuous interpersonal attraction and retraction mingle into a thrilling multi-layered performance.

Following the visual SocioPathways structure described above, each participant is assigned his or her own bubble-shaped avatar that is projected on the floor. These visuals are able to flow and change shape, yet they remain recognizable via specific colours and patterns. As described above, the avatars form a liquid social network graph that maps the connections between the participants. The graph consist of two parts: the actual location of the participant and the location that most reflects the participant's connection to his companions. Different movements and new correlations in the streams of biodata shift, distort, and fracture the graph around and in between the participants.

Participants are assigned their own signature audiostream. Together, these adiostreams create a dynamic and rich soundscape that evolves as visitors synchronize their brainwaves, heart rate, and gestures. Each participant's movement through space leaves an audio-visual trace, amplifying any disparity between the group and the individual. The intensity and melody reflect the level of synchrony between participants: an increase in synchrony results immediately renders the audiostream more harmonious. For example, when participants mimic each other's gestures, their individual sounds become more harmonized and adapt to the same rhythm. Individuals who are 'hubs' attract the traces of the other individuals, as if they were sucking the others' energy into their body. Finally, if group synchrony increases, the colors and sounds merge toward a solemn, pixelated black-and-white pattern, swallowing the individual voices and colors into ripples on the surface of a still, mesmerizing lake. Participants can choose to blindly follow the visualizations and sonifications of synchrony as they lead them through the space or to step out of their 'bubble' and rebel against the algorithms' interpretation of their connection to others.

Harmonic Dissonance (Fig. 6.12) has so far been executed at Studio Mellius (Amsterdam, 2018) and the Deutsches Filminstitut in Frankfurt (2019).

Fig. 6.12 Harmonic Dissonance. Harmonic Dissonance at Studio Mellius, Amsterdam (video stills from: plplpl/hd)

Harmonic Dissonance: Synchron(icit)y (2018–Ongoing)

DIKKER + OOSTRIK with Suzan Tunca, Sedrig Verwoert and Arad Inbar of ICK Amsterdam; David Medine of Brain Products, and Arnoud Traa of the Auditieve Dienst

In Harmonic Dissonance: Synchron(icit)y (Fig. 6.13) two dancers perform in a double duet: as their bodies dance together, so do their brainwaves, projected as *FlowTools* animations behind them. They can choose to go with the flow, coming closer—merging into each other—as their brain synchrony increases, and stepping away from each other with more dissonant brainwaves. But they can also choose to be their own agents, to move against the FlowTools tide or stand still in the stream. As such, a third duet emerges, one between their physical and mental worlds. In contrast to the other works proposed, Harmonic Dissonance: Synchron(icit)y is not an interactive installation piece but rather intended to be performed in front of an audience (following in the tradition of Oostrik's Mirror of Broken Time and Hidden Features; see www.matthiasoostrik.com).

The work invites the dancers to explore how motion synchrony (coordination, mirroring, unison, anticipation, resonance, play, etc.) may lead to a sense of togetherness, to **synchronicity** of the mind. It is realized as a performative research presentation based on fragments of repertoire from Emio Greco and P. C. Scholten, artistic directors of the International Choreographic Arts Center ICK Amsterdam. Two dancers test and highlight seven variations on *Synchrony* in dance via choreographed fragments that are intended to be performed aiming at *Synchronicity*, and via improvisation based on pre-choreographic elements[1] derived from dance notation research. As laid out in the Introduction, in the paradigm of the artistic signature of Emio Greco | PC, *Synchronicity* is defined as a "dual utopia", an aim, an ideal, whereby a qualitative relation of being "one body" is strived for, in a state and dimension of being above and beyond dancing merely in synchronized unison. The term "synchron(icit)y" is used here to indicate the emerging transdisciplinary zone between Dikker and Oostrik's conceptual decomposition of "synchrony" and the movement intention and resulting quality aiming at embodying "synchronicity" characteristic of Greco and Scholten's choreographic oeuvre, interpreted by two dancers (more on this in the Discussion). The seven scenes elaborated during a first workshop session in 2018 that highlight variations on synchron(icit)y are:

I. "Mr. Hide" (from repertoire piece Double Points: Two). The dancers move in a relation of "sameness" whereby one dancer tries to hide behind the other by aiming at performing the movements in exact "synchronicity", as "one body". The visualizations respond to the intensity of the movements. Then the

---

[1]Pre-choreographic Elements refer to the pre-phase of choreography, where the movement material is being created, shaped and tested. They are not yet part of the selection and ordering process of choreography The pre-choreographic elements are the roots of choreographic statements and constitute the conditions from which the choreographic elements can emerge. A selection of 20 pre-choreographic elements has been identified through dance notation research and is currently being investigated with regard to their generative potential in movement language creation.

Fig. 6.13 **Harmonic Dissonance: Synchron(icit)y. Top**: Stills from performance at Buitenplaats Doornburgh, the Netherlands (*Credit* Nina & Nikki Fotografie – Nikki Schuurman). **Bottom**: Theater de Meervaart (bottom, video still from: https://vimeo.com/288711606)

roles reverse: the dancers respond to the visualizations by incorporating their awareness of these visualizations into the dynamics of acceleration inherent to the dance phrase.

II. "Around Shoulder" (Pre-choreographic element as departure point for improvisation). Synchron(icit)y is established between organic movements initiated from the idea of transforming the texture of the dancer's physicality by playing with the outward and inward rotation of the shoulders, and with the resulting visualization. It is a relation of synchron(icit)y between one dancer and "the other world" of the visualization where the dancer both responds to it while at the same time lets his/her movement quality and texture be influenced by the visualization.

III. "Undulation" (movement materials based on Pre-choreographic element "Undulation" in repertoire piece "Ziel/RUOH"). The dancers play mainly with the acoustic output generated by the data derived from their motions. A sense of "dancicality" (musicality embodied in dance) is strived for, where the dancers begin to extend the impact of their movement and occasional events of synchron(icit)y may occur. The dancers combine choreographed movement material with an improvisational approach based on instant movement choices related to sensing each other's motions and in creative dialogue with the acoustic output generated by their movements. The dance unfolds based on a playful approach to "dancicality" . A "feedback loop" occurs between the sounds generated by the dancers and the dance.

IV. "Fire walks with me" (movement material from repertoire piece "Conjunto di Nero"). The dancers perform the sequence ideally in perfect "synchronicity", mirroring each others movements while moving forwards side by side. They strive towards performing the choreographed movements in "synchronicity" without being able to see each other. The sequence is performed twice with increasing scope and intensity. The dancers' quest for "synchronicity" via synchronized movement and ideally mutually aligned intention is amplified by the visuals: The visual echo of each movement pulls the dancers toward each other, to the one body that lies between them.

V. "Rhythm on 2" with exits into improvised dialogue in "Rhythm on 2" (choreography and improvisation based on Pre-choreographic element). Here the dancers are in a relation of synchrony between an internal and an external rhythm. They create a resonance between their internal models of rhythm and the externalization of rhythm via the visuals and the acoustics.

VI. Improvised dialogue: The *synchrony* relation between the dancers is here mainly on the level of *coordination*, tuning into each other's presence, dialoguing with each other via freely improvised movements, responding creatively to each other, to the acoustics and to the visualizations.

VII. Dancing bodies in synchron(icit)y without external technology.

In 2018, Harmonic Dissonance: Synchron(icit)y was performed at Buitenplaats Doornburgh and Ballet National de Marseille. In the former case, the interactive FlowTools screen remained on display for the general audience: the performance, in addition to a public lecture, contextualized the interactive piece. The performance at the Ballet National De Marseille, in contrast, served as a feasibility/pilot experiment for a performance EEG research project: EEG was collected from both the dancers and the audience. There are many challenges to conducting such a project, ranging from limiting the amount of bluetooth devices in a single space to motion artifacts in the dancers' EEG signal. The latter limits the choreographic choices. Specifically, 'usable' EEG data will most likely be collected during moments of stillness in the dancers' bodies, or at least where head motion is restricted. This need not be a limitation per se: for example, as discussed in the Introduction, synchronous movement generation (and perception) may be dependent on joint prediction/attention *prior* to the actual movement (Dikker and Pylkkänen 2013; Dikker et al. 2014 among many others, see also Introduction), which can be measured in a still period leading up to the initiation of the next movement sequence. Similarly, periods of silence *following* a movement sequence (labeled *resonance* below), can serve to measure unison in mental simulations of the completed movement. This illustrates how limitations and challenges can at times in fact be a benefit to research projects. In sum, the work not only aims at testing the feasibility a large-scale dance-audience synchrony experiment, but it also inspires theory formation. To give another example, discussed in more detail below: a fine-grained taxonomy of synchrony and synchronicity can serve to inspire hypotheses for neuroscientific exploration. The fact that dancers may at times move in unison without the subjective experience of joint action, for instance, leads to the question if neural responses that track "togetherness" can be empirically distinguished from those that track "simultaneity". Such questions, which are discussed in more detail in the Discussion, are relevant beyond performative contexts and can be asked e.g., in laboratory experiments investigating how the brain supports joint action.

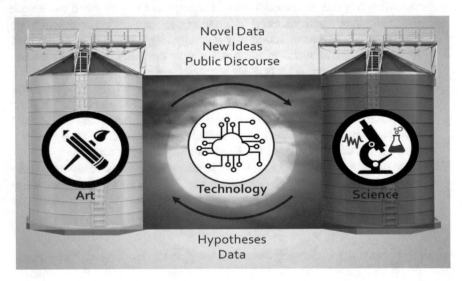

Fig. 6.14 Art + Science Synergy. Multiplicative synergy can be created by taking a transdisciplinary approach to art and science. Utilizing scientific data and hypotheses it's possible to develop installations that provide immersive aesthetic experiences. These public installations can generate novel datasets to test hypotheses and generate new ideas as well as engage public discourse. Image credits: Alexander Gruzdev, Jacopo Bonacci, Asgeir Visir, Pool party, Aiden Icons, from the Noun Project

6.3 Discussion

6.3.1 Art + Science Synergy

Art and science fundamentally seek the same truth via different perspectives. While scientists toil in the laboratory to reduce questions into testable hypotheses and artists seek to turn questions into experiences that resonate with their audience, at their best both attempt to peer into essence our brief but spectacular blip of existence. While each perspective has immense value on its own, we believe that value is multiplied by taking a transdisciplinary approach to art and science. To take a very simple metaphor illustrated in Fig. 6.14: Like colors, ideas and information multiplies. If you start with the colors red and yellow and keep them siloed, you have just 2 colors, but if you mix them suddenly you have the beautiful palette of a rising sun. And similarly, by taking a transdisciplinary approach to art and science, we can create new perspectives and ideas, generate novel datasets and methods to test those ideas, and engage public discourse in a personally moving way. As we've shown with the works described above, it's possible to bring questions and hypotheses out of the lab to create an immersive aesthetic experience that gives the viewer an opportunity for reflection and at the same time generates novel data to test those hypotheses. We believe this transdisciplinary approach creates a multiplier effect for art and science to increase curiosity, ask new questions and develop new approaches to answer those questions (for similar views, see Kerfeld 2009; Eldred 2016; Prusinkiewicz 1998; Jeffries 2011; Stevens and O'Connor 2017; Miller 2014).

Our work is closely tied to a large-scale interdisciplinary research program aimed at understanding the relationship between brainwave synchrony and communicative success. By bringing together experts from science, technology, education and art in a crowd-sourcing neuroscience approach, our projects aim to advance scientific insights into interpersonal dynamics in real-world environments, while at the same time providing visitors with both an educational and aesthetic experience. In these works the audience participates as viewers and experimental subjects at once, actively contributing to a scientific goal while exploring the intangible notion of human connectedness through artistic inquiry. Our work contributes to a myriad of fields, including software engineering, scientific design and analysis techniques, science outreach and education, and therapeutic contexts. Below we address in more detail each of the 'collateral benefits' already touched upon in the Introduction (technological innovation, real-world and 'crowdsourcing' neuroscience, neurofeedback, neuroscience outreach and education), and follow with a section devoted to the (bidirectional) conceptual enrichment enabled by art-science collaborations.

6.3.2   Technological Innovation—Hardware and Software: Bridging Commerce, BCI Art, and Science

Technology has become an important mediator between the art and science worlds in our work. The past decade has seen a sharp increase in the number of DIY and commercial-grade devices that are capable of measuring EEG and other biometrics. In contrast to standard laboratory equipment, EEG devices like the EMOTIV© EPOC and EPOC+ are portable and low-maintenance, and yet to date the EMOTIV© EEG headset has primarily been used in commercial applications. Similarly, OpenBCI offers a configurable multi-channel headset that transmits EEG data wirelessly to a computer. Data acquired from our projects are proving themselves to be extremely valuable for neuroscientists to test the applicability of the hardware and data quality for scientific purposes. The successful implementation of wearable mobile EEG recording headsets in scientific research can have significant impact on future neuroscience research. For example, they may provide a valuable tool for scientific research outside of the lab (e.g. theatres, schools, and navigating traffic; e.g., Debener et al. 2012; Zink et al. 2016) with populations that are otherwise difficult to reach (e.g. children, patients, the elderly). For applications where fewer channels are needed, the Muse and Neurosky (and formerly Zeo) devices offer a simpler process to take the headsets on and off, which can be advantageous for exhibits with lots of participants.

In addition to EEG, there are a number of additional biometrics that can potentially serve to create a brain-computer interface (Montgomery and Laefsky 2011). Tapping into the sympathetic and parasympathetic nervous systems, heart rate and other biometrics can reveal aspects of a person's psycho-physiological state (see Emergence above; Cacioppo et al. 2000; Lisetti and Nasoz 2004; Cai and Lin 2011). Termed "affective computing" (Picard 1995), by reading a constellation of biometrics including heart rate, blood pressure, electro-dermal activity, local body temperature, etc., it's possible to reliably predict changes in emotions with 60–80% accuracy (Healey and Picard 2005; Rigas et al. 2012; Collet et al. 2009; Mehler et al. 2012; Baek et al. 2009; Akbas 2011). In the last decade there has been an explosion of wearable devices on the market that measure one or more biometrics that tap into the sympathetic/parasympathetic nervous system.

The rise of wearable commercial devices that capture EEG and other biometrics has substantially lowered the barrier to entry for use in both artistic and scientific contexts. Being able to easily put on and take off wearable biometric sensors allows people to participate in BCI art works without requiring lengthy a sensor application process. Furthermore, having sensors that can communicate wirelessly with a local computer or mobile device enables novel experimental paradigms and aesthetic experiences that were previously unthinkable. While there remain challenges including obtaining open access to device data, accurately synchronizing the data across devices, scientifically validating the data with known gold standards, and the here-today/gone-tomorrow startup world (Lisetti and Nasoz 2004), the growing number of commercial-grade options continues to open up more possibilities for synergy in science and art.

Developments in open-source software and data infrastructure are also fueling greater synergy at the intersection of art and science. The custom software developed by Matthias Oostrik for the projects described above, for example, records brain activity (EEG) from multiple people simultaneously, analyzes and compares these data between individuals, and translates it into an audio-visual output in real time. This software, built on the open-source project openFrameworks (OF, https://openframeworks.cc/) and shared via github (https://github.com/), is continuously evolving and easily adaptable for different contextual uses and has proven extremely valuably in our scientific research experiments (Dikker et al. 2017; Bevilacqua et al. 2018), in addition to real-time art installations. Similarly, tools that were originally developed for the laboratory, like Lab Streaming Layer (LSL, https://github.com/sccn/labstreaminglayer) have become invaluable tools for aggregating diverse data streams in artistic contexts. As we look forward, platforms to collaboratively share data, methods and tools like Open Science Framework (OSF, https://osf.io/) may generate additional synergy at the intersection of art and science.

6.3.3 Real-World and 'Crowd' Neuroscience

To date, the study of the human brain has relied on one very fundamental assumption: laboratory-based research findings can be extrapolated to real world situations. Although this approach has proven invaluable in advancing our understanding of the human mind, researchers do at times hit a wall: realistic human interactions are much more complex and richer than can be captured in canonical laboratory experiments. By collecting data in non-laboratory settings, and by focusing on real-time and real-life dynamic social interactions, we aim to capture the brain basis of real-world everyday social interactions. This is made possible by a unique combination of recent developments: the technological innovations discussed in the previous section, the rise of hyperscanning approaches (see Introduction), and the growing interest in cross-disciplinary collaborations, which has enabled us to form partnerships with art and science institutions. This allows us to collect neuroscientific data from vast numbers of participants under circumstances that much closer resemble naturalistic human communication than could ever be achieved in a laboratory environment (Bhattacharya 2017; Parada and Rossi 2017 for short opinion pieces).

We have collected biometrics from tens of thousands of participants across the art installations described here (by comparison: most laboratory neuroscience experiments do not exceed 20 participants). These large datasets massively increase the statistical power by which to extract novel relationships in biometric data. In addition, by sampling from a more diverse group than a typical undergraduate participant pool, the effects under study are more likely to apply to the populace at large. And furthermore, combined with participant demographic data it may be possible to account for differences across the population that could lead to disparate conclusions in more traditional research settings (see e.g., Gandomi and Haider 2015 for a discussion on 'big data').

6.3.4 Improving Communication via Neurofeedback

While most existing neurofeedback applications are designed for the interaction of single individuals with technology, our projects add a dimension of human inter-action to BCI applications. As discussed above, findings from the Mutual Wave Machine suggest that explicitly providing people with neurofeedback about their brain synchrony enhances the sense of connectedness (Dikker et al. in revision). A neurofeedback tool that tracks human interaction at the neural level has multiple practical applications, including potential clinical applications in the diagnosis and treatment of social cognition disorders. For example, persons with autism may respond better to explicit cues via technological interfaces than to human cues. An easy-to-use neurofeedback application that tracks communicative success may also be useful in therapy and conflict management. For example, we are collaborating with autism researchers to develop a neurofeedback tool that can be used as a possible social training tool in high-functioning autism (see also Pineda et al. 2008), and in programs aimed at bridging intergroup conflict (e.g., involving Palestinian and Israeli youth; Levy et al. 2016).

6.3.5 Science Outreach and Education

The technology described above makes it possible create environments where issues relating to interpersonal communication are put under the looking glass in a unique, reciprocal collaboration between art, science and education. In addition to being art/science works, our projects have a clear educational mission: Scientific research findings can be ambiguous and in many cases generate more questions than they answer, even if the contrary is often suggested in media outlets. In a similar vein, the caution expressed by scientists in interpreting their findings is often misunderstood to indicate that their results don't mean anything (the discussion on human influences on climate change is one example of the kinds of misunderstandings that can arise between the general public and scientists). First-hand experience with the scientific process may help the audience better understand and reflect on research findings as they are presented to them in popular media. As such, our projects emphasize their status as research in progress instead of demonstrations of well-established scientific findings. For example, people often ask "how well they did" after participating, to which the researchers will respond in a manner that at first seems unsatisfactory: that it is currently still an open scientific question whether "more" brainwave syn-chronization means "better", and if so, better in what way? Additionally, during the setup phase in each project, participants can look at their own brainwaves presented

Fig. 6.15 Synchrony and Synchronicity. Conceptual 'cloud' of synchron(icit)y, derived from interdisciplinary art/science discussions, that can be used as a future tool for (neuro)scientific study as well as choreography

in real time on a computer screen and ask questions about their own brains and neuroscience in general. The operators can explain how EEG works and what you can and cannot tell from the ongoing EEG signal about what's going on inside people's heads (i.e. that this isn't a "mind-reading device"). We are developing a hands-on neuroscience curriculum (BrainWaves) for high school students, specifically aimed to increase the participation of underserved students in professions of science, technology, engineering, and math.

6.3.6 Conceptual Enrichment at the Art-Science Interface

In addition to the interdisciplinary transfer of 'tools' and methods between science, art, and education, we hope to show that the interdisciplinary collaboration can also lead to conceptual enrichment in each field. For example, Measuring the Magic of Mutual Gaze stemmed from a convergence of interests between artists (Marina Abramovic was interested in exploring the brain basis of the "receptive meditative mental state" that allowed her to connect to her audience), psychotherapists (e.g., Daniel Stern, co-author of the work, focused on intersubjectivity from a psychoanalytic perspective; e.g., Stern 1985), and neuroscientists, which led to one of the first studies to directly investigate the relationship between mutual gaze, social connectedness, and brain-to-brain synchrony during face-to-face social interaction.

As already briefly discussed above, Harmonic Dissonance can be considered as an other example to illustrate the *conceptual enrichment* that can be achieved by art-science collaborations. As discussed above, in the summer of 2018, we held a three-day workshop at ICK Amsterdam (dance researcher Guido Orgs was also a participant of this workshop). One of the workshop goals was to define synchrony during joint action and action observation, and decompose it into its primary parts. These notions will then be used for further explorations into the "dual utopia" of synchronicity. Our discussions led to a 'word cloud' of synchron(icit)y, illustrated Fig. 6.15, which includes a non-exhaustive list of notions, intentions, movements and other synchrony tools: (a) **physical units/objects of synchrony** (e.g., synchronizing the body to sound [Howlin et al. 2017], synchronizing the body to the mind, the body to another body [e.g., von Zimmermann et al. 2018 found that coupled movement but not simultaneous movement predicted pro-social behavior, emphasizing the importance of synchrony in group affiliation], the audience to the performer [Bachrarch et al. 2015; Orgs et al. 2016], etc.); (b) **mental/physical realizations/forms of synchrony** (e.g., leader-follower [Konvalinka et al. 2010]; together-simultaneous, etc.); (c) **mental/physical tools that are used to achieve synchrony** (mental/physical closeness vs. distance [Dikker et al. under review], anticipation of movement vs. resonance, etc.). While there are no breakthrough 'discoveries' to be found in this cloud (many scholars before us in both the sciences and the arts have used these terms, and others, to describe synchrony and synchronicity), it has been tremendously useful for both the researchers and the dancers, as it crystallizes the research goals and leads to novel research questions. For example, take the notions of "anticipation | resonance": Does synchrony during the silence leading up to a movement sequence (anticipation) predict the extent to which the movement sequence is completed? And what about the 'resonance' stage, *after* the movement sequence is completed? Can we distinguish the neural processes underlying movement sequences that are performed simultaneously versus those that are performed *together*, with the intention to *share* an experience? These questions do not apply solely to performance research, but extend to joint action more broadly. Working with a performance ensemble such as ICK Amsterdam, who think about joint action, synchronicity and (in)visible movement at a professional level, provides alternative, and more refined, angles to addressing questions that are not directly related to dance. For example, in Dikker et al. (2017), we found higher interbrain synchrony during class for those students who had engaged in silent eye contact prior to class. Is this effect due to 'cued attention' or to 'resonance'?

6.3.7 Artistic Interpretation: Objectivity Versus Subjectivity

At the crux of the intersection between art and science lies the duality between subjectivity and objectivity. While science strives toward observations that can be repeated without influence from experimenter desires, the experience of art fundamentally lives inside the viewer's own mind. Telephone Rewired pushes this duality to an extreme, revealing a broad divergence in the viewer's subjective descriptions in response to a simple flashing light. Interestingly, the works described here could help to narrow the gap between subjective and objective to explore the meaning of real time biometric data streams captured in the context of audience subjective experience. We ask ourselves to what extent these streams (brainwaves, heart rate, movement) tell us something about how connected we feel to each other ("when we feel in sync with someone, are our brainwaves literally on the same wavelength?"). With this information, we first and foremost want to create work that stands on its own as an immersive, interactive aesthetic experience, allowing visitors to gauge and explore their own interactions in a visceral, intuitive way. But we also want to raise critical questions about our often unsatisfactory quest to connect to ourselves and each other: via face-to-face interactions, scientific inquiry, tech-based communication tools, big data; about the exclusionary nature of groups, both in the real and virtual world. By challenging its audience to either obey or disregard the patterns of light flowing through space, our projects explore the clash and merger of multiple systems.

Harmonic Dissonance seeks out the friction between **social interdependence and individual agency**. This is achieved via multiple dimensions. Naturally, the social network structure inherently emphasizes this dichotomy: some individuals are strongly connected to all the others, while some remain on the outside of the circle. This is further emphasized by a careful hovering between individualized color- and soundscapes and joint A/V landscapes. Further, the BCI environments present participants with a constant internal conflict between **human-human and human-computer interaction**: The computer-generated audio-visual feedback provides a persistent distractor from face-to-face interpersonal contact, just like our everyday devices and social media platforms. Further, while the works are incomplete without collaborative effort, participants will inherently be drawn instead to focus on themselves and their sonification/visual reflection. Third, the projects explore the relationship between the **mind and the body**: Our psyche feeds on a constant battle between intuition and rationality. But is there a true dichotomy between feelings and thought, between the mind and the body? Or is this a false premise? In fact, neuropsychological evidence suggests that the body-mind dichotomy is perhaps not a biologically valid concept. The notion of 'free will' is also under attack from multiple directions, including computer science and neuroscience. In many of our projects, participants freely interact with the algorithmic interpretation/visualization of their interconnectedness, and their choices have consequences for visualizations and sonifications. In Harmonic Dissonance, participants are allowed to physically distance themselves from the social network that is drawn by their brain data, they are seduced into asking themselves questions about free will, and about the dichotomy between their brain and their body. In Telephone Rewired and Hive Mind participants' brainwaves are hijacked by the flashing lights of the installation in a manner that alters cognitive processing and subjective experience, leading to questions about the limits of free will and relationship between mind and body. Broadly speaking, our work invites the viewer to ask questions about how we're connected to one another and the world around us. How are the electrical impulses reflecting our thoughts and emotions throughout our brains and bodies fundamentally different from the electrical impulses that course through the internet? As we learn more about the brain computer interface, how can that change the way we think, learn, communicate, collaborate, and empathize? And as our biometric signals are increasingly uploaded to the cloud and artificial intelligence algorithms grow in complexity and sophistication, what possibilities exist for a world in which we can use neurofeedback to toggle the switches of our own—or other people's (Jiang et al. 2018)—cognition?

6.4 Conclusion

In sum, we aim for our work to stand on its own as interactive art, allowing visitors to gauge and explore their own interactions in a visceral, intuitive way. But we also want to raise critical questions: about the technological communication interfaces in our everyday lives; about our often unsatisfactory quest to understand ourselves and each other, either through scientific inquiry or through direct face-to-face communication; and about the exclusionary nature of groups. Our projects are situated at the intersection of immersive art and neuroscience at both the conceptual and logistic level. All the EEG data that is collected with our installations are used for scientific analysis, which in turn feeds back into the design of our installations, and further informs our research questions that derive from conversations at the art/science intersection. In communicating our work to the general public, we hope to demonstrate the richness that can result from this cross-disciplinary dialogue, and that scientific progress, like artistic expression, is often driven not by answers, but by finding the right questions to ask.

Acknowledgements Dikker and Oostrik's work is supported by Stichting Niemeijer Fonds, The Netherlands Organization for Scientific Research (VENI grant 275-89-018), Creative Industries Fund NL, TodaysArt, MAI, de Hersenstichting, Lowlands Science, Utrecht University, NEON. Montgomery's work is supported by: ISEA, Science Gallery at Trinity College Dublin, Harvestworks, National Endowment for the Arts, Rockefeller Foundation, Daejeon Museum of Art, Center for Life, RIXC Art Science Festival, OUTPOST Artist Resources, Ende Tymes Festival.

References

Abramović (2010) Marina Abramović: the artist speaks. Inside/Out. A MoMA/MoMA PS1 blog. https://www.moma.org/explore/inside_out/2010/06/03/marina-abramovic-the-artist-speaks/

Akbas A (2011) Evaluation of the physiological data indicating the dynamic stress level of drivers. Sci Res Essays 6(2):430–439

Anderson KL, Ding M (2011) Attentional modulation of the somatosensory mu rhythm. Neuro 180:165–180

Arnal LH, Giraud AL (2012) Cortical oscillations and sensory predictions. Trends Cogn Sci 16:390–398

Aron A, Aron EN, Smollan D (1992) Inclusion of other in the self scale and the structure of interpersonal closeness. J Person Soc Psych 63(4):596

Assaneo MF, Ripolles P, Orpella J, Ming-Lin, W, de Diego-Balaguer R, Poeppel D (2019) Spontaneous synchronization to speech reveals neural mechanisms facilitating language learning. Nat Neurosci

Atmaca S, Sebanz N, Knoblich G (2011) The joint flanker effect: sharing tasks with real and imagined co-actors. Exp Brain Res 211(3–4):371–385

Atmanspacher H, Fuchs CA (2014) The Pauli-Jung conjecture and its impact today. Andrews UK Limited

Babiloni F, Astolfi L (2014) Social neuroscience and hyperscanning techniques: past, present and future. Neurosci Biobehav Rev 4:476–493

Babiloni F, Cincotti F, Mattia D, Mattiocco M, Fallani FD, Tocci A, Bianchi L, Marciani MG, Astolfi L (2006) Hypermethods for EEG hyperscanning. In: Engineering in Medicine and Biology Society 28th annual international conference of the IEEE 30 Aug 2006, pp 3666–3669

Babiloni F, Astolfi L, Cincotti F, Mattia D, Tocci A, Tarantino A, Marciani MG, Salinari S, Gao S, Colosimo A, Fallani FD (2007) Cortical activity and connectivity of human brain during the prisoner's dilemma: an EEG hyperscanning study. In: Engineering in Medicine and Biology Society, EMBS 29th annual international conference of the IEEE, 22 Aug 2007 pp 4953–4956

Bachrach A, Fontbonne Y, Joufflineau C, Ulloa JL (2015) Audience entrainment during live contemporary dance performance: physiological and cognitive measures. Front Hum Neurosci 9:179

Baek HJ, Lee HB, Kim JS, Choi JM, Kim KK, Park KS (2009) Nonintrusive biological signal monitoring in a car to evaluate a driver's stress and health state. Telemed e-Health 15(2):182–189

Bar M (2007) The proactive brain: using analogies and associations to generate predictions. Trends Cogn Sci 11:280–289

Bašnáková J, Weber K, Petersson KM, van Berkum J, Hagoort P (2013) Beyond the language given: the neural correlates of inferring speaker meaning. Cereb Cortex 24(10):2572–2578

Bastiaansen M, Hagoort P (2006) Oscillatory neuronal dynamics during language comprehension. Prog Brain Res 159:179–196

Beauchene C, Abaid N, Moran R, Diana RA, Leonessa A (2017) The effect of binaural beats on verbal working memory and cortical connectivity. J Neural Eng 14(2):026014

Benford S, Greenhalgh C, Crabtree A, Flintham M, Walker B, Marshall J, Koleva B, Rennick Egglestone S, Giannachi G, Adams M, Tandavanitj N (2013) Performance-led research in the wild. ACM TOCHI 20(3):14

Bevilacqua D, Davidesco I, Wan L, Oostrik M, Chaloner K, Rowland J, Ding M, Poeppel D, Dikker S (2018) Brain-to-brain synchrony and learning outcomes vary by student–teacher dynamics: evidence from a real-world classroom electroencephalography study. J Cogn Neurosci 1–11

Bhattacharya J (2017) Cognitive neuroscience: synchronizing brains in the classroom. Curr Biol 27(9):R346–R348

Böckler A, Knoblich G, Sebanz N (2012) Effects of a coactor's focus of attention on task performance. J Exp Psychol: Hum Percept Perform 38(6):1404

Brenninkmeijer J (2013) Neurofeedback as a dance of agency. BioSocieties 8(2):144–163

Buzsáki G (2006) Rhythms of the brain. Oxford University Press, New York, NY, US

Cacioppo JT, Bernston GG, Larsen JT, Poehlmann KM, Ito TA (2000) The psychophysiology of emotion. In: Handbook of emotions. The psychophysiology of emotion. Guilford, pp 173–191

Cai H, Lin Y (2011) Modeling of operators' emotion and task performance in a virtual driving environment. Int J Hum Comput Stud 69:571–586

Canolty RT, Edwards E, Dalal SS, Soltani M, Nagarajan SS, Kirsch HE, Berger MS, Barbaro NM, Knight RT (2006) High gamma power is phase-locked to theta oscillations in human neocortex. Science 313(5793):1626–1628

Collet C, Clarion A, Morel M, Chapon A, Petit C (2009) Physiological and behavioural changes associated to the management of secondary tasks while driving. Appl Ergon (England) 40:1041–1046

Davis M (1996) Interpersonal reactivity index. In: Empathy: a social psychological approach, pp 55–56 de Graaf TA, Gross J, Paterson G, Rusch T, Sack AT, Thut G (2013) Alpha-band rhythms in visual task performance: phase-locking by rhythmic sensory stimulation. PLoS ONE 8(3):e60035

Debener S, Minow F, Emkes R, Gandras K, De Vos M (2012) How about taking a low-cost, small, and wireless EEG for a walk? Psychophysiology 49(11):1617–1621

Desimone R, Duncan J (1995) Neural mechanisms of selective visual attention. Annu Rev Neurosci 18(1):193–222

Dikker S, Pylkkänen L (2013) Predicting language: MEG evidence for lexical preactivation. Brain Lang 127(1):55–64

Dikker S, Rabagliati H, Pylkkänen L (2009) Sensitivity to syntax in visual cortex. Cognition 110(3):293–321

Dikker S, Rabagliati H, Farmer TA, Pylkkänen L (2010) Early occipital sensitivity to syntactic category is based on form typicality. Psychol Sci 21(5):629–634

Dikker S, Silbert LJ, Hasson U, Zevin JD (2014) On the same wavelength: predictable language enhances speaker–listener brain-to-brain synchrony in posterior superior temporal gyrus. J Neurosci 34(18):6267–6272

Dikker S, Wan L, Davidesco I, Kaggen L, Oostrik M, McClintock J, Rowland J, Michalareas G, Van Bavel JJ, Ding M, Poeppel D (2017) Brain-to-brain synchrony tracks real-world dynamic group interactions in the classroom. Curr Biol 27(9):1375–1380

Dikker S, Michalareas G, Oostrik M, Kahraman H, Serafimaki A, Struiksma ME, Poeppel D (in revision) Crowdsourcing neuroscience: brain synchrony during face-to-face interaction outside the laboratory

Doelling KB, Poeppel D (2015) Cortical entrainment to music and its modulation by expertise. Proc Natl Acad Sci USA 112(45):E6233–E6242

Donoghue JP, Sanes JN, Hatsopoulos NG, Gaal G (1998) Neural discharge and local field potential oscillations in primate motor cortex during voluntary movements. J Neurophysiol 79:159–173

Dumas G, Nadel J, Soussignan R, Martinerie J, Garnero L (2010) Inter-brain synchronization during social interaction. PLoS ONE 5(8):e12166

Eldred SM (2016) Art–science collaborations: change of perspective. Nature 537(7618):125–126

Enns JT, Lleras A (2008) What's next? New evidence for prediction in human vision. Trends Cogn Sci 12:327–333

Federmeier KD (2007) Thinking ahead: the role and roots of prediction in language comprehension. Psychophysiology 44:491–505

Fox NA, Bakermans-Kranenburg MJ, Yoo KH, Bowman LC, Cannon EN, Vanderwert RE, Ferrari PF, van IJzendoorn MH (2016) Assessing human mirror activity with EEG mu rhythm: a meta-analysis. Psychol Bull 142(3):291–313

Friston K (2003) Learning and inference in the brain. Neural Netw 16:1325–1352

Gandomi A, Haider M (2015) Beyond the hype: Big data concepts, methods, and analytics. Int J Inf Manag 35(2):137–144

Garrod S, Pickering MJ (2009) Joint action, interactive alignment, and dialog. Trends Cogn Sci 1(2):292–304

Goede MNM (2017) Narrating the brain through neurofeedback art. Master's thesis, University of Amsterdam, Amsterdam, the Netherlands. http://www.scriptiesonline.uba.uva.nl/en/scriptie/624073

Goldstein P, Weissman-Fogel I, Dumas G, Shamay-Tsoory SG (2018) Brain-to-brain coupling during handholding is associated with pain reduction. Proc Natl Acad Sci USA 201703643

Golumbic EMZ, Ding N, Bickel S, Lakatos P, Schevon CA, McKhann GM, Goodman RR, Emerson R, Mehta AD, Simon JZ, Poeppel D, Schroeder CE (2013) Mechanisms underlying selective neuronal tracking of attended speech at a "cocktail party". Neuron 77(5):980–991

Gomez-Ramirez M, Kelly SP, Molholm S, Sehatpour P, Schwartz TH, Foxe JJ (2011) Oscillatory sensory selection mechanisms during intersensory attention to rhythmic auditory and visual inputs: a human electrocorticographic investigation. J Neurosci 31(50):18556–18567

Gruber T, Tsivilis D, Giabbiconi CM, Müller MM (2008) Induced electroencephalogram oscillations during source memory: familiarity is reflected in the gamma band, recollection in the theta band. J Cogn Neurosci 20(6):1043–1053

Haegens S, Golumbic EZ (2017) Rhythmic facilitation of sensory processing: a critical review. Neuro Biobehav Rev

Hari R, Kujala MV (2009) Brain basis of human social interaction: from concepts to brain imaging. Physiol Rev 89(2):453–479

Hari R, Himberg T, Nummenmaa L, Hämäläinen M, Parkkonen L (2013) Synchrony of brains and bodies during implicit interpersonal interaction. Trends Cogn Sci

Hasson U, Frith CD (2016) Mirroring and beyond: coupled dynamics as a generalized framework for modelling social interactions. Philos Trans R Soc B 371(1693):20150366

Hasson U, Ghazanfar AA, Galantucci B, Garrod S, Keysers C (2012) Brain-to-brain coupling: a mechanism for creating and sharing a social world. Trends Cogn Sci 16(2):114–121

Healey JA, Picard RW (2005) Detecting stress during real-world driving tasks using physiological sensors. IEEE Trans Intell Transp Syst 6:156–166

Hipp JF, Hawellek DJ, Corbetta M, Siegel M, Engel AK (2012) Large-scale cortical correlation structure of spontaneous oscillatory activity. Nat Neurosci 15(6):884–890

Howlin C, Orgs G, Vicary S (2017) The impact of soundtrack congruency on the aesthetic experience of contemporary dance: exploring aesthetic interaction in terms of arousal and enjoyment ratings in three audio settings. Age 1000:29–6

Jeffries S (2011) When two tribes meet: collaborations between artists and scientists. The Guardian. http://www.theguardian.com/artanddesign/2011/aug/21/collaborations-betweenartists-and-scientists. Accessed 26 Nov 2016

Jiang L, Stocco A, Losey DM, Abernethy JA, Prat CS, Rao RP (2018) BrainNet: a multi-person brain-to-brain interface for direct collaboration between brains. arXiv:1809.08632

Jirakittayakorn N, Wongsawat Y (2017) Brain responses to a 6-Hz binaural beat: effects on general theta rhythm and frontal midline theta activity. Front Neurosci 11:365

Jung CG (2012) Synchronicity: an acausal connecting principle (from vol 8 of the Collected Works of C. G. Jung). Princeton University Press, pp 114–115

Keitel C, Quigley C, Ruhnau P (2014) Stimulus-driven brain oscillations in the alpha range: entrainment of intrinsic rhythms or frequency-following response? J Neurosci 34(31):10137–10140

Kerfeld CA (2009) When art, science, and culture commingle. PLoS Biol 7(5):e1000100

Keysers C, Gazzola V (2009) Expanding the mirror: vicarious activity for actions, emotions, and sensations. Curr Opin Neurobiol 19(6):666–671

Kinreich S, Djalovski A, Kraus L, Louzoun Y, Feldman R (2017) Brain-to-brain synchrony during naturalistic social interactions. Sci Rep 7(1):17060

Knoblich G, Butterfill S, Sebanz N (2011) 3 psychological research on joint action: theory and data. Psychol Learn Motiv Adv Res Theory 54:59

Konvalinka I, Vuust P, Roepstorff A, Frith CD (2010) Follow you, follow me: continuous mutual prediction and adaptation in joint tapping. Q J Exp Psychol 63(11):2220–2230

Lakatos P, Shah AS, Knuth KH, Ulbert I, Karmos G, Schroeder CE (2005) An oscillatory hierarchy controlling neuronal excitability and stimulus processing in the auditory cortex. J Neurophys 94(3):1904–1911

Lakatos P, Karmos G, Mehta AD, Ulbert I, Schroeder CE (2008) Entrainment of neuronal oscillations as a mechanism of attentional selection. Science 320(5872):110–113

Levy J, Goldstein A, Influs M, Masalha S, Zagoory-Sharon O, Feldman R (2016) Adolescents growing up amidst intractable conflict attenuate brain response to pain of outgroup. Proc Natl Acad Sci USA 113(48):13696–13701

Lisetti CL, Nasoz F (2004) Using noninvasive wearable computers to recognize human emotions from physiological signals. EURASIP J Adv Signal Process 2004:1672–1687

Lumsden J, Miles LK, Richardson MJ, Smith CA, Macrae NC (2012) Who syncs? social motives and interpersonal coordination. J Exp Soc Psychol 48:746–751

Marsh KL, Richardson MJ, Baron RM, Schmidt RC (2006) Contrasting approaches to perceiving and acting with others. Eco Psychol 18(1):1–38

Marsh KL, Isenhower RW, Richardson MJ, Helt M, Verbalis AD, Schmidt RC, Fein D (2013) Autism and social disconnection in interpersonal rocking. Front Integr Neurosci 7

Mathewson KE, Prudhomme C, Fabiani M, Beck DM, Lleras A, Gratton G (2012) Making waves in the stream of consciousness: entraining oscillations in EEG alpha and fluctuations in visual awareness with rhythmic visual stimulation. J Cogn Neurosci 24(12):2321–2333

Matusz PJ, Dikker S, Huth AG, Perrodin C (2018) Are we ready for real-world neuroscience? J Cogn Neurosci

Mehler B, Reimer B, Coughlin JF (2012) Sensitivity of physiological measures for detecting systematic variations in cognitive demand from a working memory task: an on-road study across three age groups. Hum Factors (United States) 54:396–412

Meyer L, Henry MJ, Gaston P, Schmuck N, Friederici AD (2016) Linguistic bias modulates interpretation of speech via neural delta-band oscillations. Cereb Cortex 27(9):4293–4302

Miller AI (2014) Colliding worlds: how cutting-edge science is redefining contemporary art. WW Nort Comp

Montgomery SM, Laefsky I (2011) Bio-sensing: hacking the doors of perception. Make Magazine 26:104–111

Mullen T, Khalil A, Ward T, Iversen J, Leslie G, Warp R, Whitman M, Minces V, McCoy A, Ojeda A, Bigdely-Shamlo N (2015) MindMusic: playful and social installations at the interface between music and the brain. In: More playful user interfaces. Springer, Singapore, pp 197–229

Neuper C, Pfurtscheller G (1999) Motor imagery and ERD. In: Pfurtscheller G, Lopes da Silva FH (eds) Event-related desynchronization, revised edition Handbook. Electroencephalogr Clin Neurophysiol vol 6. Elsevier, Amsterdam, pp 303–325

Nijholt A (2015) Competing and collaborating brains: multi-brain computer interfacing. In: Brain-computer interfaces. Springer, Cham, pp 313–335

Nolte G, Bai O, Wheaton L, Mari Z, Vorbach S, Hallett M (2004) Identifying true brain interaction from EEG data using the imaginary part of coherency. Clin Neurophysiol 115(10):2292–2307

Orgs G, Caspersen D, Haggard P (2016) You move, I watch, it matters: aesthetic communication in dance. Shar Represent: Sensorimotor Found Soc Life 627–654

Parada FJ, Rossi A (2017) Commentary: brain-to-brain synchrony tracks real-world dynamic group interactions in the classroom and cognitive neuroscience: synchronizing brains in the classroom. Front Hum Neurosci 11:554

Paxton A, Dale R (2012) Frame-differencing methods for measuring bodily synchrony in conversation. Behav Res Methods 1–15

Peelle JE, Davis MH (2012) Neural oscillations carry speech rhythm through to comprehension. Front Psychol 3:320

Pfurtscheller G, Lopes da Silva FH (1999) Event-related EEG/MEG synchronization and desynchronization: basic principles. Clin Neurophys 110(11):1842–1857

Pfurtscheller G, Neuper C (1994) Event-related synchronization of mu rhythm in the EEG over the cortical hand area in man. Neuro Lett 174(1):93–96

Picard RW (1995) Affective computing. M.I.T Media Laboratory Perceptual Computing Section Technical Report No. 321

Pickering MJ, Garrod S (2013) An integrated theory of language production and comprehension. Behav Brain Sci 36(4):329–347

Pineda JA, Brang D, Hecht E, Edwards L, Carey S, Bacon M, Futagaki C, Suk D, Tom J, Birnbaum C, Rork A (2008) Positive behavioral and electrophysiological changes following neurofeedback training in children with autism. Res Autism Spectr Disord 2(3):557–581

Powers SR, Rauh C, Henning RA, Buck RW, West TV (2011) The effect of video feedback delay on frustration and emotion communication accuracy. Comput Hum Behav 27(5):1651–1657

Prusinkiewicz P (1998) In search of the right abstraction: the synergy between art, science, and information technology in the modeling of natural phenomena. Na

Ramnani N, Miall RC (2003) A system in the human brain for predicting the actions of others. Nat Neurosci 7(1):85–90

Richardson MJ, Marsh KL, Isenhower RW, Goodman JR, Schmidt RC (2007) Rocking together: dynamics of intentional and unintentional interpersonal coordination. Hum Mov Sci 26(6):867–891

Richardson DC, Dale R, Shockley K (2008) Synchrony and swing in conversation: coordination, temporal dynamics, and communication. Embodied Commun Hum Mach 75–93

Rigas G, Goletsis Y, Fotiadis DI (2012) Real-time driver's stress event detection. IEEE Trans Intell Transp Syst 13:221–234

Sänger J, Lindenberger U, Müller V (2011) Interactive brains, social minds. Commun Integr Biol 4(6):655–663

Schroeder CE, Lakatos P (2008) Low-frequency neuronal oscillations as instruments of sensory selection. Trends Neurosci 32(1):9–18

Sebanz N, Bekkering H, Knoblich G (2006) Joint action: bodies and minds moving together. Trends Cogn Sci 10(2):70–76

Shteynberg G (2010) A silent emergence of culture: the social tuning effect. J Pers Soc Psychol 99(4):683

Spaak E, de Lange FP, Jensen O (2014) Local entrainment of alpha oscillations by visual stimuli causes cyclic modulation of perception. J Neurosci 34(10):3536–3544

Stephens GJ, Silbert LJ, Hasson U (2010) Speaker–listener neural coupling underlies successful communication. Proc Natl Acad Sci USA 107(32):14425–14430

Stern DN (1985) The interpersonal world of the infant: a view from psychoanalysis and developmental psychology. Karnac Books

Stevens C, O'Connor G (2017) When artists get involved in research, science benefits. The Conversation. theconversation.com/when-artists-get-involved-in-research-science-benefits-82147

Stivers T, Enfield NJ, Brown P, Englert C, Hayashi M, Heinemann T, Levinson SC (2009) Universals and cultural variation in turn-taking in conversation. Proc Natl Acad Sci USA 106(26):10587–10592

Sun X, Lichtenauer J, Valstar M, Nijholt A, Pantic M (2011) A multimodal database for mimicry analysis. In: International conference on affective computing and intelligent interaction, October 9. Springer, Berlin, Heidelberg, pp 367–376

Tognoli E, Lagarde J, DeGuzman GC, Kelso JA (2007) The phi complex as a neuromarker of human social coordination. Proc Natl Acad Sci USA 104(19):8190–8195

Van Berkum JJA (2013) Anticipating communication. Theor Lin 39(1–2):75–86

Van Berkum JJ, Brown CM, Zwitserlood P, Kooijman V, Hagoort P (2005) Anticipating upcoming words in discourse: evidence from ERPs and reading times. J Exp Psychol Learn Mem Cogn 31(3):443

Van Berkum JJA, Holleman B, Nieuwland MS, Otten M, Murre J (2009) Right or wrong? The brain's fast response to morally objectionable statements. Psychol Sci 20:1092–1099

Van den Brink D, Van Berkum JJA, Bastiaansen MCM, Tesink CMJY, Kos M, Buitelaar JK, Hagoort P (2012) Empathy matters: ERP evidence for inter-individual differences in social language processing. Soc Cogn Affect Neurosci 7:173–182

Van Essen DC, Smith SM, Barch DM, Behrens TE, Yacoub E, Ugurbil K, Wu-Minn HCP Consortium (2013) The WU-Minn human connectome project: an overview. Neuroimage 80:62–79

Verbeke WJ, Pozharliev R, Van Strien JW, Belschak F, Bagozzi RP (2014) "I am resting but rest less well with you." The moderating effect of anxious attachment style on alpha power during EEG resting state in a social context. Front Hum Neurosci 8:486

von Zimmermann J, Vicary S, Sperling M, Orgs G, Richardson DC (2018) The choreography of group affiliation. Top Cogn Sci 10(1):80–94

Watson D, Clark LA (1994) The PANAS-X: manual for the positive and negative affect schedule-expanded form. Uni Iowa, Ames

Williams JH (2001) Frequency-specific effects of flicker on recognition memory. Neuroscience 104(2):283–286

Williams J, Ramaswamy D, Oulhaj A (2006) 10 Hz flicker improves recognition memory in older people. BMC Neurosci 7:21

Wilson EO (2012) The social conquest of earth. WW Nort Comp

Wu CH, Chang HC, Lee PL, Li KS, Sie JJ, Sun CW, Yang CY, Li PH, Deng HT, Shyu KK (2011) Frequency recognition in an SSVEP-based brain computer interface using empirical mode decomposition and refined generalized zero-crossing. J Neurosci Methods 196(1):170–181

Yun K (2013) On the same wavelength: face-to-face communication increases interpersonal neural synchronization. J Neurosci 33(12):5081–5082

Zamm A, Debener S, Bauer AKR, Bleichner MG, Demos AP, Palmer C (2018) Amplitude envelope correlations measure synchronous cortical oscillations in performing musicians. Ann N Y Acad Sci

Zhang D, Maye A, Gao X, Hong B, Engel AK, Gao S (2010) An independent brain–computer interface using covert non-spatial visual selective attention. J Neural Eng 7(1):016010

Zink R, Hunyadi B, Van Huffel S, De Vos M (2016) Mobile EEG on the bike: disentangling attentional and physical contributions to auditory attention tasks. J Neural Eng 13(4):046017

Suzanne Dikker is a research scientist at Utrecht University and New York University. Her research, funded by a Netherlands Organisation of Scientific Research VENI grant, merges cognitive neuroscience, education, and performance art in an effort to understand the brain basis of human social interaction. As part of the art collective **DIKKER + OOSTRIK**, she works with interactive media artist Matthias Oostrik (and friends) on a series of arts/BCI projects that are simultaneously art installations and social neuroscience experiments.

Sean Montgomery is a technologist, educator and new-media artist in New York City. Using research methodologies combined with emerging technologies, Sean takes a trans-disciplinary look at the human condition, using technology to enhance our understanding of ourselves and create new ways for people to interact with one another and the objects around them.

Suzan Tunca is a dance researcher at ICK Amsterdam. She is developing and implementing an artistic research curriculum for dancers at CODARTS University of the Arts, is member of the performing arts research group DAS research THIRD! and Ph.D. candidate at PhDArts Leiden University. With her work as a performing artist and as dance researcher in professional and educational contexts she aims to contribute to the continuous regeneration and advancement of dance as an autonomous art form and as an invaluable source for embodied knowledge.

Chapter 7
EEG KISS: Shared Multi-modal, Multi Brain Computer Interface Experience, in Public Space

Karen Lancel, Hermen Maat and Frances Brazier

Abstract Can shared intimate experience of social touch be mediated through multi-brain-computer interface (Multi-brain BCI) interaction in public space? Two artistic EEG KISS orchestrations, both multi-modal, multi-brain BCIs, are shown to create novel shared experiences of social touch in public space. These orchestrations purposefully disrupt and translate known forms of face-to-face connection and sound, to re-orchestrate unfamiliar sensory syntheses of seeing, hearing, touching and moving, connected to data-visualization and audification of brain activity. The familiar sensory relations between 'who you kiss and who is being kissed, what you see and what you hear' are captured in a model of digital synaesthetics in multi-modal multi brain BCI interaction for social touch. This model links hosted self-disclosure, witnessing, dialogue and reflection to intimate experience in public space through syntheses of the senses. As such, this model facilitates the design of new shared intimate experiences of multi modal multi brain BCI interaction through social touch in public space.

Keywords Social engagement · Interactive digital art · Shared experience · Intimate touch · Multi-modal multi-brain computer interface · Public space

7.1 Introduction

Can shared intimate experience of multi-brain-computer interface interaction be mediated through social touch? Artistic orchestrations of multi-modal BCI mediated interaction have been shown to create novel shared experiences (Abramovic et al.

K. Lancel (✉) · F. Brazier
Delft University of Technology, Delft, The Netherlands
e-mail: lancel@xs4all.nl

F. Brazier
e-mail: F.M.Brazier@tudelft.nl

K. Lancel · H. Maat
Artists duo Lancel/Maat, Amsterdam, The Netherlands
e-mail: maat@xs4all.nl

© Springer Nature Switzerland AG 2019
A. Nijholt (ed.), *Brain Art*,
https://doi.org/10.1007/978-3-030-14323-7_7

Fig. 7.1 EEG KISS (EEG KISS. Promotion Photo. © Lancel/Maat 2014)

2014; Dikker et al. 2016; Mori 2005; Novello 2016; Sobell 1974, 2001). They purposefully disrupt and translate known forms of face-to-face connection and sound, to re-orchestrate unfamiliar sensory syntheses of seeing, hearing and moving, connected to data visualization and audification of brain activity, often in playful exploration (Lysen 2019; Prpa and Pasquier 2019). This paper extends insights gained in such multi-modal BCI mediated orchestrations focusing explicitly on the effects of design choices for shared intimate experience of multi-brain multi-modal BCI interaction through social touch. Digital synaesthetic (Gsöllpointner et al. 2016), shared intimate social touch experience is explored and an integrated model of multi-modal, multi-brain BCI interaction for social touch is proposed. Two BCI mediated artistic orchestrations performed internationally by Lancel/Maat[1] are analysed to this purpose. In these artistic orchestrations, participants are invited to feel, see, touch and share an intimate kiss experience. The familiar relation between 'who you kiss and who is being kissed, what you see and what you hear' is purposefully disrupted and explored for a new, shared sensory synthesis (Fig. 7.1).

[1]Lancel/Maat are artistic partners whom have been working together since 1998. Their works include the artistic performance installation EEG KISS discussed in this paper. Their works have been presented in Venice Biennial 2015—China Pavilion; Ars Electronica Linz 2018; ZKM Karlsruhe; Transmediale Berlin; Stedelijk Museum Amsterdam; Rijksmuseum Amsterdam; World Expo Shanghai 2010; HeK Haus for Electronic Art Basel; ISEA2016 Hongkong; ISEA2011 Istanbul; Banff Center Canada; RIXC Riga; V2-Institute Rotterdam; Beall Center for Art + Technology USA; BCAC Beijing; 2nd TASIE Art Science exhibition at Millenium Museum Beijing; Third TASIE Art Science exhibition at Science & Technology Museum Beijing; Public Art Lab Berlin. http://www.lancelmaat.nl/work/e.e.g.-kiss/.

7.2   Related Work

Brain Computer Interfaces are being explored by a growing community of international artists (Lysen 2019; Prpa and Pasquier 2019). In interactive performances, installations, cinema, multi-user game and theatre, artists are exploring new types of BCI interaction that are often not primarily anchored in scientific understanding of physiological data (Delft University of Technology 2015). These installations often focus on aesthetics, ethics and affective experience (Gürkök and Nijholt 2013; Roeser et al. 2018).

Brain Computer Interfaces (BCI) enable direct communication between brain activity (the input) and control of (internal or external) devices (the output). Often, BCIs process and combine representations of brain activity with other audio, visual and haptic information. BCI interfaces combined with virtual reality (VR) and augmented reality (AR) technologies have, for example, been designed to enhance realistic, immersive experiences, using haptic sensors, motor imagination and feedback based on action visualization, for art, entertainment, training, therapy, sex, gaming, robotics (Gomes and Wu 2017; Lupu et al. 2018; Nijholt and Nam 2015; Ramchurn et al. 2019). In other works, functional Near Infrared Spectroscopy (fNIRS) has been used in BCI Interfaces to explore arousal of shared engagement (Bennett et al. 2013; Lancel/Maat and Luehmann 2017).

Research of direct brain-to-brain communication between humans, or between humans and robots enhanced with Artificial Intelligence technologies, using EEG to record electrical activity in the brain and transcranial magnetic stimulation (TMS), is still in an early stage, but is promising (MIT 2018).

BCIs are designed for single users, or for multiple users in multi-brain interfaces (Multi-brain BCI) (Nijholt 2015). Multi-brain BCIs process brain activity of two or more participants as input for shared experience of joint (parallel or sequential) brain activity.

In some cases, output is based on 'spontaneous' (Gürkök and Nijholt 2013) participant input (Sobell 1974; De Boeck 2009; Casey 2010). In other cases, output is based on 'controlled' (Pike et al. 2016) or 'directed' input (Mori 2005; SPECS 2009). Sobell (1974) for example, explores the influence of different augmented representations of joint brain activity as output, predominantly based on 'spontaneous' participant input. In contrast, Dikker et al. (2016) and Gabriel (1993) have designed systems in which individual participants purposively influence their individual input (e.g. altering between level of arousal) to collectively 'direct' the output of the multi-brain BCI. In other systems, the threshold between 'spontaneous' and 'directed' is mixed (Novello 2016; Rosenboom 1990; Sobell 2001). In other artistic orchestrations, multi-brain BCIs have been used to direct brain activity synchrony in coordinated social interaction to explore empathy and connectedness (Dikker et al. 2016).[2]

---

[2]Note, that although many of these orchestrations combine co-located forms of interaction sometimes these are networked from different locations (Casey 2010). Multi-user game orchestrations, which are not subject to this chapter, are, for example, often networked from different locations.

Fig. 7.2 EEG KISS Orchestration 2 at Stedelijk Museum Amsterdam, 'Stedelijk Statements' Series 2017 and UvA University of Amsterdam 'Worlding the Brain Conference' 2017. © Lancel/Maat

However, although current BCI research includes hedonic and affective touch experience, intimate touch communication is not yet well understood (Björnsdotter et al. 2014). fMRI research shows that tactile experiences of slow (1–10 cm/s), gentle stroking (caressing) of the skin and the system is associated by participants with affection. This is in line with research that shows that intimate touch provides a means to share empathic, intimate emotions (Van Erp and Toet 2015), for which (a) vulnerability and self-disclosure, (b) physical proximity and (c) witnessing and responsibility are essential (Lomanowska and Guitton 2016) (Fig. 7.2).

Designing for witnessing (Nevejan 2007) and embodied vulnerability as an 'intimate aesthetic' (Loke and Khut 2014) through touch, is better understood with respect to shared experiences in hosted performance art with participants and spectators (Benford et al. 2012; Cillari 2006–2009; Clark 1963–1988; CREW 2016; Lancel et al. 2018; Osthoff 1997; Vlugt 2015).[3] In such artistic orchestrations, intimate touch is used to evoke embodied and cognitive reflection (Kwastek 2013).

In interactive digital art, components of intimate experience have been orchestrated based on aesthetic principles of disruption, unfamiliarity, risk and unpredictability in digital synaesthetic (Gsöllpointner et al. 2016) orchestrations. These aesthetic principles incite ambivalence and immersion, as essential conditions for engagement and reflection to emerge (Benford and Giannachi 2012; Kwastek 2013). Such orchestrations explore relations between brains, bodies, personal embodied

---

[3]Live Art, http://www.thisisliveart.co.uk/, last accessed 2018/10/20.

knowledge and perception, technologies and the surrounding environment (Gill 2015; Lysen 2019).

However, multi-modal, multi-brain orchestrations, for shared intimate experiences, through orchestration of social touch, are, to the authors' knowledge, not yet explored.

7.3 Artistic Motivation

This chapter explores the effects of multi-modal, multi-brain BCI designs of two artistic orchestrations on shared engagement through intimate touch. This section describes these two artistic orchestrations of the multi-modal multi-brain BCI interface EEG KISS (Lancel/Maat 2014–2018) from the perspective of the artists Lancel and Maat.

7.3.1 Introduction to the Artists' Research on Mediated Touch

Lancel and Maat have explored new approaches to interfacing mirroring affective touch. In their artworks, the person touching and being touched does not have to be the same (telematically present) person to whom the haptic connection is attributed (Lancel et al. 2018). Paradoxically, in these orchestrations, participants are requested to touch or caress themselves to haptically relate with other participants. For example, in the artwork Saving Face (Lancel/Maat 2012), based on use of face recognition technologies, participants are invited to caress their own faces to visually connect with others on an electronic screen. In the artwork Tele_Trust (Lancel/Maat 2009), participants wear a smart textile, full body covering 'data-veil'. Participants are invited to caress their bodies to connect with others through smart phones. In both artworks participants experience novel haptic connections from an individual to others in public, digitally distributed environments.

This chapter describes two artistic orchestrations that explore new ways to similarly create a unique sense of communal haptic connections with others in the network. These two orchestrations have been designed by the artists to be both 'expressive' and 'magical', as defined by Reeves et al. (2005). As an 'expressive' interface, it has been designed to attract people, inviting them to participate, when encountered with its 'magical' and secretive nature. 'Spontaneous' EEG interaction between participants is part of the design. Brain signals during intimate interaction of kissing are made explicit to facilitate a shared experience of social touch through reflection and dialogue.

7.3.2   EEG KISS: Short Description of Two Artistic Orchestrations

In two EEG KISS, multi-modal, multi-brain BCI orchestrations (performance installations), shared intimate experience of social touch is explored. Participants are
invited to feel, see, touch and share an intimate kiss to incite both an aesthetic and a
sensory experience.

Both orchestrations combine one-to-one and multi-user participation. Members
of the public are invited to kiss while wearing EEG headsets, as Actors in the orchestration. The 'kissing' brainwaves of the Actors, are measured and are made visible
as EEG data, shared with surrounding Spectators. Spectators, in turn, are invited to
watch the kiss. In the second orchestration, the EEG data of the kiss are translated
real-time to a floor projection that encircles the kissers with their real-time streaming
EEG data. The same data are 'translated' into a music score generated by the brain
computer interface (based on a novel algorithm design to this purpose).

These artistic orchestrations have been designed to provide an immersive, engaging environment for intimate experiences in shared multi-modal, multi-brain computer interaction through social touch. A Host facilitates this process.

7.3.3   Design of Two EEG KISS Orchestrations

This section describes the technical, spatial and social design of two EEG KISS
orchestrations, depicted in Fig. 7.3. The visual familiarity with aesthetics of medical
BCI representations is purposefully deployed by the artists to evoke reflection, on
current expectations towards scientific validations of intimate interaction.

Fig. 7.3 Spatial model of digital synaesthetics in multi-modal multi brain BCI interaction for social
touch: Artistic Orchestration 1 © Lancel/Maat and Studio Matusiak (2015)

Technically, the orchestrations consist of 2 EEG headsets[4] with four contact points on the skull (A) of which three positioned in the motor cortex. The brain activity is translated to a multi-modal, data visualization on two individual screens (2b) or as a floor projection that encircles the kissers in Orchestration 2 (2b in Fig. 7.9).

Spatially, two chairs are positioned across from each other, central stage, together forming 'a love seat', for Actors to take place and for Spectators to gather around.

Socially, people are invited to participate in various roles: that of Actor or Spectator. The staged acts of Actors kissing (1), have been designed to be 'performative'.[5] Spectators (3) view from a distance (and can become Actors themselves). They watch the kissing and the EEG data.[6] In Orchestration 2, they data audification is added. Aspects of Hosting, data visualization and data audification are described in the sections below.

7.3.3.1 Host

A Host (C), performed by the artists or by volunteers, is part of the interface design. The performance procedure is depicted in Fig. 7.4.

In Phase 1 Actors interact with the Host. The Host explains that the artistic orchestration explores and studies social engagement through mediated touch and performative interaction, using words such as 'online kissing', 'digital touch', 'kissing online', 'brain technology', 'share', 'privacy' that are internationally understood. Body language is used to visualize 'kissing' and 'being close'. The explanation serves both as a spoken manual and as contextualization indicating that EEG data primarily measure muscle tension and that scientific interpretation of EEG data from intimate kissing is not possible in this artistic orchestration.

Actors are asked to firstly close their eyes before kissing, secondly take all the time they need to kiss and thirdly, to keep their eyes closed when they feel the kiss is ending and to remember how the kiss felt. The Host then places EEG headsets on the Actors' heads and asks them to close their eyes, to concentrate and to reflect on the experience to come.

In Phase 2, the Host determines when to proceed, based on observations of the Actors' embodied behaviour and movements as well as the visual EEG data

---

[4]The wireless EEG headsets (IMEC 2014) are instrumented with dry electrodes. They measure at four contacts points on the skull [cz, pz, c3, c4 (Teplan 2002)]. Measuring emotional arousal is not the focus of these headsets, as the locations primarily associated with emotions [(pre)frontal cortex] are not measured. https://www.elektormagazine.com/news/wireless-activeelectrode-eeg-headset, last accessed 2018/12/17.

[5]Instead of referring to the notion of *performance* as a form of 'role-playing', *performativity* (Butler 1990) is, in this context, considered to be a repetitive act designed for public spaces, to share reflection on social engagement.

[6]Spectators can only participate as Actors by giving verbal consent for recording all EEG data non-anonymously, by adding their first names. Adding their names also serves a second purpose, namely to identify their own contribution, to be able to engage with their replayed data visualization at a later date.

Phase 1 *(left, right)*: Interaction with the Host who explains and places the headsets.

Phase 2 *(left)*: Brain acitivity is measured with eyes closed, of Actors concentrating.
Phase 3 *(right)*: Brain acivity is measured while kissing.

Phase 4 *(left)*: Brain acitivity is measured with eyes closed, of Actors concentrating.
Phase 5 *(right)*: Host in dialogue with Actors on experience of kissing.

Phase 5 *(left, right)*: Host in dialogue with Actors relating experiences of kissing
to EEG recordings.

Fig. 7.4 Performative actions in both artistic EEG KISS orchestrations of Actors and Host

sequences. Once the Host observes that the participants are sitting quietly and the EEG data sequences depict low frequencies, the Host softly tells them they can start kissing.

In Phase 3, the Host then witnesses the Actors' performativity of kissing from a short distance, ensuring a feeling of safety.

In Phase 4, after kissing, the Actors keep their eyes closed to remember how the kiss felt. When necessary, the Host reminds Actors to keep their eyes closed and remember what they have experienced.

In Phase 5, the Actors again interact with the Host. Once the Host observes that Actors have started to talk and look around, their headsets are removed. The Host then mediates reflection through an open-ended dialogue, with questions such as: (1) How did your kiss feel and how did your kiss feel in EEG data? (2) Did you hear the sound during kissing and did it affect your kiss? (3) Did you feel the audience around? (4) How is this kiss different from your other kisses? (5) How intense did you experience the presence of Spectators, artificial system and data audification while kissing? (6) Can your kiss be measured? On a scale of 1–10: how intimate was this kiss? (7) Would you agree to save your kisses in a database to be used by others?[7] The Host, in fact, mediates between physical and virtual presence, between experience of kissing and representing datafication, between public space and intimate space. The Host mediates the multi-modal and multi-brain feedback processes between Actors, Spectators, data visualization and data audification.

7.3.3.2   Data Visualization

The EEG data visualization (2b in Fig. 7.3) emerges real-time from acts of kissing, in two different orchestrations. The first orchestration shows individual data of two kissing persons on separate screens.[8]

In the second orchestration (2b in Fig. 7.9), both data sequences are integrated into one visualization. The separate data sequences are visually placed on top of each other[9] in different colours, to both compare individually and merge. Spatially, the combined data sequences are projected real-time around the Actors kissing as 'Dancing Data', as a floor-projection designed to function as a dynamic stage, bridging and isolating Spectators and Actors in communal patterns and flow.[10]

The data sequences in both orchestrations are derived directly from the four electrodes and are shown as separate 'lines' on the screen. Top-down, the first three 'lines'

---

[7]Note that these conversations are an essential part of the artwork and are not recorded.

[8]In this visualization, the feedback of starting and ending of kisses enable Spectators to synchronize the Actors kissing to the visualization of EEG data. The markers are activated by the Host, based on observing participants starting and ending their acts of kissing.

[9]The data-sequences are visually placed on top of each other without fusing them previously.

[10]In both orchestrations, data sequences differ in each performative phase (Fig. 7.4).

When Actors close their eyes from Phase 1 to Phase 2, waves become smaller. However, sometimes, interestingly, they seem to synchronize and 'flow.' In those cases, visual sequences move like waves that cross each other rhythmically.

are the data from the channels C3, Cz and C4, showing measurements from the motor cortex, including measurements of sensory and motor functions (Teplan 2002),[11] and weak motory intention (mu rhythms) (although more activities are reflected in C3 and C4 in comparison to Cz). In measurements of all positions (including Pz), motor artefacts (such as of neck, face and tongue muscles) measurements and alpha rhythms (due to the participants having their eyes closed while kissing and reflecting) and cognitive relaxation are measured. In the visual feedback, the measured arousal activity is not separated from the measured motor cortex activity. As a consequence, the data visualization predominantly shows motor intention and body movement of kissing.

7.3.3.3 Data Audification

In the second orchestration, a sound based sensory feedback module has been added[12] to enable Actors and Spectators to share multi-modal neurofeedback of the act of kissing 'digital synaesthesia'. The sound of each kiss is unique.

Technically, the algorithm on which the sound is based makes use of pre-defined combinations and averages of both of the participants' EEG data signals, to generate sound patterns.[13] The algorithm adapts to the various performative phases (Fig. 7.4), with different sound patterns, separated manually by the Host.

A 'sound flow' is acquired by crossfading separate sound patterns, as 'spheres', based on artistic choices.[14] In phases 1 and 5, the algorithm generates soft, 'ticking and crackling' sound patterns (by electric disturbances of the 50 Hz system). In phases 2 and 4, 'water bubbles tickling' sound patterns are dominant (based on low tones). In phase 3 (during kissing), the sound of phase 2 is combined with sparks of bells tingling (achieved through soft high tones).

7.4 Disruption for Engagement

These two multi-modal, multi-brain orchestrations focus on intimate touching, of kissing in public. Design of engagement for participants is based on disruption, as an aesthetic principle to orchestrate experiences of unfamiliarity, unpredictability and risk, to evoke ambivalence, immersion and reflection. Sensory perception of seeing and touching while 'intimately kissing in public' is disrupted. A new interaction between Spectators, Actors and EEG data is re-orchestrated, in three interdependent

---

[11]These functions relate to processing touch and sensation as well as keeping track of the location of body parts (proprioception).

[12]The algorithm and sound were designed in collaboration with Tijs Ham (STEIM Amsterdam). In his artworks he applies programming, live-electronics techniques and system design. https://www.soundlings.com/staff/tijs-ham/, last accessed 2019/1/30.

[13]In the data processing, EEG signals are translated via OSC to Super Collider.

[14]This research does not focus on soundtrack valences in relation to emotion elicitation.

forms of digital synaesthetics, starting from the kiss as source, in which feedback is defined, as depicted in Fig. 7.5.

The three models in Fig. 7.5 show interdependent, cross-modal feedback processes in both orchestrations. The first shows audification feedback, the second visualization feedback and the third model shows social feedback.

(1) In the audio-feedback, the direct sensory intimate connection between Actors is disrupted through amplified, EEG data audification, for kissing Actors and Spectators to share. The Actors' brain activity affects the BCI input and the brain activity itself is affected by the BCI output, and as a consequence, in fact input and output inform a loop (Gürkök and Nijholt 2013).

(2) The visual connection Actors and Spectators is disrupted when Actors close their eyes to kiss. The kissing Actors' brainwaves are translated to a EEG data visualization for Spectators to watch. They watch and compare an aesthetic, ambivalent orchestration, of both physical acts of kissing and an abstract, digital data visualization that represents the kiss.

(3) Social feedback builds on the witnessing Spectators in relation to the self-disclosure of Actors and discussions with the Host on embodiment and cognitive reflection with both Actors and Spectators.

The fourth model (Fig. 7.6) combines all models in an integrated model of digital synaesthetics in multi-modal, multi brain BCI feedback processes for shared intimate experience of social touch.

7.5   EEG KISS: Two Artistic Orchestrations

This section analyses the two artistic orchestrations described above to answer the question: Can shared intimate experience of social touch be mediated through multi-brain-computer interface (multi-brain BCI) interaction? These orchestrations are evaluated based on interaction between the components described above and depicted in Figs. 7.5 and 7.6.

7.5.1   Research Method

The effects of the artistic design choices behind these two artistic orchestrations are analysed on the basis of (1) observations (by the Host) of Actors and Spectators' actions and reactions; (2) thick descriptions of open ended interviews with Actors and Spectators. The analysis is further based on (3) photo and short video documentations that support these observations, when available.

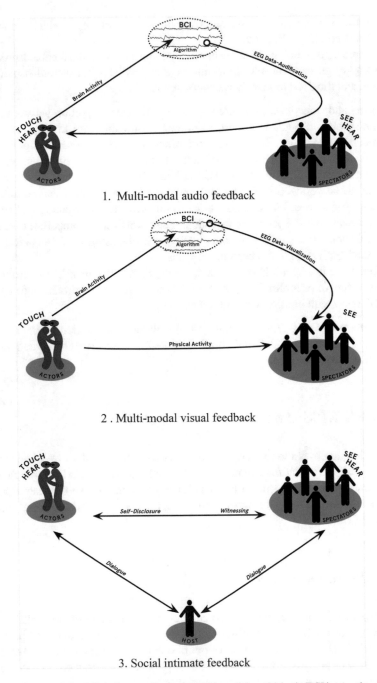

1. Multi-modal audio feedback

2 . Multi-modal visual feedback

3. Social intimate feedback

Fig. 7.5 Three models of digital synaesthetics in multi-modal, multi-brain BCI interaction of social touch

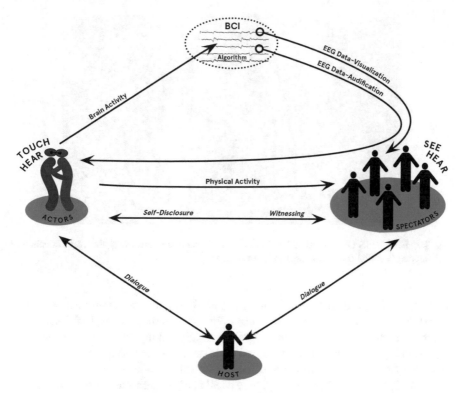

Fig. 7.6 Integrated model of digital synaesthetics in multi-modal, multi-brain BCI interaction of social touch (based on the three models depicted in Fig. 7.5)

7.5.2 EEG KISS: Artistic Orchestration 1

Orchestration 1 was held during the exhibition Reality Shift, during the Discovery Festival at Tolhuistuin in Amsterdam in 2014.[15] 14 Couples and 300 Spectators participated. This orchestration explores the reactions of the public to disrupted connections, and new sensory and social connections and data visualization.

7.5.2.1 Results

Many Spectators stop to see the orchestration, seemingly attracted by people kissing. The Host observes that many Spectators become immersed in the orchestration once they are told that they too can become participants and see that others are also participating. While Actors are kissing, the Host almost always observes that Spectators

---

[15]Orchestration 1 at Tolhuistuin Amsterdam. Discovery and Transnatural Festival 2014: Beyond Biennial Exhibition 2014 © Lancel/Maat.

Fig. 7.7 EEG KISS Orchestration 1 at Tolhuistuin Amsterdam. Discovery Festival/Transnatural and Beyond Biennial Exhibition. 2014 © Lancel/Maat

are immersed in a disrupted, twofold gaze, shifting between kissing acts and data representations on the screens. They turn to the data on the screens, look back at the kissing act and back to the screens again, seemingly linking the kissing gestures and the emerging data traces. Their focus remains on the Actors after the act of kissing, when the Actors eyes are still closed, even if this phase takes up to 5 min.

Couples, friends and strangers, people of all ages, kiss. Initial reactions expressed to the Host such as: "All of those people who are watching!" often include indications of shyness, nervousness and discomfort, but also enthusiasm.

As indicated in Fig. 7.7, Actors are asked to firstly close their eyes before kissing, secondly take all the time they need to kiss and thirdly, to keep their eyes closed when they feel the kiss is ending and to remember how the kiss felt. Some couples start kissing right away while others wait a few minutes to seemingly overcome shyness. The duration of kissing is between 20 s and 2 min. Different ways of kissing are observed and interpreted by the Host to vary between still, silent, tender, dynamic and expressive. Giggling a little before or during the kiss is not unusual. The Host observes that if Actors do not close their eyes and reflect on their kiss, both Actors and Spectators do not concentrate on the act of kissing.

When Actors open their eyes after the kiss, different reactions are perceived by the Host: some Actors express exaltation, others express tension, others are silent, perceived by the Host to be opening their eyes as if awakening, needing time to find words and staring in mid-space. Expressions include "I feel disoriented" or "I forgot where I was".

Not all but many Actors state that while kissing, they are first aware of the surrounding Spectators, but that after some time they lose touch with the Spectators, expressed e.g. as "I felt fear at the beginning but soon forgot all around us. Our kiss was all that mattered". Nevertheless, during some performances, the Host observes that the Actors' hands dwell towards sexually arousing parts of the body but stop at that.[16] One Actor expressed the role of Spectators as that they "come and go" in his mind "like waves".

Comparably, in most cases, Actors state that they 'forget about visualization of the data'. However, ambivalently, the Host observes that Actors only start kissing if they have seen the data visualization before kissing. In the few cases that Actors started kissing without Spectators being present, they indicated to the Host that they experienced their act of kissing as instrumental to digital data production and interpretation, and not as an intimate act. A few of these Actors also expressed concern about what data visualisation of their kiss may be "giving away". One these Actors stated "I am concerned that these data are judged by others."

Actors are always interested in the data visualisation of their kiss. Although at forehand Actors have been told that scientific interpretation of EEG data from intimate kissing is not possible in this artistic orchestration, and not the focus of this research, almost all participants seem to be convinced that the artistic orchestration reveals information about their kiss, their 'kiss-qualities', and the quality of the Actors' relationships (as expressed to the Host).

Although the individual data sequences are visible on two different screens, Actors often talk about both sequences as a composed representation of the kiss and often refer to the combination of sequences as "the portrait of our kiss", as an act of co-creation. When discussing the data in dialogue with the Host, Actors' expressions include 'an enigmatic mirror of their kisses': "It leaves sense making to ourselves" and "Only we know what these traces mean", interpreting data as depicting their experience of intenseness ('on fire'), concentration ('like waves of a river"), or the feeling of togetherness during their kiss ("This reminds me of the intimate moment we just had together."). In some cases, Actors are observed to silently gaze at the data, smile and seem to lose all sense of time with expressions interpreted by the Host to indicate tenderness, disbelief and curiosity.

Spectators express attribution of meaning to the data sequences. Example of such attributions are: "I can clearly see from the data sequences that one of the persons kissing was more passionate than the other." Or "I love how these data-lines move together and many times I could see whose line belongs to who from the way they kiss." One of the Spectators stated "I could see the kiss being mirrored in the data visualization. Although in fact I don't know what I was seeing, I felt I could see it." Others expressed other experiences related to emergence: "I could see they were passionate and I could see that feeling in the data too.", "You really see them going in the data", and "I love the data emerging. Of course, I knew they were emerging from kissing." Some Spectators indicate the importance of synchronization between the beginning and end of physical kissing and its visualization of EEG data, for their experience. Often lively discussions start about data interpretation, intimacy in public and issues of privacy, in relation to the information value of EEG data.

---

[16]Visible in video documentation.

7.5.2.2   Discussion

Couples, friends and strangers of various ages and diverse cultural backgrounds have participated in these experiments, some for hours. Ambivalence, both purposefully designed and emergent, is shown to be essential to evoke engagement and immersion of Actors and Spectators in the shared orchestration.

Firstly, Kissing Actors need to ambivalently trust the Host while simultaneously risking judgement of their vulnerable act of kissing and the resulting data by Spectators. For immersion in intimate experience, it has shown to be important for Actors to 'semi-lose touch' with Spectators. However, ambivalently, they also have expressed the need to have confirmation that Spectators are present.

Secondly, ambivalently, individual Spectators have expressed the need to witness the Actors' physical (intimate kissing) gestures and simultaneously give meaning to the emerging abstract data visualization. In this process, seeing the Actors kissing gestures is shown to be needed to 'feel' the data visualization as being intimate.

Thirdly, to individually interpret the EEG data visualization as an expression of intimacy, Spectators and Actors have indicated the need to be confirmed of each other's presence during the kiss. Ambivalently, they express the need of a shared experience to interpret individually. Spectators also express the need to be able to witness other Spectators. Actors have shown the need the presence of Spectators witnessing their emerging BCI data, to appropriate the data visualization in retrospect as 'their portrait' of shared intimacy.

Importantly, shared intimate experience is only reported if reflection is facilitated, for all to share and co-experience.

7.5.3   EEG KISS: Artistic Orchestration 2

Orchestration 2 took place at the Frascati Theaters Amsterdam, 2016. 11 Couples and 43 Spectators participated. In this adapted orchestration Actors are surrounded by sound and by a visual, abstract, streaming EEG data floor projection witnessed by Spectators, depicted in Fig. 7.9. This orchestration explores whether spatial data visualization and data audification enhance shared engagement for BCI mediated intimate connections, as described above. Note, that in this second orchestration, Actors and Spectators behaviour observed by the Host was comparable to Orchestration 1. Only new aspects are described Fig. 7.8.

7.5.3.1   Results

When asked by the Host, Actors indicate that they experience kissing the other person as both familiar and unfamiliar. Actors follow the Host's invitation to close their eyes, to listen to the sound, and to immerse in each other's kiss.

Fig. 7.8 EEG KISS Orchestration 2 at Frascati Theaters, Amsterdam. 2016. © Lancel/Maat

As in the previous orchestration, in phase 5, Actors are asked to reflect about the kiss in dialogue with the Host. Actors refer to the impact of sound with words such as: "The sound made my kiss more intense and more focused. The tickling sound, that emerged from my brain activity, made me imagine electric rain drops that enhanced and merged with my experience of electrified kissing." and "It felt like our kiss was being borne by the music". A few Actors, who indicated that they tried to control the sound through different ways of kissing, referred to their kiss as 'fun' rather than as intimate.

Spectators are observed by the Host to be more concentrated and immersed in the circular data environment and data audification, in comparison to Orchestration 1. Both Actors and Spectators express for example: "This situation is weird but feels strangely safe.", "The sound makes the space reflective", "This feels like a kind of trance" or "I could stay here forever." Some stay for hours, talking quietly with each other. More often than in Orchestration 1, the Host observes that Spectators encourage each other to become Actors and kiss. Even strangers kiss. While the average time for Actors to start to kiss is between 1 and 30 s, strangers starting to kiss can take up to 5 min. These 5 min are reported by all Actors and Spectators to be experienced as being very intense. Furthermore, in this second orchestration, the duration of kissing is longer (between 20 s and 10 min) in comparison to orchestration 1.

The circular data visualization is designed for Spectators to stand around the Actors. A few Actors indicate that they experience the circular, emerging data visualization as 'a radar.' Spectators are observed to never enter the floor projection while Actors kiss. Most Actors and Spectators describe their experience of the data visualization as immersive, indicated by for example "Can I step into it?", "Is this a sort of brain data space?", "These lines here are moving more wildly than those lines over there" (while observed to be pointing at the projection on the ground proximate to their bodies)", or "Am I staying in their brain activity?".

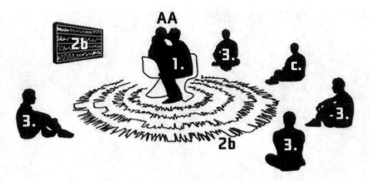

Fig. 7.9 Spatial model of digital synaesthetics in multi-modal multi brain BCI interaction for social touch: Artistic orchestration 2. © Lancel/Maat and Studio Matusiak (2016) (Lancel et al. 2018)

The Host observes that Actors seem more comfortable entering the staged space and to being exposed, in comparison to Orchestration 1. Furthermore, after kissing, Actors take more time to talk about their kisses with the Host, to explore their memories of kissing. They are, in general, seen to be more comfortable, talking while watching the streaming data around them, indicated by one Actor to be both 'beautiful and strange'.

7.5.3.2 Discussion

Orchestration 2 shows that circular, streaming visual data and data audification, of the multi-brain BCI enhances focus and immersion for all participants. Ambivalently, spatially, the data visualization distances Actors from the Spectators, while at the same time bridging them. Furthermore, ambivalently, Actors have shown to need the data visualization to experience and remember an intimate kiss, but most Actors express forgetting the data visualization while kissing. For Actors, the shared data audification is perceived by many to be shared feedback of their performativity, merging with their kissing experience. The combination of shared sound and spatial data visualization has been observed to increase embodied, immersive experience of the BCI data for Spectators. This multi-modal BCI orchestration with spatial visualization and shared sound has shown to increase participants' feeling of safety and involvement, both in time and in intimate connection with each other.

7.6 Conclusion and Future Research

This chapter explores design syntheses for artistic orchestrations of shared intimate experiences, of multi-brain, multi-modal BCI interaction, through touch. Digital synaesthetic, shared intimate social touch experience is explored through disruption of familiar relations between 'who you kiss and who is being kissed, what you see and what you hear'. A model of digital synaesthetics in multi-modal, multi-brain BCI interaction for social touch is proposed.

Two orchestrations show that for engagement in shared, intimate experiences mediated by multi-brain BCI, sensory connections and feedback processes through seeing, hearing, and touching need to be disrupted. They need re-orchestration into multiple ambivalent connections, such as connections between participants (Actors and Spectators), senses, actions and connections between physical presence and virtual, spontaneously emerging BCI representations. Audification of BCI data of touch, enhances the Actors and Spectators' experience of feedback. Spatial data visualization provides an embodied, immersive relation to the BCI data, both isolating and bridging Actors and Spectators. The combination of spatial BCI data visualization and audification increases focus, concentration, immersion and feelings of safety. Central stage intimate touch is essential to shared experience and reflection.

Meaningful shared intimate experience mandates an orchestration that includes vulnerable self-disclosure, witnessing, dialogue and reflection, embedded in individual and shared interpretation, in co-presence with all participants. The mediating role of the Host is a crucial element of the design. The proposed model in which the role of Hosting is defined, is based on shared perception of a social, sensory synthesis.

This model of digital synaesthetics facilitates the design of new shared intimate experiences of social touch mediated by multi modal multi brain BCI interaction in public spaces.

Current and future research extends the BCI interface to include brain activity of both Actors (touching) and Spectators (mirror-touching) for a Multi Brain BCI visualization (Lancel/Maat 2018).

The correlation of EEG and fNIRS data visualization to experiences of connectedness (synchronization), in a social-technical performative synthesis, is currently being explored. These data visualizations are studied from an aesthetic perspective, for a new approach to understanding data patterns. This approach is explored in dialogue with participants in the artistic orchestrations, for an emotionally intelligent machine learning system for intimate experience (Figs. 7.10 and 7.11).

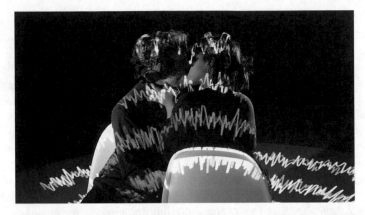

Fig. 7.10 EEG KISS Orchestration 2 at HeK, Haus for Electronic Art, Basel, 2018 (EEG KISS Orchestration 2, HeK Haus for Electronic Art, Basel, 2018 © Lancel/Maat 2018.)

Fig. 7.11 Kissing Data, Ars Electronica Linz, 2018 (Kissing Data, Ars Electronica Linz, 2018 © Lancel/Maat 2018.)

Acknowledgements The authors thank Caroline Nevejan, the EU's Horizon 2020 *Hack the Brain* programme partners (Waag Society Amsterdam, *Science Gallery Dublin, TU Delft and TU Berlin*), TNO (the Netherlands Organisation for Applied Scientific Research), the Holst Center, Fourtress Eindhoven, Phillips Lab Eindhoven, Baltan Laboratories ('Hack the Body Group'), EMAP (European Media Art Platform;) University of Applied Sciences Vienna ('Digital Synaesthesia Group'), TASML Art-Science-Media Lab and Tsinghua University Beijing and the 'Dutch Touch Group' for their contributions to joint research and development of EEG KISS. The authors are also very grateful for financial support from Mondriaan Fund, NWO KIEM The Netherlands Organisation for Scientific Research, FWF Wissenschaftsfonds Austria, BEAM Lab/BEAM Systems Amsterdam and Eagle Science Amsterdam.

References

Abramovic M, Dikker S, Oosterik M (2011) Measuring the magic of mutual gaze. http://www. artbrain.org/marina-abramovic-measuring-the-magic-of-mutual-gaze. Accessed 19 Oct 2018

Benford S, Giannachi G (2012) Interaction as performance. Interactions 19(3):38–43

Benford S, Greenhalgh C, Giannachi G, Walker B, Marshall J, Rodden T (2012) Uncomfortable interactions. In: Proceedings of the SIGCHI conference on human factors in computing systems. ACM, pp 2005–2014

Bennett RH, Bolling DZ, Anderson LC, Pelphrey KA, Kaiser MD (2013) fNIRS detects temporal lobe response to affective touch. Soc Cogn Affect Neurosci 9(4):470–476

Björnsdotter M, Gordon I, Pelphrey KA, Olausson H, Kaiser M (2014) Development of brain mechanisms for processing affective touch. Front Behav Neurosci 8(2014):24

Butler J (1990) Gender trouble, feminism and the subversion of identity. Routledge, New York

Casey K (2010) Global mind project. http://www.globalmindproject.com. Accessed 19 Oct 2018

Cillari S (2006–2009) Se Mi Sei Vicino. http://www.li-ma.nl/site/catalogue/art/sonia-cillari/se-mi-sei-vicino-if-you-are-close-to-mc/9774. Accessed 19 Oct 2018

Clark L (1963–1988). http://en.wikipedia.org/wiki/Lygia_Clark. Accessed 19 Oct 2018

Crew (2016) C.a.p.e. Drop_Dog. http://www.crewonline.org/art/project/704. Accessed 19 Oct 2018

De Boeck C (2009) Staalhemel. http://imal.org/en/more/tangible-feelingsexhibition. Accessed 19 Oct 2018

Delft University of Technology (2015) BrainHack: bringing the arts and sciences of brain and neural computer interface together. Horizon 2020 Project, part of EU framework programmes for research and innovation. https://cordis.europa.eu/project/rcn/199028_en.html. Accessed 19 Oct 2018

Dikker S, Abramovic M, Oosterik M (2016) Measuring the magic of mutual gaze. http://www. suzannedikker.net/art-science-education#mwm. Accessed 19 Oct 2018

Van Erp JB, Toet A (2015) Social touch in human–computer interaction. Front Digit Hum 2(2015):2

Gabriel U (1993) Terrain 02. https://bci-art.tumblr.com/post/163479016197/terrain-02. Accessed 19 Oct 19

Gabriel U (1996) Barriere. https://bci-art.tumblr.com/post/163479268367/barriere. Accessed 19 Oct 2018

Gill SP (2015) Tacit engagement. In: Tacit engagement. Springer, Cham

Gomes LM, Wu R (2017) Neurodildo: a mind-controlled sex toy with E-stim feedback for people with disabilities. In: International conference on love and sex with robots. Springer, Cham, pp 65–82

Gsöllpointner K, Schnell R, Schuler RK (eds) (2016) Digital synesthesia: a model for the aesthetics of digital art. Walter de Gruyter GmbH & Co KG

Gürkök H, Nijholt A (2013) Affective brain-computer interfaces for arts. In: Humaine Association conference on affective computing and intelligent interaction (ACII), 2013. IEEE, pp 827–831

Kwastek K (2013) Aesthetics of interaction in digital art. MIT Press

Lancel K, Maat H, Brazier FM (2018) Kissing data, distributed haptic connections through social touch. In: Acoustic space volume No 17. Riga's Center for New Media Culture RIXC and Art Research Laboratory of Liepaja University

Lancel/Maat (2000–2018). http://www.lancelmaat.nl/work/. Accessed 19 Oct 2018

Lancel/Maat (2009) Tele_Trust. http://www.lancelmaat.nl/work/tele-trust/. Accessed 19 Oct 2018

Lancel/Maat (2012) Saving face. http://lancelmaat.nl/work/saving-face/. Accessed 19 Oct 2018

Lancel/Maat (2014–2018) EEG KISS and digital synaesthetic EEG KISS. http://www.lancelmaat. nl/work/e.e.g-kiss/. Accessed 22 Jan 2019

Lancel/Maat (2018) Kissing data. At: Ars Electronica Linz, RIXC Riga and EMAP European Media Art Platform. https://ars.electronica.art/error/en/kissing. Accessed 22 Jan 2019

Lancel/Maat, Luehmann A (2017) EEG KISS II (fNIRS). At: Stedelijk Museum Amsterdam and Worlding the Brain conference 2017. https://www.metropolism.com/nl/features/33557_ worlding_the_brain. Accessed 22 Jan 2019

Loke L, Khut GP (2014) Intimate aesthetics and facilitated interaction. In: Interactive experience in the digital age. Springer, Cham, pp 91–108

Lomanowska AM, Guitton MJ (2016) Online intimacy and well-being in the digital age. Internet Interv 4(2016):138–144

Lupu RG, Irimia DC, Ungureanu F, Poboroniuc MS, Moldoveanu A (2018) BCI and FES based therapy for stroke rehabilitation using VR facilities. Wirel Commun Mob Comput

Lysen F (2019) Kissing and staring in times of neuromania: the social brain in art-science experiments. In: Borgdorff H, Peters P, Pinch T (eds) Artful ways of knowing, dialogues between artistic research and science & technology studies. Routledge Advances in Art & Visual Studies series, forthcoming in 2019

Martin D (ed) (2018) Mirror-touch synaesthesia: thresholds of empathy with art. Oxford University Press

MIT Technology Review (2018) The first 'social network' of brains lets three people transmit thoughts to each other's heads. https://www.technologyreview.com/s/612212/the-first-social-network-of-brains-lets-three-people-transmit-thoughts-to-each-others-heads/. Accessed 22 Jan 2019

Mori M (2005) Ufo wave. https://bci-art.tumblr.com/post/163513717167/ufo-wave. Accessed 19 Oct 2018

Nevejan C (2007) Presence and the design of trust. University of Amsterdam

Nijholt A (2015) Multi-modal and multi-brain-computer interfaces: a review. In: 10th international conference on information, communications and signal processing (ICICS) 2015. IEEE, pp 1–5

Nijholt A, Nam CS (2015) Arts and brain-computer interfaces (BCIs). Brain Comput interfaces 2(2–3):57–59. https://doi.org/10.1080/2326263X.2015.1100514

Novello A (2016) (Un)focussed. https://instrumentsmakeplay.nl/album/unfocussed/. Accessed 19 Oct 2018

Osthoff S (1997) Lygia Clark and Hélio Oiticica: a legacy of interactivity and participation for a telematic future. Leonardo 30(4):279–289

Pike M, Ramchurn R, Benford S, Wilson ML (2016) # scanners: exploring the control of adaptive films using brain-computer interaction. In: Proceedings of the 2016 CHI conference on human factors in computing systems. ACM, pp 5385–5396

Prpa M, Pasquier P (2019) Brain computer interfaces in contemporary art: a state of the art and taxonomy. In: Nijholt A (ed) Brain art: brain-computer interfaces for artistic expression. Springer human-computer interactions series, forthcoming in 2019

Ramchurn R, Martindale S, Wilson ML, Benford S, Chamberlain A (2019) Brain controlled cinema. In: Nijholt A (ed) Brain art: brain-computer interfaces for artistic expression. Springer human-computer interactions series, forthcoming in 2019

Reeves S, Benford S, O'Malley C, Fraser M (2005) Designing the spectator experience. In: Proceedings of the SIGCHI conference on Human factors in computing systems. ACM, pp 741–750

Rodil LF (2014) R2 (Braid 2). https://vimeo.com/104188040. Accessed 19 Oct 2018

Roeser S, Alfano V, Nevejan C (2018) The role of art in emotional-moral reflection on risky and controversial technologies: the case of BNCI. Ethical Theory Moral Pract 21(2):275–289

Rosenboom D (1990) Extended musical interface with the human nervous system: assessment and prospectus. International Society for the Arts, Sciences and Technology

Sobell N (1974) Brainwave drawing. http://www.ninasobell.com/. Accessed 19 Oct 2018

Sobell N (2001) Thinking of you. http://www.ninasobell.com/. Accessed 19 Oct 2018

SPECS Synthetic, Perceptive, Emotive and Cognitive Systems (2009) Brain orchestra. http://news.bbc.co.uk/1/hi/sci/tech/8016869.stm. Accessed 19 Oct 2018

Teplan M (2002) Fundamentals of EEG measurement. Meas Sci Rev 2(2):1–11

Vlugt M (2015) Performance as interface | Interface as performance. IT&FB Amsterdam

von Lühmann A, Soekadar S, Müller KR, Blankertz B (2017) Headgear for mobile neurotechnology: looking into alternatives for EEG and NIRS probes. In: Proceedings of the 7th Graz brain–computer interface conference 2017, Graz. Verlag der TU Graz, Graz University of Technology

Chapter 8
New Ways of Knowing Ourselves. BCI Facilitating Artistic Exploration of Our Biology

Laura Jade and Sam Gentle

Abstract As rapidly advancing technologies become more widely available, having access to tools that collect biometric data and in particular BCI technology, is providing artists with new ways of exploring our biological selves as well as creating new modes of audience interaction. Brainlight is a large illuminated interactive sculpture that integrates biology, lighting design and BCI technology to explore the hidden aspects of our minds. The installation is controlled with a wireless EMOTIV EPOC+ EEG headset that detects live neural activity which is translated into a light display within the brain sculpture. In real time it visualises the brain frequencies of Theta (3.5–7.5 Hz) as green light, Alpha (7.5–13 Hz) as blue light, and Beta (16–32 Hz) as red light. Previously, in more traditional art, when an audience views an artwork their own psychological process would be a passive, hidden, private experience. The aim of Brainlight is to harness the brain as the creator of an interactive art experience where no physical interplay is required except for the electrical activity of the mind. The project exposes some key developments in the use of BCI technology for artistic purposes, such as how to accurately collect and process EEG data aesthetically, and what license the artist can take with this data in order to facilitate meaning or allow space for the audience to bring their own meaning to the work. This chapter will explore these developments and outline the collaborative process behind the research and development of the work and the contexts in which it has subsequently been exhibited and used by the public.

Keywords Electroencephalography (EEG) · Brain-computer interface (BCI) · Brain data visualisation · Neurofeedback · Interaction design · Mind-controlled art · Illuminated brain sculpture · Interactive art · Biofeedback art · Art-science · Audience

L. Jade (✉) · S. Gentle
Sydney, Australia
e-mail: laura@laurajade.com.au

S. Gentle
e-mail: sam@samgentle.com

8.1   Introduction

8.1.1   Technology: An Artist's Tool

Technology and art have always been bound. Art, like technology, shapes and is shaped by the social and cultural environment in which it is created. In recent years, as rapidly advancing technologies become democratised, access to tools that collect biometric data and in particular brain-computer interface (BCI) technology, have expanded the boundaries of what are considered artists' tools. When methods are appropriated from other disciplines (such as science), new artistic mediums are generated. Biosensors that collect heart rate, breath or skin conductance are turning our biological data into mediums for artistic exploration. When brain-computer interfaces (BCI) become the tool, the brain and consciousness itself become the medium for inquiry.

8.1.2   Hybrid Artists and Biologically-Driven Interactive Artworks

In our globalised world, contemporary artists increasingly have access to research outside their field, which is facilitating cross-disciplinary and interdisciplinary practice. This is creating a new ecology of hybrid artists, who use scientific tools to harness the body's bio-rhythms and generate live artistic interactions. Artists creating mind-driven interactive artworks are exploring the capabilities of augmented BCI technologies to interface directly with the brain. The most common technique applied in these devices, is Electroencephalography (EEG)—a recording of small electrical currents along the scalp generated by the synchronous activity of neurons in the cortex—the outer layer of the brain.

One of the earliest and most well-known examples of sonifying brain activity, was conducted by Adrian and Matthews (1934), who listened to human brain alpha oscillations; the sound of which they dubbed the "Berger Rhythm" (after Hans Berger who invented electroencephalography and discovered the Alpha wave in 1924). Subsequently, artists have been among the pioneers of EEG use outside clinical settings, designing situations and applications for EEG use in "real-life contexts" since the 1960s. The composer Alvin Lucier applied Adrian and Matthew's idea in his 1965 work *Music For Solo Performer,* where he amplified his alpha rhythms through percussive instruments for a live audience (Lucier 1976).

Fifty years later, the creative potential of BCI in contemporary art practices has only increased due to the availability of affordable, easy to use EEG technologies. In the last decade in particular, BCI has proliferated across a wide variety of artistic practices. Many artists using BCI are curious to see how their inner mental states can inform their art practice and offer new forms of expression. Sculptural work such as Ian Popian's *Mental Fabrications* (2014) translates EEG data into 3D printed topographical sculptures, exploring how our emotional responses could inform architectural design practices. Random Quark's *The Art Of Feeling* turns the collected EEG data of a person's emotional memory, (such as "the birth of my son"), into digital paintings (Papatheodorou and Chambers 2018). *Neuro-knitting* is a project which translates EEG activity into textiles (Guljajeva et al. 2018). Media artist Refik Anadol's *Melting Memories* (2017) turns EEG recordings focused on long and short term memory into mesmerising large-scale projections and animations. Lisa Park gives the invisible energies of her brain an auditory and visible form in her work Eunoia II (2014) where real-time EEG signals translate Park's changing brain activity into sound vibrations that manipulate 48 pools of water. When describing her work, Park says "I wanted to make a connection that our brainwaves, feelings and sound waves are all frequencies of energies… my work attempts to embody this idea of giving the invisible a physical form to create an external representation of myself" (2014).

These artists are demonstrating how electrical data generated by the brain can be transformed artistically into a wide variety of cross-modal sensory experiences. According to Gsöllpointner, experiencing your brain's electrical activity modified into an artwork can result in an altered perception of the self by inducing "digital synesthesia" (2016). Synesthesia is a phenomenon of perception where sensations experienced in one sensory domain are translated and expressed in another, such as sounds experienced as colours. Digital synesthesia is produced when a sensory stimulus is transferred across other sensory domains by way of a digital interaction. For instance, brain data collected from an EEG can be translated via software into visual images or sounds allowing you to experience the activity of your mind through visual or auditory sensory channels which alters the way you naturally sense yourself.

The virtue of this practice in art lies in its capacity to reveal biological systems that are otherwise imperceptible, offering revelations on aspects of humanness that come from the extension or alteration of the self through technology. Artists working with BCI are contributing to the convergent field of practice that seeks to explore the juncture between art, technology and the mind; a framework that has been defined by Roy Ascott as the "technoetic arts". Ascott says that "the body is no longer a solid biological entity but a technologically connected or enhanced cyborg". Ascott calls "cyberception", "the emergent human faculty of technologically augmented cognition and perception," which acts not only as an extension or enhancement of the senses, but as a unification and distribution of the mind, producing new human faculties (Ascott 1999). Artistic research of this nature may offer new pathways within the field of human computer interaction by introducing novel sensory methods of interfacing with computer systems that aim to amplify human qualities (Vygandas 2018).

8.2 Brainlight

8.2.1 Introduction to Brainlight

Brainlight is an artwork that explores how technology can aesthetically interface with the mind. It integrates biology, lighting design and BCI technology into an interactive brain sculpture, lasercut from transparent perspex and engraved with neural networks. The installation is controlled with a wireless EMOTIV EPOC+ EEG headset which detects and outputs live neural activity, translating electrical signals from the user's brain, into a vivid and dynamic light display within the brain sculpture. In real-time Brainlight visualises the brain frequencies of theta (3.5–7.5 Hz) as green light, alpha (7.5–13 Hz), as blue light and beta (16–32 Hz) as red light (Fig. 8.1).

   The project highlights some key developments in the use of BCI technology for artistic purposes, such as how to collect and process EEG data in an artistic context, how to translate it into a live interaction that communicates the data aesthetically (explored in Sect. 8.3), how the work has been experienced in various contexts (Sect. 8.4), and what license the artist can take with this data in order to allow space for the audience to bring their own meaning to the work (Sect. 8.5). Further developments of the work are explored in Sect. 8.6, and evaluation methods and future directions are explored in Sect. 8.7.

8.2.2 Artist Aims

Various methods for exploring the mind have been used throughout human history. Yet most of us live with very little understanding of the underlying processes within our own minds. Consciousness continues to be one of the more enigmatic problems for both the natural sciences and philosophy. One of its most perplexing properties is that it materialises as an intimate, subjective, experiential sense of self (Menon et al. 2014).

Fig. 8.1 Brainlight and the illuminated colours that represent each brain frequency

What can interactive art tell us about the self? According to Rokeby, the interactive artist holds up a mirror to the spectator, resulting in a shifting reflection. These "transformed reflections are a dialogue between the self and the world beyond. The echo operates like a wayward loop of consciousness through which one's image of one's self and one's relationship to the world can be examined, questioned and transformed" (Rokeby 1995).

While all art engenders a relationship between the audience and the work, in Brainlight's case, the audience also enters a relationship directly with the self. The artwork transfers neuro-feedback therapy, a technique used to teach self-regulation of brain activity (e.g. Hammond 2007; Peper et al. 1979), from a clinical setting to an artistic one by creating a sculpture that aesthetically embodies a live visualisation of brain activity, allowing a participant to have an intimate and unique interaction with their inner selves—to "meet their own mind"—externally. The work aims to facilitate a curiosity to know and sense oneself more intimately, while at the same time explores the creative potential of BCI technologies.

Wadeson et al. (2015) identifies four types of user control of artistic BCI's: passive control, selective control, direct control and collaborative control. Brainlight can be classified as a 'selective control' BCI as users can intentionally control their brain activity through emotion, relaxation or excitement etc. in order to influence the artworks pre-determined parameters. The artwork is partly an extension of the user, however the relationship between the user and the work is externally defined by myself the artist.

Experiencing Brainlight as an audience member invites not just a dialogue with their own mind, but also invites them to question and engage with their experience of BCI technology. Brainlight's visual simplification of the brains complexity through coloured light, makes tangible only a small glimpse of the true reality of the brain's electrochemical processes. The work inevitably reveals a tension between our desire for self-reflection and the inexplicable gap between the physical brain and the ethereal mind.

8.2.3 Communicating Emotion

There is considerable neuroscience research into understanding how humans best communicate with one another (e.g. Sherry 2015). Particularly important appears to be empathetic communication and the transfer of feelings and emotions. The field of affective computing aims to bridge the gap between human emotions and computational technology (Heanue 2018). One approach to communicating emotion via a computer, without using language, is by using pattern recognition algorithms to pick up facial expressions (Ekman 1994) or body gestures (Kleinsmith and Bianchi-Berthouze 2013). Other techniques for non-verbal emotional measurement include, heart rate, blood pressure, temperature, electro-dermal responses and respiration (Molina et al. 2009).

Since the early experiments with EEG on humans in the 1920s, the use of EEG in the study of the brain has been mainly focused on clinical diagnostics and trying to understand neurological processes and functions in a research laboratory environment (Maskeliunas et al. 2016). Only recently has EEG received specific interest for its potential to be harnessed as a communication channel for BCI. The advantage of having access to real-time brain activity with EEG means that a person's current emotional state can become a passive or active method for BCI control (Molina et al. 2009; Mühl et al. 2014). As Gürkök and Nijholt suggest, if art is a way to express emotion (emotions we might not yet understand), then BCI generated art could even help us understand the emotions we are experiencing (2013).

The universal struggle to express our innermost feelings led me to the question of what it might be like to be able to transfer internal states and emotions to one another through BCI communication. As an artistic exploration, Brainlight uses BCI technology to tune in as best as possible to the unspoken, subtle forms of communication of the electrical activity that produces our thoughts and emotions. Despite the complexity of emotions and the limitations of EEG, I was curious to see if a simplification of live brain activity, symbolically visualised through colour, could communicate a sense of a person's inner reality to an audience and generate a meaningful experience. In doing so, the artwork asks the audience to imagine a future where technology may be able to enhance our ability to capture and share inner qualities that are innately human, and inevitably ask themselves whether or not this would be desirable.

8.2.4 Light, Art and the Brain

One of the most fascinating biological relationships is between our bodies and light. We depend on light for all kinds of important metabolic functions, such as vitamin D and melatonin production and maintaining healthy circadian rhythms. Beyond this, light is also our connection to the universe. Through light we can observe distant galaxies, nebulas and look back at the beginning of existence itself.

There is also an interesting connection between light frequencies and brain frequencies. Light is a photon travelling through space in an electromagnetic wave. The visible light spectrum is the particular electromagnetic frequencies that interact with our visual system in order to stimulate the perception of colour. Because colour does not actually exist in nature—it is all generated in our mind—our own brain is essentially collaborating with light in order to perceive the world around us.

Brainlight takes this idea a step further by harnessing BCI interaction to create a neuro-feedback loop of this process of perception. The artwork is activated by your brain's electrical activity; it transforms these electrical frequencies into light waves, which are then re-translated by your brain into colour. The colour which your brain is now seeing, is a visual representation of this very perceptual process, meaning you as viewer bear witness to a real-time loop of your brain transforming the image of your brain transforming the image etc. This is no different to the constant input-output mechanism performed by the visual cortex, with the exception that the input is now also the output. The recursive nature of the feedback demonstrates the potential of artistically devised interactive technologies to bring us closer to the pure process of perception itself.

8.2.5 Communicating with Colour

Brainlight displays, as coloured light, a live stream of dominant brain frequencies, creating a neuro-feedback loop between the artwork and the viewer. Red, blue and green (RGB) light was chosen to represent the brain states beta, alpha and theta respectively, in order to make use of the artistic symbolism associated with each colour.

Because of connotations with speed, fire, heat and intensity, the colour red was used to represent beta (16–32 Hz) frequencies, which have a higher energy and can signify states of alertness and intense emotions such as excitement and stress (Alonso et al. 2015; Ray and Cole 1985). Dominant alpha oscillations (7.5–13 Hz) have been correlated with calm, meditative and relaxed states, particularly in the occipital channels when the eyes are closed (Ahani et al. 2014; Chiesa and Serretti 2009; Khare and Nigam 2000; Lutz et al. 2007), and so the colour blue was chosen due to its association with peace, introspection and tranquility. Green, symbolic of nature, was used to represent Theta (3.5–7.5 Hz) which has been linked to a large number of cognitive processes, such as integrating affective and cognitive sources of information in working memory tasks and action monitoring (Cavanagh et al. 2011; Kawasaki et al. 2010; Klimesch 1999), heightened expressiveness and creativity (Gruzelier 2008; Gruzelier et al. 2014) and deep meditation and present-moment awareness (Cahn and Polich 2006) to name just a few.

As RGB are the three primary colours of light, they have the added benefit of being visually distinct, allowing each dominant brain state to be communicated clearly.

The electrode positions on the EEG headset were mapped to corresponding positions on the brain sculpture via the projected light. The dominant frequencies in each electrode could then be individually visualised, creating a dynamic multi-coloured array of light displaying the rhythm and movement of the dominant electrical activity emanating from the brain.

Although the software doesn't explicitly mix the colours in the sculpture, early in the process we were surprised to notice additive secondary colours emerging when neighbouring regions of the brain were displaying different dominant frequencies. This allowed for interesting subjective meanings to be created by audience members. One example, at Illuminate, a light festival in Wagga Wagga, Australia, an 8-year-old girl was asked to imagine what made her most happy, after which the entire brain sculpture radiated a warm magenta light. She told us she was thinking about her guinea pig, which she loved very much. The magenta was created because her brain produced equal amounts of calm alpha frequencies (blue light) and excited beta frequencies (red light), and the coincidental mixing of the two states appeared to communicate a "loving" state of mind to the audience.

The nature of this subjective interpretation opened up interesting questions in relation to public interpretation of the data and my role as an artist facilitator in allowing subjective meaning to be created during an individual's personal experience with the Brainlight. (This is explored in more depth in Sect. 8.5.)

8.2.6 Limitations of EEG

It should be mentioned that analysis and interpretation methods for EEG are still limited and a consensus on the relationships between complex cortical dynamics sensed through an EEG is not apparent in the literature. It is well established that EEG is best suited for sensing fast temporal dynamics, which makes it ideal for interactive artworks which rely on fast and responsive feedback in order to facilitate a perceivable interaction with an audience. In this respect, EEG works well for studying responses to stimuli by showing real-time changes in regular brain activity (Zioga et al. 2014).

In contrast to high temporal resolution, a significant limitation of EEG is poor spatial resolution. EEG is most sensitive to the electrical activity produced in the outer layers of the cortex which means that the activity produced by deeper structures inside the brain contribute far less to the EEG signal. Because of this we can only observe how the outer brain signals change in response to various types of activities or stimuli and then make inferences about the brain processes involved in such situations. Indeed, it is very unlikely that a single cerebral rhythm is associated with a specific cerebral function, particularly when it has been shown that even single neurons have the ability to oscillate at multiple frequencies (Mantini et al. 2007).

In an article by Herrmann et al. (2016) the authors state that "almost every cognitive process has been associated with an event-related EEG oscillation. However, there are many more different cognitive processes than the five different well-established frequency bands (delta, theta, alpha, beta, and gamma). Therefore, it is obvious that one cannot establish a 1:1 mapping between cognitive processes on the one side and EEG oscillations on the other side. It is more likely that EEG oscillations contribute to different cognitive functions depending on where in the brain and with what parameters (amplitude, frequency, phase, coherence) they occur."

In light of this, Brainlight's visualisation of theta, alpha and beta frequencies and the associations with particular conscious states referred to throughout this chapter, are based on the most frequently replicated and widely accepted findings within the literature.

Another limitation that should be mentioned is that EEG recordings can easily pick up noise and non-brain artifacts such as signals produced by muscular movement, heart activity or other exterior disturbances that interfere with the purity of the signal. Commercial grade EEG headsets are particularly prone to this and therefore offer only a rudimentary accuracy. According to Stamps and Hamam (2010) the EMOTIV EPOC+ is the most usable low cost EEG device and Maskeliunas et al. (2016) show that it performed better in attention/meditation tasks than other devices of similar value on the market. Duvinage et al. (2013) demonstrated that the EPOC's performance is above random and is therefore suitable be used for gaming or for communication for the disabled. It is for these reasons that we chose to use EMOTIV.

It is also worth noting that EEG frequencies in humans vary widely according to the brain anatomy of the person, stress, mood, age, neurological diseases, memory performance, therefore any specific analysis of EEG must be interpreted with caution (Klimesch 1999).

It seems unavoidable that BCI technologies will continually need a high level of processing and human decision making in order to interpret the raw data and extract meaning from it. Unlike heart monitors (ECG) or electrodermal sensors, which access a more direct expression of a person's biological inner life, EEG devices may never accurately "read our minds" in a pure sense due to the distortion inherent in interpreting EEG signals. Despite this, it seems likely that future advances in our interactions with computers through BCI will become ever richer as we increase our understanding of the brain's inner states (Mühl et al. 2014).

8.3 Context and Collaboration

8.3.1 University of Technology, Sydney and Culture at Work

Brainlight was created as a research project for a Masters of Design in Lighting at the University of Technology, Sydney (UTS). Two professors were of particular influence in their unique approach to technology and lighting design, Michael Day, head of the Lighting Studio, and Bert Bongers, leader of the Interactivation Lab, both in the faculty of Design, Architecture and Building.

Alongside UTS, Brainlight was created through Culture at Work's (CAW) 2015 art + science residency program. CAW is a non-for-profit organisation that connects art and science through artist residencies, educational programs and exhibitions. The residency provided mentorship, curatorship, and a studio space for four weeks, followed by a two-week exhibition at CAW's Accelerator Gallery.

The values of both these institutions, in terms of encouraging experimental cross-disciplinary practice as well as the conditions they provided, such as access to new research, were highly influential in Brainlight's creation. It is important that environments such as these continue to encourage and engage in innovative methods of working, where experimental, cross-disciplinary collaborative projects can be nurtured and explored.

8.3.2 The Collaborative Process

An interdisciplinary approach was important in the research and development of Brainlight. The team included software engineer Sam Gentle, neuroscience and medical innovation researcher Peter Simpson Young, industrial designer Neill Wainwright and electronics engineer Sami Sabik.

One of the most important aspects of collaborating with an interdisciplinary team is learning how to communicate. In order to understand each other we needed to spend time learning each other's disciplinary approach; grasping new vocabulary, language and terminologies in the process. As the artist, being able to communicate aesthetic ideas and conduct experiments with the lighting visualisations required grasping the possibilities and limitations of the software architecture that Sam was designing. In turn, he had to learn to work within an iterative artistic framework by designing software that allowed for versatility.

A large part of my role was synthesising everyone's input in order to realise an over-all creative vision, balancing what was inherently important to each discipline and cultivating alternative directions to tackle obstacles. Different components of the project also moved at different speeds and here communication was particularly important for remaining on schedule and maintaining momentum.

Brainlight required three key components to come together:

1. The design of the physical artwork, including a brain sculpture, portable base and lighting system.
2. The science of electroencephalography (EEG), data collection and processing.
3. A soft and hardware architecture to transform EEG data into a light display.

Neill Wainwright assisted with the 3D development and design of the physical sculpture. Using an MRI scan of a 35-year-old healthy male we translated the 3D model into a slice-form which was laser cut out of clear 5 mm perspex (see Fig. 8.2). The brain sculpture consists of 25 vertical slices that slide into a central

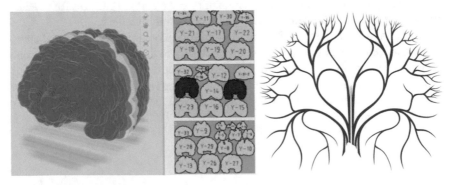

Fig. 8.2 Slice-form brain sculpture (left) created from an MRI brain scan and neural network design (right) laser etched into each brain slice

spine which delineates the right and left hemispheres. Each slice is etched with neural pathways creating the illusion of a three-dimensional neural network within the sculpture (Fig. 8.2).

Peter-Simpson Young guided us on the anatomical and neuronal organisational structure of the cerebral cortex (and its various sensory, motor and cognitive functions) and then advised us on current neuroscience research into EEG technology and its capacity for capturing the electrical activity associated with different brain states. We tested several low-cost commercial EEG headsets, which had between one and four channels but decided that their resolution and feedback potential was too limited, instead choosing to work with an EMOTIV EPOC+ which provided a better EEG resolution via its 14 channels.

Sam Gentle developed the custom software and hardware architecture (detailed in Sect. 8.3.3). A Raspberry Pi was used to run the software which processed the raw EEG data received wirelessly through a USB. The visual output was sent over HDMI to a 5000 lm data projector which was positioned above the brain sculpture.

An important quality of the collaboration was the mutual gain that emerged from learning about each other's disciplines, as well as extending our skills in our own practices. Seeing one discipline through the lens of another can offer valuable perspectives and new insights.

After our collaboration it is interesting to note how the payoffs for each collaborator varied. As Brainlight is primarily an artwork originally designed to be experienced in an art gallery, the recompense for myself as the artist are the rewards of the art industry; exhibitions, invitations for artistic performances and conferences within that field. While these were valuable to myself as an artist by way of career development, these rewards were less valuable to the other collaborators. However, indirect benefits for the team emerged later in the form of further collaborations and career opportunities. Sam became creative technologist in residence at CAW, moving away from commercial software development to develop his own artworks. Peter and Sam worked together on a BCI sound project titled "Mind Music" for Spotify, and Neill and Peter are currently developing a non-invasive brain stimulation device within the Science Of Innovation Lab at UNSW, Sydney.

8.3.3   EEG Processing

EEG signals were acquired from the 14 channels of the EMOTIV EPOC+ at 128 Hz, 14-bit resolution. We used a 5th-order Butterworth filter to remove frequencies below 3 Hz which are more vulnerable to noise and artifacts. We further improved signal clarity by applying a Blackman-Harris window function and a fast Fourier transform at 0.5 Hz resolution to deconstruct the time domain of the EEG signals into the frequency domains of Theta (3.5–7.5 Hz), Alpha (7.5–13 Hz), Beta (16–32 Hz).

8.3.4 Software Architecture

Creating software and hardware systems for artworks requires flexibility and stability. To support creativity, a system must be able to respond to its artist, to iterate and improve as the work takes shape. However, an exhibition demands a system that is reliable and predictable. Neither audiences nor artists want a canvas that reboots to apply updates, or suddenly fails during a show and cannot be replaced.

Achieving both stability and flexibility in a single system is difficult: if it is easy to change on purpose, it also tends to be easy to change accidentally. It is possible to overcome this problem by creating one larger system out of a number of smaller systems, known as modules.

In our system, the modules were completely independent, so that editing one component of the software did not require a complete overhaul of the entire code. For example, the "epoc" module, responsible for acquiring the EEG data, hardly changed at all during the design process. The "freqs" module, which did the bulk of the data processing, changed more often as we experimented with time-delays and neuro-feedback potentials. While the "vis" module, which displayed the visualisation, was in a state of constant flux until the final artistic output was chosen. This independence allowed each individual module to have its own trade-off between flexibility and stability. Key to this was the knowledge that changes in one module could not cause problems in another.

Despite changes and tweaks to the system right up until opening night, it worked without failure or further modification, not just during that exhibition, but through years of subsequent exhibitions both locally and internationally.

8.3.5 Visualisation

The visualisation needed to achieve the aesthetic goals of the artwork, while communicating the brain's activity as accurately as possible, while working within the limitations of the hardware.

To make the visualisation fast and responsive, we tuned our signal processing to achieve a compromise between accuracy and speed, but even so it took 2 s of EEG data to produce 1 frame of visual output, an unavoidable limitation of the Fourier transform. To work around this, we used pipelining: multiple overlapping transforms running simultaneously. For example, one transform could run from 0:00 to 0:02, another from 0:01 to 0:03, 0:02 to 0:04, and so on. The processing still takes 2 s, but because of the overlap there is an update every second. In fact, we overlapped 8 transforms so the data would update 4 times per second.

Although 4 updates per second is quite fast by EEG standards, it is slow for animation, where rates of 25, 30 or even 60 frames per second are common. It became clear as we worked with the perspex sculpture, that it looked best having rapid changes in brightness, colour and movement which caused the light to twinkle

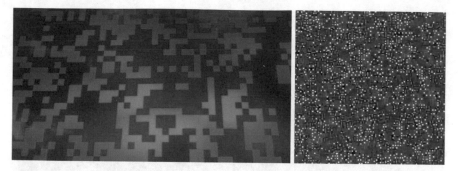

Fig. 8.3 Still images of cellular automaton patterns used to animate *Brainlight's* visualisation of EEG data

and reflect between its interior surfaces. To bridge this gap, we added an additional layer of animation based on cellular automata.

A cellular automaton is a grid of cells, where each cell obeys rules governing its interactions with its neighbours. By carefully tuning the rules, it is possible to achieve a variety of organic patterns. In our case, the grid nature of cellular automata matched quite naturally to pixels on a projector, and the organic movement complemented the behaviour of the EEG data.

The final visualisation consisted of a cellular automaton animation layer (Fig. 8.3), which created organic patterns of light and dark using colours derived from the EEG frequency bands layered on top. This layering technique allowed the animated visuals to make the best aesthetic use of the perspex brain while still providing a clear representation of the underlying EEG data.

8.3.6 Visual Experimentation

During the four-week residency at CAW we developed interactive interfaces which translated the electrical data from the EEG headset into various forms of visual communication, including projections (Fig. 8.4), prototype brain maquettes (Fig. 8.5) and the final Brainlight sculpture (Fig. 8.6).

8.3.7 Testing Prototypes

During the CAW residency we had several opportunities to engage with the public and test the prototypes we had created for visualising brain activity. In May 2015, during Pyrmont festival, we presented Cerebral Orb (see Fig. 8.4), a circular light projection that allowed participants wearing the EEG headset to observe their dominant brain frequencies (mapped to corresponding locations within a circular light

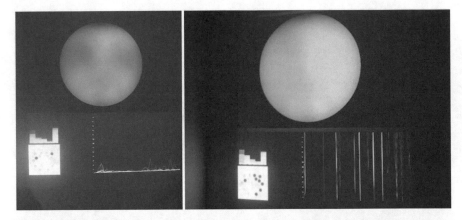

Fig. 8.4 *Cerebral Orb*, interactive light projection and EEG frequency graph

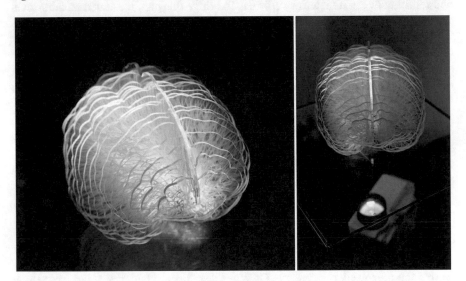

Fig. 8.5 *Cerebral Nebula*, laser-cut and hand-etched perspex

display) change colour depending on their state of mind. A second projection displayed a graph showing the frequency ranges as well as the connectivity level of the EEG electrodes to the scalp in order to present a measure of the connectivity of each electrode and the purity of the signal.

As interactive artworks invite audience participation there is a level of unpredictability in how they will be used. In the case of Brainlight, in order to observe as wide a range of natural interactions as possible, I gave very little guidance to the public other than a basic understanding of what the headset was capturing and which colours correspond to which frequencies. There was substantial variation in how the public approached the artwork. Some were nervous about what might be

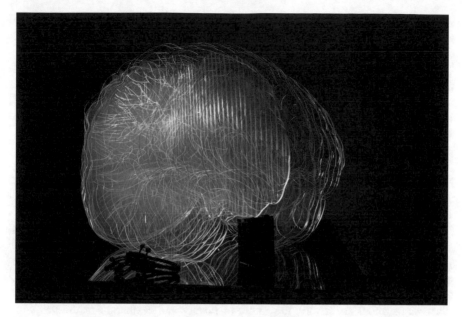

Fig. 8.6 *Brainlight*, laser-cut perspex, 120 cm × 120 cm × 110 cm

revealed, leading to cautious interactions, others were more outgoing and curious, eager to experiment and stimulate their own brain in order to activate colour changes. Some gained a level of control over their mind by using the visual neurofeedback to practice retaining particular states, while others found it challenging to gain any sense of control over the work.

This test period was useful for two reasons. First, it enabled us to see a much wider spectrum of brain activity with noticeable variation from person to person in terms of rhythm and frequency dominance. Second, it gave me a chance to observe audience behaviour, which informed the way I facilitated future interactions with the work. It allowed me to develop a sensitivity towards different participant's temperaments and to test a range of emotion or memory based "triggers" in the form of questions that stimulated people to access different brain states in order to more easily see the resulting colour change (discussed in Sect. 8.5).

8.4 Exhibition Journey

Since the launch in 2015, Brainlight has been experienced in a wide range of contexts; from science museums, art galleries, festivals and conferences, to universities, schools, corporate offices and private homes. Each of these settings have their own cultural codes and conventions which influence the interaction, interpretation and response to Brainlight, affecting the aesthetic impact of the work and its subjective meaning and significance.

In Australia, Brainlight has exhibited at a number of Sydney's cultural institutions, including The Museum of Contemporary Art, The Museum of Applied Arts and Sciences (an institute focused on the impact of technology, engineering, science and design), The Australian Museum (the oldest museum in Australia dedicated to anthropology and natural history) and Vivid Light Festival Ideas Conference on neuroscience and creativity.

Internationally there have been further opportunities to exhibit Brainlight in contexts that specifically celebrate the nexus of art, science and technology, including Hoy Es Diseno (Design of the Future) in Cali, Colombia, Ars Electronica Festival, Austria, Athens Digital Art Festival, Greece, Starmus festival, Canary Islands and GOGBOT (AI and Robotics festival), Netherlands.

An important characteristic of Brainlight has been its ability to create links from one disciplinary context into another by generating integrative dialogues with the public. Within the domain of art, audience members have often asked me more questions about the scientific or technological aspect of the work than the artistic. Surprisingly, I have found the converse to also be true and have ended up in numerous discussions about art history, the nature of beauty and aesthetic sensibilities with neuroscientists and computer programmers.

8.4.1 Brainlight and the Sydney Art Quartet

Playing and listening to music is a multi-sensual experience involving numerous higher order, motor and sensory areas of the brain which stimulates emotions, memories, and drives reward centers (Chanda and Levitin 2013). There is even neurochemical and physiological evidence to suggest music may have played a central role in the evolution of the modern human mind (Cross 2006; Harvey 2018). In light of this, collaborations between BCI and musical performers have provided opportunities to explore the links between music and the brain.

Since Lucier's sonification of brain signals in 1965, a wide variety of experimental brain-driven interfaces for musical expression have been created. Among these, many use EEG signals as a trigger for music generation, such as the MoodMixer by Leslie and Mullen which composes new music based on the combined EEG signals of multiple participants (2011). To a lesser extent, brain-driven interfaces have been created in order to show or study music's effect on the brain (Mullen et al. 2015). One example is Ringing Minds (2014), a collaborative installation by David Rosenboom, Tim Mullen and Alexander Khalil which uses the collective brain responses of multiple audience members listening to music to influence a live music composition.

In 2017, Brainlight was invited to perform with the Sydney Art Quartet (SAQ) in a series of three evening concerts titled *Light Fantastic: Music + Neuroscience + Light* (see Fig. 8.7). The performances focused on the concept of "genius" in music; bringing to life five to five-hundred-year-old compositions, from Beethoven

Fig. 8.7 *Light Fantastic: Music + Neuroscience + Light*, Brainlight performance with Sydney Art Quartet

to Bach, Dvorak to Tool. The music was chosen to contrast the historical with the contemporary as well as the technical with the emotive. The role of Brainlight was to visualise the brain activity of the musicians performing in order to compare it to the brain activity of the audience listening.

Across three evenings, Brainlight exposed some remarkable differences between the brain patterns of the musicians and the audience. Lead cellist (and SAQ director) James Beck wore the EEG headset for the first two compositions. His performance was both technically complex and highly emotive, yet rarely did Brainlight flicker into the upper frequencies of beta or alpha (associated with highly analytical and calm states of mind respectively) remaining instead in a dominant state of theta.

Similarly, when the headset was placed on violinists Anna Albert and Thibaud Pavlovic-Hobba, their brain activity remained in a steady state of theta throughout their performance with very little dynamic fluctuation (Fig. 8.8).

The audience members who subsequently wore the EEG headset displayed a remarkable contrast; exhibiting unstable fluctuations of predominately higher frequencies (alpha and beta) suggesting a larger, more constantly changing range of brain activity and emotions. Visually this contrast was obvious, having remained bright green while worn by the musicians, when the headset was passed to the audience Brainlight began cycling quickly through red and blue, often displaying all three colours simultaneously, visualising the varying emotional and physiological reactions to the music's own fluctuations in mood and rhythm.

Fig. 8.8 SAQ and Brainlight performance featuring violinists Anna Albert (wearing the EEG) and Thibaud Pavlovic-Hobba

The collaboration between SAQ and Brainlight demonstrated how BCI technologies can visualise underlying neurological phenomena in musical performance, adding a new layer of interest for the audience by revealing the hidden differences between performing and listening to music.

8.4.2 Brainlight and Science Communication

Science exhibitions and museums are increasingly employing tangible interactive technologies in order to provide a higher engagement with information. As Wellington has previously illustrated, "one of the achievements of hands-on science centres has been to relate science and technology to the things that most people see and use" (1990).

In 2018, the University of Nottingham's "Quantum Sensing the Brain" exhibit at the Royal Society's Summer Science Exhibition in London included an immersive "brain room" where visitors could wear an EEG headset and perform some basic actions to influence an illuminated installation to learn how the brain works (Brookes 2018). Exhibitions such as this demonstrate how interactive models are playing an important role in a learners' investigation of complex phenomena (Fleck and Simon 2013).

Brainlight has proven to be a captivating medium for science communication and public engagement with neuroscience. ABC's Catalyst (an Australian national science communication television series) used Brainlight in an episode titled "Brain Stimulation" featuring scientists (including Brainlight collaborator Peter Simpson Young) discussing brain-enhancing devices. In a follow up episode titled "Sleep

Matters" they used Brainlight's colour coded light display to communicate brain activity cycling through the stages of sleep.

The work has also acted as a creative stimulus in more unusual scientific encounters. In 2017, I was invited by Professor Avi Schroeder to the Technion, Israeli Institute of Technology, to present Brainlight to his research team. Schroeder's research lab is focused on nanotechnology for targeted medicine, creating miniature medical devices that can couple diagnosis and therapy, called theranostic devices. These drug-loaded nanoparticles can be remotely triggered with ultrasound to release an anti-cancer chemotherapeutic inside tumours (Schroeder 2018). Schroeder's interest in Brainlight was to take his researchers out of the traditional scientific realm in order to encourage them to think laterally and creatively about the brain's natural frequencies and how they might be harnessed for nanomedicine. Utilising Brainlight, Schroeder asked his lab researchers to consider the possibility of using neurofeedback training to trigger a medical therapeutic device targeted to a tumour site within the brain.

Brainlight has also recently been commission by Dr. Adrian Ivanescu, assistant professor of Anatomy and Embryology at the University of Medicine and Pharmacy in Târgu Mureş Romania. Ivanescu is planning to use the artwork as an educational tool for his students and has since presented it at their annual NEURON conference (Neuron 2018).

8.4.3   Experiential Learning and Education

Brain activity is hard to understand because it cannot be sensed like other bodily systems, for instance respiration, therefore it needs to be conceptualised (Frey et al. 2014). BCI technologies are creating interesting opportunities for experiential learning and education by creating new methods of conceptualising the brain.

Teegi (Tangible EEG Interface), for example, is a project that uses a tangible character to visualise and analyse a user's brain activity in real-time through various EEG filters, such as motor, vision and mediation (Frey et al. 2014). The project also enables users to better understand the kinds of brain activity that can be detected in EEG signals in order to demystify BCI technologies.

Brainlight provides a similar interactive educational experience and has been developed into a workshop for primary and secondary students. In the Brainlight workshop, students are invited to interact with the artwork and partake in stimulus activities such as puzzles, maths problems, memory games, and emotive role play in order to predict and observe the changes in brain activity. In the process, students are introduced to the concepts of neuroscience, brain anatomy, emotional intelligence, mindfulness, data visualisation and BCI technologies.

Both Teegi and the Brainlight workshop demonstrates how biologically-driven interactive models can offer students tangible, memorable and novel opportunities for self-discovery by stimulating a unique engagement with their own bodies and minds.

8.5   Brainlight and the Audience

Brainlight's interactive method of using a person's mind to co-create the work, challenges the traditional role of an audience member. Typically, the viewer's psychological process is a private experience, hidden from the rest of the audience. Brainlight harnesses this passive experience, transforming the electrical activity that defines it, into the artwork itself, on display for all to see.

Since 2015, approximately 3200 people have used the artwork as a participant and an estimated 20,000 people have viewed it in exhibitions around the globe. On average, people tend to spend about 10 min with Brainlight, but in settings outside the gallery, away from crowds, it's not uncommon for people to spend an hour or more with the work.

During public showings of Brainlight, I often experiment with emotional cues by asking the participant to imagine scenarios relating to their lives, or to relive a memory. For instance, getting the participant to re-imagine the feelings they experienced during the birth of their first child; thinking of something that brings them a sense of peace; imagining a stressful scenario. As they settle into the feeling associated with the memory, the brain activity associated with that feeling affects the colour of the sculpture, allowing onlookers to witness an externalized embodiment of the participant's emotional state.

On many occasions participants have been able to identify emotional states that emit strong dominant frequencies permitting them to cycle through the different colours at will. Typically, this is achieved when the participant spends a few minutes experimenting with thoughts, memories or mental challenges until they find one that stimulates the desired colour. With practice, some people are able to hold on to that particular brain state and maintain the brain sculpture in the desired colour. Perhaps the most common state I've found people—(including children)—are able to maintain, are Beta waves (the highest level of active cognition, shown as red light) when attempting to solve a challenging mathematical problem.

Because interactivity in art, particularly with works that employ EEG, is a relatively new concept for some people, some approach the work with skepticism, requiring proof of its interactivity. As Rokeby suggested, the proof that will most easily satisfy the audience is 'predictability' (i.e. if one makes the same action twice, the work will respond identically each time) (1995). This test only works with Brainlight if a person's brain produces the same signal on their command, which is not a straightforward task. As Rokeby further observes, "the complexity of this relationship is, in this case, not so much a function of the complexity of the system, but of the complexity of the participants themselves" (1995). Whether Brainlight is seen as interactive or not is therefore highly dependent on the quality of the behaviour of the audience.

8.5.1 Emotion and Cognition

Emotion is core to the appreciation of art, "from ancient to modern times, theories of aesthetics have emphasized the role of art in evoking, shaping, and modifying human feelings" (Silvia 2005). The field of neuroaesthetics has shown complementary neurological pathways work in tandem to create both conscious and unconscious aesthetic response; "the cortical pathway, which leads to recognition and conscious thought; and the thalamo–amygdala pathway, which gives emotional colour and meaning to all information that passes through our senses" (Barry 2006).

Experiencing art nourishes our psychological needs which elicits a certain intrinsic pleasure. It provides a "sensory anchor" for our thoughts and emotions by inviting personal involvement with its affective impact (Perkins 1994). In artistic applications, BCIs can satisfy our psychological needs by having 'influence' over our affective state as well as giving users new creative abilities to express emotions (Gürkök and Nijholt 2013).

Brainlight's interactive BCI creates a relationship between a viewer's inner state and their influence on the artwork. Placing the audience at the center of a live neuro-feedback interaction not only challenges the participant to witness and confront their own emotional state and aesthetic response, more interestingly, it challenges their ability to have agency over it, inviting them to experiment with methods of emotional regulation and control.

While the idea of control was not the initial intention of Brainlight (rather the intention was to have an encounter, whatever that encounter may be for one person or another) I have noticed that the artwork can provide a person with a sense of power when they gain control over the interaction. The opposite is also true, when someone does not gain control of the work, it can elicit a feeling of being powerless or out of control.

Our aesthetic response is deeply connected to the universal drive for pattern recognition. As Barry explains, "the brain is a meaning-seeking mechanism, and this suggests that recognition of pattern is at the heart of all perception, the process by which we make meaning from both stimuli from the outer world and prior experience stored in memory" (2006).

Our innate search for synthesis seeks to reduce complexity into its simplest form in order to understand it. In the same way, Brainlight simplifies the electro-chemical activity of the brain into a colour coded visual experience. The audience then attempts to make sense of patterns generated by a complex mix of brain activity which includes perceptual responses, memory and emotion. What becomes interesting is how audiences attempt to create a narrative and recognise a pattern that signifies personal meaning for them. Their desire for pattern recognition happens at the same time as their mind, emotions and subjective interpretation continually influence the artwork.

8.5.2 Games and Competition

For many participants, once they realise they can affect change within the artwork, the ability to control the colours on display quickly becomes the focus of the experience, rather than to explicitly attain awareness and control over their thoughts. Often audience members become competitive, trying to see how quickly they can get the artwork to respond, how long they can retain a colour and how easily they can go back and forwards between colours at will. This focus on controlling the colours may be partly due to the expectations I set up when explaining the artwork to participants, or it may be because controlling the colours is the most obvious, novel and satisfying outlet for creative interaction with the work.

Amusingly, audience interaction tends to become most gamified around the calm and meditative blue alpha state, which in a busy gallery, is usually the most difficult to accomplish. In order to experience this state, users need to contend with the many obstacles within the exhibition environment: eager onlookers, the brightly illuminated sculpture, the novelty of the situation, exposure to other sensory stimuli (such as noise), as well as their own state of mind and how comfortable they are being the spectacle in the room. Interestingly, when a participant does manage to attain the blue state, they must master their excitement in order to prevent the brain from switching immediately to red. This effect happens the other way around as well, and it is perhaps one of the most interesting manifestations of the feedback loop. Witnessing one's current state of mind, so often changes it.

This scenario is the focus in Hjelm and Browall's Brainball project (2000) where two players wearing EEG's must remain calm in order to win a competitive game. The tricky part of the game is that the players must master their ability to relax at the same time as competing. When a player gets close to winning they get excited and so a considerably excited player will be at a disadvantage.

Alongside competition, one of the reasons Brainlight has been so popular with the public may be attributed to our natural curiosity for self-knowing. I have observed that most viewers want to learn if they have "good" brains, they want to test their level of self-control; many seem to use it to demonstrate their "meditation prowess" to their peers. I have noticed though, that the more time someone spends with the work, the more they move beyond a superficial appreciation, to genuine curiosity about the inner workings of their mind.

8.5.3 Spectatorship and Surveillance

Trust is a central dimension in the relationship between human beings and technologies (Moritz 2017). Many of us have concerns about how we actively share our data and our ability to retain a level of control over what we choose to share. When people are faced with an interactive artwork, in the process of interacting they reveal something about themselves (Rokeby 1995). Brainlight is an example of a technology that "quantifies the self" by providing a level of self-reflection in exchange for participatory public surveillance in an exhibition context. During a Brainlight interaction an observer wearing the EEG becomes a "performer" and their mind becomes a public spectacle.

The spectatorship dynamic between audience and user means that some people feel a need to self-regulate their behaviour in order to try and retain a level of privacy. They do this by attempting to regulate their thoughts and emotions, as well as by limiting how much they verbally share about themselves to myself and the surrounding public. For an audience member to agree to participate in the experience they need to have a level of trust in the process of the collection, and display, of their personal data and how that data is situated in the view of the crowd's subjective gaze.

I have noticed a participant's level of trust is usually proportionate to their understanding of what EEG technology is capable of detecting and displaying about them. People who are not familiar with the limitations of EEG often believe that the artwork is revealing more about their inner self than it actually is. Because of this, people's fears and anxieties occasionally come to the fore—perhaps their mind would appear "abnormal" and embarrassing, perhaps intimate details about their emotional state or the type of person they are would be revealed to strangers. On many occasions these people have displayed a greater sense of awe towards the experience, viewing the artworks ability to "read one's mind" as an almost "mystical" quality of the technology.

Research has shown differences in art appreciation among those with artistic training and expertise compared to those with no expertise, revealing its influence on aesthetic engagement, interpretation and judgment (Else et al. 2015). In the case of Brainlight, I noticed that the viewers level of scientific expertise played a key role in influencing their response in relation to fears of surveillance. People who had neuroscience expertise, an understanding of EEG technologies or a higher level of scientific literacy, could appreciate the work for its novel visualisation of brain activity, tempered by an understanding of the current limitations of the technology.

While the inherent limitations of EEG mean we can't yet decode complex thoughts, we can already make assumptions about a person's mood, and it's possible that we may succeed in understanding more of the brain's complexity over time. An emerging neuroethical debate is starting to permeate the BCI research community about the possible misuse of BCIs in the future (Tamburrini 2009). Much of the debate is focused on ethical concerns regarding BCI as a medical intervention for locked-in patients, assistive therapies or BCI controlled prosthesis. However, as BCI is becoming more prevalent in popular culture, the perceived risks that relate more to the general public are also being explored (Nijboer et al. 2011). Social issues such as mind-reading and privacy, mind control, selective enhancement and social stratification are just a few examples.

An interactive theatre performance titled Noor: A Brain Opera addresses these ideas more directly by asking the question "is there a place in human consciousness where surveillance cannot go?" (Pearlman 2017). Similar to Brainlight's setup, a performer wears an EMOTIV EEG and as their brain state changes so does abstract video footage. The changing colours signify different emotions: yellow for excitement, pink for interest, turquoise for meditation, and red for frustration. The performance encourages the audience to consider a future where mental surveillance is possible.

Dunn states that "art can open us up to new ideas and beliefs, and artists can make a massive impact as role models, either in a positive or a negative manner. Because art communicates with us on so many different levels, and appeals to our senses, emotion, reason, language and imagination it inevitably affects us more than other areas of knowledge" (2013). The value and impact of art is highly determined by what the public bring to it with their prior knowledge in combination with the subjective meaning they generate from it. Works such as Noor and Brainlight do not provide concrete answers or positions, rather they provide space for audiences to formulate their own questions.

Interactive artworks that are highly engaged with by an audience, like Brainlight, are facilitating "interactive literacy", allowing the public to experiment with possible future relationships with technology. Having interacted with technologies at the experiential level, audiences may then have a better understanding and a deeper engagement with global issues around emerging technologies. Because these issues are likely to become more complex in the future, "understanding autonomy and feedback and permeability and transparency and internalization of tech and externalization of self are all things we need to become literate in if we are to make good decisions" (Ekman and Rokeby 2014).

8.5.4   Metacognition, Subjectivity and Intimacy

An encounter with Brainlight permits people to have a moment of self-reflection, occasionally providing a level of higher self-analysis that results from seeing their "mind" as an entity outside their heads. This third person perspective stimulates people to have a conversation with their own mind, whereby Brainlight acts as a symbolic "oracle" that embodies their own process of metacognition. This process has often stimulated personal insights or created new meaning around a participant's own thoughts and memories.

Indeed, in this context, the subjectivity of a person's interpretation becomes the work itself. Time spent with Brainlight allows a person to establish a personal identity with the work which becomes a reflection of their thoughts, feelings and presence.

The ambiguity involved in the audience's interpretation of their own brain data through the artwork raises interesting questions about accuracy versus subjectivity in "science inspired" art, and how bio-sensing technologies can mediate our subjective identities (Moritz 2017).

I realised early on that this blurring of personal interpretation and scientific accuracy would become a feature of Brainlight when I noticed participants were often eager to furnish the interpretations of their brain signals with their own subjective speculation about what might be driving their mind's behaviour.

Audience members revealed stresses they were under at work, stories of heartbreak and grief, aspirations and fantasies; stories unique to each person, but common to the human condition. Many of the interpretations participants brought to the work were deeply personal and unlikely to have been volunteered to a room full of strangers under ordinary circumstances. The nature of these intimate exchanges demonstrates how art and art spaces allow people to explore vulnerabilities that might remain concealed in other public social settings (Khut 2006).

This vulnerability raised further questions about my role as an artist in facilitating the interpretation of the artwork: how much should be explained and how much space should remain for people to generate their own meaning? While Brainlight uses traditional scientific tools to explore the mind, the artistic and aesthetic framework attempts to strike balance between providing a level of scientific validity and leaving room for ambiguity, uncertainty, subjectivity, imagination and emotional response.

8.6 Further Evolutions of Brainlight

Two further iterations of the original Brainlight have been created since 2015, which I'll briefly describe here.

8.6.1 Mini Brainlight

As the large perspex sculpture is cumbersome to transport I decided to create a miniature version of Brainlight (see Fig. 8.9) that packs down into a briefcase for easy portability. The "mini" sculpture is the size of a human brain and is illuminated by 190 individually programmed LEDs mapped to corresponding electrode positions and illuminated according to the EEG data received from the headset. The work was developed in collaboration with creative technologist and electronics engineer, Sami Sabik, who designed the custom printed circuit board, Sam Gentle, who programmed the software to interact with the LEDs and Neill Wainwright, who assisted in the design of the customised base.

8.6.2 Projection Mapping the Mind

Brainlight has also been re-configured as a projection mapping system, allowing any object, room or building to be illuminated with an interactive live visualisation of brain activity (Fig. 8.10). Created in collaboration with Sam Gentle, the original EEG processing software was adjusted to run on a "master" Raspberry Pi which communicates data to four secondary Raspberry Pi's over a network. Each secondary Raspberry Pi can be connected to a data projector which displays the correspond-

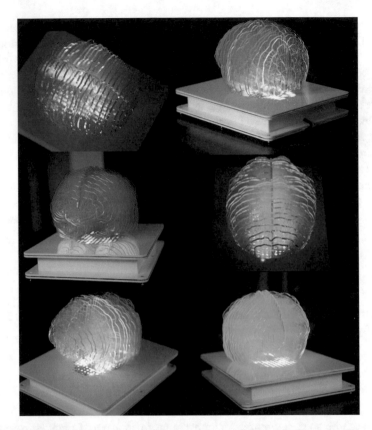

Fig. 8.9 *"Mini" Brainlight*, perspex sculpture with custom base and LEDs

Fig. 8.10 Projection mapped car using live audience brain activity for the launch event of MG3 × ELLE 2018

ing visualisation of the associated region of the brain. When the four projections are unified it creates a "complete brain" (frontal lobe, left and right hemispheres and occipital lobe). The animated visualisation was created through the cellular automata program used in the original Brainlight, with some adjustments to the size and movement of individual pixels, creating the impression of an electrically charged web of vibrant neural networks.

8.7 Discussion

As I have illustrated in this chapter, Brainlight unites art, BCI technology and the audience to realise new ways of exploring our biological selves, as well as creating new modes of audience interaction. The work contributes to an international community of artists, designers and technologists who have been utilising BCI technologies for artistic expression during the last 50 years. These artistic BCI projects are important to both research communities as well as the public because they explore, question and reveal new relationships with technologies and facilitate creative methods of connecting to each other and to ourselves. This discussion examines other artworks which utilise BCI technology in a similar way to Brainlight to explore our specific contributions to the artistic BCI field. I also discuss further methods of evaluating Brainlight and outline potential creative directions for the future.

Artists have been among the pioneers of EEG use outside clinical settings, designing situations and applications for EEG use in "real-life contexts" since the 1960s. One of the earliest interactive works was Nina Sobell's 1973 BrainWave Drawings, which paved the way for audience involvement in EEG generated biofeedback loops.

Akin to Brainlight's aim of exploring and sharing subtle non-verbal forms of communication, Sobell is interested in revealing a "universal mental language" (Sobell 2019). During a BrainWave Drawing session, two participants watch their brain activity changing in real time displayed over a closed-circuit video of themselves, creating a joint visual "drawing" of their silent communication. When describing her work, Sobell says, "in these projects I see myself as a facilitator or vehicle" (2019). As with Brainlight, Sobell facilitates the audience's experience and observes how people interact and improvise with the work in the moment. As with Brainlight, the social dynamics between participants, and the relationship between audience and artist become equally as important as the installation.

Another work with objectives and outcomes comparable to Brainlight in terms of "mirroring the self" for neurofeedback, educational, and entertainment purposes is the Mind-Mirror (Mercier-Ganady et al. 2014). This work enables the experience of seeing "inside your own head". The system uses a semi-transparent mirror positioned in front of a screen, allowing users to see a virtual display of their brain activity in different colours superimposed on their own reflection. Both Mind-Mirror and Brainlight facilitate an audience to "meet their own mind" externally by using realism. Brainlight uses an anatomically inspired brain sculpture and Mind-Mirror uses a literal mirror to visualise brain activity inside the skull.

While Mind-Mirror's technology is certainly advanced in terms of computer graphic capabilities, the work remains similar to classical 2D, screen-based, visual feedback, providing a single viewpoint for a single user which they must navigate by rotating their head while keeping their eyes on the mirror. Sobell's work is also largely screen based, using oscilloscopes and closed-circuit video. One clear distinction between these works is that Brainlight is a sculptural model with no screens or monitors involved. The advantage of Brainlight's three dimensional design is that the work can be experienced spatially, allowing large groups of people to view it from multiple angles. In combination with a wireless EEG headset, the participant is able to walk around the brain sculpture freely, allowing them to experience a tangible, 360° view of their changing brain activity.

In addition to creating live feedback loops, some artists are also documenting or recording audience experience in ways which can be later used to evaluate the work.

In George Khut's "Behind Your Eyes, Between Your Ears: Neurofeedback portrait project" the artist is able to capture and record an audiences subjective experience more tangibly through a series of brain-wave controlled video portraits (Khut 2015). A participant's face is overlaid with a colour projection and an electronic soundscape that is controlled via their alpha brainwave patterns. A voiceover of the participant's retrospective recollection of their experience during the EEG recording plays over the top. The artwork demonstrates a method of collecting and recording the experience of audience while using it as part of the work itself.

There are conflicting motivations when using audience experience to evaluate a work within an interdisciplinary space. Traditional methods for evaluation are very different for the fields of art, science or human computer interaction. Approaching Brainlight from a scientific standpoint, the inclination might be to rigorously test and measure its effectiveness in neurofeedback training. A study such as The Sensorium: Psychophysiological Evaluation of Responses to a Multimodal Neurofeedback Environment, which uses an immersive sound and light environment influenced by EEG and ECG signals, could be a possible model to replicate (Hinterberger and Fürnrohr 2016). The study used three phases, a mindfulness meditation, a guided body scan exercise, and a "Pseudo-Sensorium" using pre-recorded data that did not reflect the subject's own physiology, followed by a feedback questionnaire, in order to test its neurofeedback performance.

Alternatively, new research methods of evaluation may be necessary for more nuanced assessments of artworks that unify disciplines. Muller et al. describes the locus of encounter between art, science and the public as a "third space", a civic space of "trans disciplinary knowledge production, requiring new research methods that capture emergent knowledge". Evaluating an EEG sound installation of amnesic memory by artist Shona Illingworth in the Amnesia Lab exhibition in Sydney 2014, Muller et al. conducted a group-based psychosocial method of analysis—the visual matrix—designed to evaluate the transformative effects of aesthetic practice and interdisciplinary arts-science projects. The matrix allows participants to stay with the lived experience of the exhibition which enables researchers to "capture and characterize knowledge emerging in third space, where disciplinary boundaries are fluid and there is no settled discourse" (Muller et al. 2018). The visual matrix fosters a space of dialogue between scientific and artistic modes of thought without orienting to the established goals of either discipline. It also provides a way of capturing the "shared, complex, emergent and transformative aspects" of art-science exhibitions by illuminating how artistic intention is transformed into audience experience (Muller et al. 2015).

Most artists have limited ability to capture audiences affective and sensory responses to their work. Conventional studies that evaluate audience experience have the audience report on its impact after their experience. One benefit of artworks such as BrainWave Drawings, Behind Your Eyes, Between Your Ears: Neurofeedback portrait project and Brainlight, which all foster a real-time sharing of a participant's inner experience with the artist, is that by its very nature they provide a deeper insight into how the work is experienced by the audience during the moment of impact. Emulating the evaluation method outlined by Muller et al. could be a unique way to further investigate and document this process.

While it is interesting exploring ways in which technology can sense and communicate hidden inner biological states, it remains true that even the subtlest and most precise technological biosensors (both current and future) will nevertheless require a level of human processing, interpretation and analysis, as well as being subject to a decision making process of how to communicate the analysed data. Each one of these steps removes us further from the original source; thus "pure" unadulterated communication and a transferal of feelings through technology may never be possible, as it will always require a level of external human mediation.

Despite this, it remains interesting to imagine the future of these mediations and what scenario's might arise through further experimentation with technologies that are ever more sensate.

Appealing to the idea of digital synesthesia, future directions for Brainlight could be to expand brain activity from being translated through visual and auditory domain to include other senses, such as the transferal of the mind to other people's bodies through touch, vibration, or smell.

Another possibility would be to connect audiences in increasingly playful ways by further exploring the performative "spectacle" aspect of Brainlight for experiential entertainment. Audiences could be given more spectacular rewards for their ability to maintain a state of calmness. A great example is The Ascent, a "mind controlled levitation ride" which pairs an EEG headset with a 3-D theatrical flying harness, allowing users to "fly" by retaining a meditative state (Duenyas 2011). If they manage to levitate all the way to the top they trigger an explosive light and sound experience as a reward.

Multi-brain interactions are increasingly common in BCI art applications. Currently Brainlight allows a single person to explore their own mind in a general way, so there is fertile ground for expanding the work to include multiple users and to focus on more specific intimate interactions. A work such as EEG Kiss, for example, focuses on a singular intimate exchange, exploring how a kiss can be translated into data (Lancel and Maat 2014). The value of this would be to translate and share joyful human connections in new ways.

As this discussion has outlined, Brainlight has expanded on past similar work, while simultaneously there remains room for further evaluation of audience experiences and to further develop the work in new creative directions. This leaves exciting potential for Brainlight to collaborate with both the artistic and scientific communities on future research projects.

8.8 Conclusion

The Brainlight artwork, public exhibitions and audience interactions described in this chapter demonstrate how BCI technology is providing new sensory connections to our own hidden and immaterial neurobiological processes. Brainlight uses neuroscientific tools as a vehicle to explore the mind and our subjective responses in intimate, but public environments, beyond the usual clinical settings and laboratories where these technologies usually reside.

Interactive artworks such as Brainlight can encourage an increased self-awareness and self-mastery by offering new possibilities for extending or enhancing our senses through the technological augmentation of cognition. Brainlight does this through a visual display of an audience's metacognitive process, which includes the perception of cognition as well as its regulation.

In the three years of facilitating Brainlight interactions, the discussions that emerged between myself and the audience and among audience members extended to topics far beyond those directly relevant to the work and continued long after direct engagement with the work had ended. They ranged from philosophical debates about the phenomenology of the mind and the nature of consciousness to future speculation about post-humanism and how technology is redefining what it means to be human. While Brainlight uses a removable EEG device and is only a temporary experience, it stimulates discussions about humanity's relationship with technology and provides a glimpse into an imagined future where we might have more permanently integrated technological sensors, in the form of biological implants, finely calibrated to our individual bio-rhythms, which may one-day facilitate more precise non-verbal communication.

References

Adrian E, Matthews B (1934) The Berger rhythm: potential changes from the occipital lobes in man. Brain 57:355–385. https://doi.org/10.1093/Brain/57.4.355

Ahani A, Wahbeh H, Nezamfar H et al (2014) Quantitative change of EEG and respiration signals during mindfulness meditation. J NeuroEng Rehabil 11:87. https://doi.org/10.1186/1743-0003-11-87

Alonso J, Romero S, Ballester M et al (2015) Stress assessment based on EEG univariate features and functional connectivity measures. Physiol Meas 36:1351–1365. https://doi.org/10.1088/0967-3334/36/7/1351

(2014) Amnesia Lab. UNSW Art & Design. In: Artdesign.unsw.edu.au. https://artdesign.unsw.edu.au/unsw-galleries/amnesia-lab. Accessed 26 Jan 2019

Anadol R (2017) In: Refikanadol.com. http://refikanadol.com/works/melting-memories/. Accessed 13 Dec 2018

Ascott R (1999) The technoetic predicate. Leonardo 32:219–220. https://doi.org/10.1162/leon.1999.32.3.219

Barry A (2006) Perceptual aesthetics: transcendent emotion, neurological image. Vis Commun Q 13:134–151. https://doi.org/10.1207/s15551407vcq1303_2

Brookes M (2018) Giant brain to light up Royal Society Summer Science Exhibition—The University of Nottingham. In: Nottingham.ac.uk. https://www.nottingham.ac.uk/news/pressreleases/2018/june/giant-brain-to-light-up-royal-society-summer-science-exhibition.aspx. Accessed 30 Jan 2019

Cahn B, Polich J (2006) Meditation states and traits: EEG, ERP, and neuroimaging studies. Psychol Bull 132:180–211. https://doi.org/10.1037/0033-2909.132.2.180

Cavanagh J, Zambrano-Vazquez L, Allen J (2011) Theta lingua franca: a common mid-frontal substrate for action monitoring processes. Psychophysiol 49:220–238. https://doi.org/10.1111/j.1469-8986.2011.01293.x

Chanda M, Levitin D (2013) The neurochemistry of music. Trends Cogn Sci 17:179–193. https://doi.org/10.1016/j.tics.2013.02.007

Chiesa A, Serretti A (2009) A systematic review of neurobiological and clinical features of mindfulness meditations. Psychol Med 40:1239–1252. https://doi.org/10.1017/s0033291709991747

Cross I (2006) Music, cognition, culture, and evolution. Ann N Y Acad Sci 930:28–42. https://doi.org/10.1111/j.1749-6632.2001.tb05723.x

Duenyas Y (2011) The ascent mind controlled levitation ride—Yehuda Duenyas. In: Yehuda Duenyas. https://xxxyehuda.com/theascent/. Accessed 26 Jan 2019

Dunn M (2013) What is the relationship between art and ethics? In: Theory of knowledge. https://www.theoryofknowledge.net/areas-of-knowledge/the-arts/what-is-the-relationship-between-art-and-ethics/. Accessed 15 Dec 2018

Duvinage M, Castermans T, Petieau M et al (2013) Performance of the Emotiv Epoc headset for P300-based applications. BioMedical Eng OnLine 12:56. https://doi.org/10.1186/1475-925x-12-56

Ekman P (1994) Strong evidence for universals in facial expressions: a reply to Russell's mistaken critique. Psychol Bull 115:268–287. https://doi.org/10.1037/0033-2909.115.2.268

Ekman U, Rokeby D (2014) Transformations of transforming mirrors: an interview with David Rokeby. Postmod Cult. https://doi.org/10.1353/pmc.2014.0004

Else J, Ellis J, Orme E (2015) Art expertise modulates the emotional response to modern art, especially abstract: an ERP investigation. Front Hum Neurosci. https://doi.org/10.3389/fnhum.2015.00525

Fleck S, Simon G (2013) An augmented reality environment for astronomy learning in elementary grades. In: Proceedings of the 25th ICME conference francophone on l'Interaction Homme-Machine—IHM '13. https://doi.org/10.1145/2534903.2534907

Frey J, Gervais R, Fleck S et al (2014) Teegi: tangible EEG interface. In: UIST-ACM user interface software and technology symposium

Gruzelier J (2008) A theory of alpha/theta neurofeedback, creative performance enhancement, long distance functional connectivity and psychological integration. Cogn Process 10:101–109. https://doi.org/10.1007/s10339-008-0248-5

Gruzelier J (2014) EEG-neurofeedback for optimising performance. II: Creativity, the performing arts and ecological validity. Neurosci Biobehav Rev 44:142–158. https://doi.org/10.1016/j.neubiorev.2013.11.004

Gruzelier J, Hirst L, Holmes P, Leach J (2014) Immediate effects of alpha/theta and sensory-motor rhythm feedback on music performance. Int J Psychophysiol 93:96–104. https://doi.org/10.1016/j.ijpsycho.2014.03.009

Gsöllpointner K (2016) Syn-aesthetics of digital art. In: Gsöllpointner K, Schnell R, Schuler R (eds) Digital synesthesia: a model for the aesthetics of digital art, 1st edn. De Gruyter, Berlin, Bostonpp 10–28

Guljajeva V, Canet M, Mealla S (2018) NeuroKnitting | Varvara & Mar. In: Varvara & Mar. http://www.varvarag.info/neuroknitting/. Accessed 13 Dec 2018

Gürkök H, Nijholt A (2013) Affective brain-computer interfaces for arts. In: 2013 Humaine Association conference on affective computing and intelligent interaction. https://doi.org/10.1109/acii.2013.155

Hammond D (2007) What is neurofeedback? J Neurother 10:25–36. https://doi.org/10.1300/j184v10n04_04

Harvey A (2018) Music and the meeting of human minds. Front Psychol. https://doi.org/10.3389/fpsyg.2018.00762

Heanue M (2018) Group overview ‹ Affective computing—MIT Media Lab. In: MIT Media Lab. https://www.media.mit.edu/groups/affective-computing/overview/. Accessed 16 Dec 2018

Herrmann C, Strüber D, Helfrich R, Engel A (2016) EEG oscillations: from correlation to causality. Int J Psychophysiol 103:12–21. https://doi.org/10.1016/j.ijpsycho.2015.02.003

Hinterberger T, Fürnrohr E (2016) The sensorium: psychophysiological evaluation of responses to a multimodal neurofeedback environment. Appl Psychophysiol Biofeedback 41:315–329. https://doi.org/10.1007/s10484-016-9332-2

Hjelm S, Browall C (2000) Brainball-using brain activity for cool competition. In: NordiCHI, pp 177–178

Kawasaki M, Kitajo K, Yamaguchi Y (2010) Dynamic links between theta executive functions and alpha storage buffers in auditory and visual working memory. Eur J Neurosci. https://doi.org/10.1111/j.1460-9568.2010.07217.x

Khare KC, Nigam SK (2000) A study of electroencephalogram in meditators. Indian J Physiol Pharmacol 44:173–178

Khut G (2015) Behind your eyes, between your ears: neurofeedback portrait project. In: George Khut. http://www.georgekhut.com/behind-your-eyes-between-your-ears/. Accessed 25 Jan 2019

Khut G (2006) Development and evaluation of participant-centred biofeedback artworks. Doctor of Creative Arts (DCA), University of Western Sydney

Kleinsmith A, Bianchi-Berthouze N (2013) Affective body expression perception and recognition: a survey. IEEE Trans Affect Comput 4:15–33. https://doi.org/10.1109/t-affc.2012.16

Klimesch W (1999) EEG alpha and theta oscillations reflect cognitive and memory performance: a review and analysis. Brain Res Rev 29:169–195. https://doi.org/10.1016/s0165-0173(98)00056-3

Lancel K, Maat H (2014) Lancel/Maat—E.E.G. KISS. In: Lancelmaat.nl. https://www.lancelmaat.nl/work/e.e.g-kiss/. Accessed 26 Jan 2019

Leslie G, Mullen T (2011) MoodMixer: EEG-based collaborative sonification. In: Proceedings of the international conference on new interfaces for musical expression, pp 296–299

Lucier A (1976) Statement on: music for solo performer. In: Rosenboom D (ed) Biofeedback and the arts, results of early experiments, 1st edn. Aesthetic Research Center of Canada, Vancouver, pp 60–61

Lutz A, Dunne J, Davidson R (2007) Meditation and the neuroscience of consciousness: an introduction. Camb Handb Conscious 499–552. https://doi.org/10.1017/cbo9780511816789.020

Mantini D, Perrucci M, Del Gratta C et al (2007) Electrophysiological signatures of resting state networks in the human brain. Proc Natl Acad Sci 104:13170–13175. https://doi.org/10.1073/pnas.0700668104

Maskeliunas R, Damasevicius R, Martisius I, Vasiljevas M (2016) Consumer grade EEG devices: are they usable for control tasks? PeerJ 4:e1746. https://doi.org/10.7717/peerj.1746

Menon S, Sinha A, Sreekantan B (2014) Interdisciplinary perspectives on consciousness and the self, 1st edn, pp 1–8

Mercier-Ganady J, Lotte F, Loup-Escande E et al (2014) The mind-mirror: see your brain in action in your head using EEG and augmented reality. In: 2014 IEEE virtual reality (VR). https://doi.org/10.1109/vr.2014.6802047

Molina G, Tsoneva T, Nijholt A (2009) Emotional brain-computer interfaces. In: 2009 3rd international conference on affective computing and intelligent interaction and workshops. https://doi.org/10.1109/acii.2009.5349478

Moritz J (2017) Augmented humanity. Technoetic Arts 15:341–352. https://doi.org/10.1386/tear.15.3.341_1

Mullen T, Khalil A, Ward T et al (2015) MindMusic: playful and social installations at the interface between music and the brain. In: Nijholt A (ed) More playful user interfaces. Interfaces that invite social and physical interaction, 1st edn. Springer, Singapore, pp 197–229

Muller L, Bennett J, Froggett L, Bartlett V (2015) Understanding third space: evaluating art-science collaboration. In: 21st international symposium of electronic art

Muller L, Froggett L, Bennett J (2018) Emergent knowledge in the third space of art-science. Leonardo 1–11. https://doi.org/10.1162/leon_a_01690

Mühl C, Allison B, Nijholt A, Chanel G (2014) A survey of affective brain computer interfaces: principles, state-of-the-art, and challenges. Brain Comput Interfaces 1:66–84. https://doi.org/10.1080/2326263x.2014.912881

NEURON (2018) Blogul UMF » NEURON 2018. In: Blog.umftgm.ro. https://blog.umftgm.ro/tag/neuron-2018/. Accessed 10 Oct 2018

Nijboer F, Clausen J, Allison B, Haselager P (2011) The Asilomar survey: stakeholders' opinions on ethical issues related to brain-computer interfacing. Neuroethics 6:541–578. https://doi.org/10.1007/s12152-011-9132-6

Papatheodorou T, Chambers T (2018) Random quark—creative technology studio. In: Randomquark.com. http://randomquark.com/case-studies/mindswarms.html. Accessed 13 Dec 2018

Park L (2014) Lisa Park. In: Lisa Park. http://www.thelisapark.com/#/cunoia-ii/. Accessed 13 Dec 2018

Pearlman E (2017) Brain opera: exploring surveillance in 360-degree immersive theatre. PAJ: J Perform Art 39:79–85. https://doi.org/10.1162/pajj_a_00367

Peper E, Ancoli-Israel S, Quinn M (1979) Mind/Body integration: essential readings in biofeedback, 1st edn. Plenum Press, New York

Perkins D (1994) The intelligent eye: learning to think by looking at art. Getty Center for Education in the Arts, Santa Monica, Calif

Popian I (2014) iondesign | PROJECTS. In: iondesign. http://www.ionarch.com/projects. Accessed 13 Dec 2018

Ray W, Cole H (1985) EEG alpha activity reflects attentional demands, and beta activity reflects emotional and cognitive processes. Science 228:750–752. https://doi.org/10.1126/science.3992243

Rokeby D (1995) Transforming mirrors: subjectivity and control in interactive media. In: Penny S (ed) Critical issues in electronic media, 1st edn. Suny Press, pp 133–158

Schroeder A (2018) Schroeder-lab. In: Schroeder-lab. https://www.schroederlab.com/. Accessed 10 Oct 2018

Schroeder A. Avi Schroeder - הפקולטה להנדסה כימית |ח. הפקולטה להנדסה כימית. https://chemeng.technion.ac.il/avi-schroeder/. Accessed 10 Oct 2018

Sherry J (2015) Neuroscience and communication. Commun Methods Meas 9:117–122. https://doi.org/10.1080/19312458.2014.999756

Sobell N (2019) Nina Sobell: artist statement http://ninasobell.com. Accessed 25 Jan 2019

Silvia P (2005) Emotional responses to art: from collation and arousal to cognition and emotion. Rev Gen Psychol 9:342–357. https://doi.org/10.1037/1089-2680.9.4.342

Stamps K, Hamam Y (2010) Towards inexpensive BCI control for wheelchair navigation in the enabled environment—a hardware survey. In: International conference on brain informatics. Springer, Berlin, Heidelberg, pp 336–345

Tamburrini G (2009) Brain to computer communication: ethical perspectives on interaction models. Neuroethics 2:137–149. https://doi.org/10.1007/s12152-009-9040-1

Vygandas Š (2018) Humanizing technology through post-digital art. PhD, KTH Royal Institute of Technology

Wadeson A, Nijholt A, Nam C (2015) Artistic brain-computer interfaces: state-of-the-art control mechanisms. Brain Comput Interfaces 2:70–75. https://doi.org/10.1080/2326263x.2015.1103155

Wellington J (1990) Formal and informal learning in science: the role of the interactive science centres. Phys Educ 25:247–252. https://doi.org/10.1088/0031-9120/25/5/307

Zioga P, Chapman P, Ma M, Pollick F (2014) A wireless future: performance art, interaction and the brain-computer interfaces. In: Proceedings of Inter-Face: international conference on live interfaces (ICLI 2014)

Part III
Your Brain on Art: Perceiving, Understanding, and Creating

Chapter 9
Understanding Perceptual Experience of Art Using Mobile Brain/Body Imaging

Zakaria Djebbara, Lars Brorson Fich and Klaus Gramann

Abstract This chapter draws on the importance of movement for human perceptual experience and how it influences brain dynamics. By use of Mobil Brain/Body Imaging (MoBI), artists with interest in the experience of art can get insights into human cortical activity during artworks. Specifically, art that depends on action faces challenges regarding the exploration of human brain activity during their artistic acts or performances. We give an account of how architectural experience, which essentially rests on perception and movement, can be investigated using a MoBI method. We present results from studies that indicate fundamental differences in cognitive and behavioural responses when comparing active behaviour compared to passive perception. Consideration of the processes underlying movement and cognition suggests that action alters perception, which in turn alters experience. MoBI is therefore able to reveal aspects of natural cognition, which would otherwise go unnoticed highlighting the advantage of using MoBI in animate forms of art.

Keywords Mobile brain body imaging · Mobile EEG · Architectural cognition · Enactivism · Active inference

9.1 Introduction: Experience and Cognition

A plethora of philosophers have written on aesthetics and the appreciation of art (Dickie 1997). Both art and philosophy focus on a common facet shared by all forms of art: the perceptual aspect. In order to experience art, it needs to be perceivable.

Z. Djebbara (✉) · L. B. Fich
Department of Architecture, Design, Media and Technology, Aalborg University, Aalborg, Denmark
e-mail: zadj@create.aau.dk

K. Gramann
Biological Psychology and Neuroergonomics, Berlin Institute of Technology, Berlin, Germany

School of Software, University of Technology Sydney, Sydney, Australia

Center for Advanced Neurological Engineering, University of California, San Diego, CA, USA

© Springer Nature Switzerland AG 2019
A. Nijholt (ed.), *Brain Art*,
https://doi.org/10.1007/978-3-030-14323-7_9

Studies by pioneers in brain imaging and visual processing gave rise to an interdisciplinary movement, namely neuroaesthetics (Kawabata and Zeki 2004; Zeki 1998), proposing that perceptual processes of art as realized through the human brain, and the visual cortex in particular, can inform our knowledge of human judgment of beauty. By this conception, it is assumed the human brain holds a common organizational principle enabling communication through and about art (Zeki 2002). The approach has often been that of treating art comparable to the treatment of a stimulus in any neuroscientific or psychological experiment. Often fMRI has been the experimental method used, staging both the context in which art has been presented to the participants, but also the time-frame, which is often limited to only seconds. Furthermore, established experimental approaches that measure the reaction of person to the presentation of visual stimuli leaves the test-person as a passive receiver. However, from presenting a stimulus to arriving at a judgment of beauty about the very same is a complex process, moving from physical characteristics of an object to a reflected subjective judgement. The limitations of this perspective as derived from the neuroaesthetics approach lie in the assumptions that beauty, if materialized in the brain through neural activation, must be traceable through brain imaging techniques during the experience of beauty, and that beauty is a fixed reproducible phenomenon. In this chapter, instead of setting out to resolve the general puzzle of judging beauty, we take one step back, and aim at providing an account on the nature of *perception* and the emergence of *experience* of artistic expression.

The nature of experience itself reaches a level of complexity beyond the capabilities of a single field of study. This belief is shared with cognitive science, which seeks to include the perspectives offered in philosophy, experimental approaches of psychology, empirical studies by neuroscience and other fields (Miller 2003; Thagard 2009). Particularly, phenomenology—the philosophical study of underlying structures of human experience—demonstrates the wide-ranging features of experience through reflections on the experience itself, by not restricting experience to representation and perception. In fact, phenomenology teaches an important lesson on human experience by emphasizing time as a necessity for experience (Gallagher and Zahavi 2012, Chap. 4). Merleau-Ponty et al. (1968, p. 29) underlines that from feeling and vision will be retained only what animates and sustains them, meaning that perceptual experience is what enables a continuity, a passage from one moment to another. Experience as such does not become a matter of "knowing" and rationally untangling the depth of perception. Rather, experience develops from perception as "in action" and thereby, attributing perception a primacy due to its bodily relation (Merleau-Ponty and Edie 1964, pp. 12–13). Experiencing art is not a static rational process of information processing, instead it is active, embodied and temporally continuous—a property essential for any kind of experience, and thus underlying the majority of arts.

Perceptual processes, positioned at the transition from the objective physical to the subjective perceptual world, can be conceived as the preceding stepping-stone to human experience. The perceptual process then precedes the experience as *perceptual experience,* that is, the epistemic content about the world. Perceptual processes have been extensively investigated in neuroscience through various methodologies, and due to the fast temporal scale of these processes, electroencephalography (EEG) with its high temporal resolution of the recorded brain electrical signal has proven important in such investigations. Considering sensory evoked potentials, it is possible to trace the advancement from sensory to perceptual experience through cortical activation patterns (Luck and Kappenman 2011, Chap. 4). The EEG enables us to investigate perceptual and cognitive processes through carefully designed experiments. Furthermore, mobile EEG allows the subjects to move in space during experimentation, which of course is a great advantage when the experiments concerns three dimensional objects or spatial structures like architecture.

Perception and perceptual experience has lately been rediscovered in philosophy of perception, e.g. (Macpherson 2014). In cognitive science, perceptual experience has been suggested to be best understood through skilled sensorimotor behaviour (Noë 2004; Thompson 2007; Varela et al. 1991). The nature of perception is argued to be active, in the sense that it is not passively collected from the environment, but rather actively entangled with bodily action. Perception is thus driven by interaction of informed movement and sensory feedback. Essentially, proprioception stemming from movements including eye saccades, head movements, reaching, moving back and forth and other gestures, inform the cognitive system about the state of our physical structure with respect to a cognitive goal. The brain generates predictions of sensory input, via a deep multilevel cascade based on prior experiences, to infer incoming sensory data (Friston 2005). The general idea of a Bayes-optimal account of the brain, is referred to as *predictive processing* (Clark 2015). Eventually, perception is the process of a mismatch between multilevel bidirectional flows in a probabilistic model, continuously matching top-down predictions with bottom-up sensory signals. Movement thus initiates bottom-up sensory signals, which then are cancelled out if correctly predicted (Clark 2015; Hohwy 2013). According to *active inferences*, the perception and action cycle is explained as minimizing prediction-errors by either changing the predictions to explain the incoming sensory signal or by actively changing sensory signals through movement to fulfil the prediction (Brown et al. 2013). Both action and perception thus contribute to cortical responses, altering one another through active inferences. Such an account ultimately indicates the importance of action for a holistic experience.

In dealing with brain and art, cognitive neuroscience and phenomenology indicate that movement and immersion with the environment play a significant role in the reported experience and measured brain responses. Architecture and movement in particular serve as an excellent example of how neuroscience can inform architects about the experience of their designs, and in turn inform cognitive neuroscientists on the nature of cortical functioning. Here, we pave the way for a neuroscientific approach to investigate architecture as the art of communicating designed experience that necessitates active movement over time.

9.2   Architecture and Movement

Essentially, art communicates a sense, and perhaps a logic, through a medium suitable
for the message, e.g. music through instruments, dance through movement, paintings
through paint on canvas etc. Architecture, containing many art forms in itself, communicates at least also through the medium of movement (Vesely 2004, pp. 74–86).
It is a complex, dynamic, and heterogeneous art form, which releases and restricts
the body through space and time to make sense of an animate and moving world.
Ancient architectural gems indicate early awareness of the body and its movement
in architecture. Egyptians illustrated such awareness by differing spatial dimensions
of their temples in a systematic manner, while Greeks introduced natural relations
to their facades, and Romans, during the renaissance, reintroduced mathematical
relations between spaces (see e.g., Alberti et al. 1986; Fazio et al. 2008, Chaps. 1,
2, 5; Vitruvius and Morgan 1960). Common to these examples is the necessity of
movement to experience the intentions of the architect. The reason for this early
awareness, according to Robinson (2011) and Pallasmaa (Holl et al. 2006, Chap. 3),
is that traditional builders shaped their buildings with their own bodies, similar to
how a bird shapes its nest by its body. Their design was informed by how they would
approach or confront the building, how the body weight meets the weight of the door,
the eyes measure the distances, the feet measure the height of the steps etc. (Holl
et al. 2006, p. 35). Similar traces are emphasized by Bachelard (1969) through a phenomenological account of the peculiar qualities of houses. This branch of philosophy
is suitable for architects as it focuses on the embodied and holistic experience of the
body being situated in space, and understands the human being through descriptions
of the perceptual experience itself (Gallagher and Zahavi 2012). This approach has
given rise to numerous concepts developed from observations of architectural experiences, e.g., atmospheres and *genius loci* (Böhme and Engels-Schwarzpaul 2017;
Norberg-Schulz 1997; Zumthor 2006).

  Steen Eiler Rasmussen is considered an architectural phenomenologist. In particular his writing *On Experiencing Architecture*, made an immense impact on architectural theory. Rasmussen introduced the idea of *time* into architectural experience with
an approach that is strikingly similar to contemporary cognitive science. Although he
evidently appreciated Gestalt theory of perception, Rasmussen (2012/1957) aban-
doned the pictorial approach to architecture and experience, referring to the non-
temporal issue (2012, pp. 41–42). Instead, Rasmussen insisted on the importance of
movement, perspective and the active perception of an environment (2012, p. 35).
We actively move about and construct our impression of the architectural character,
which means the disembodied idea of forms, such as Gestalt and other abstract ratio-
nales, becomes obsolete in architecture. The fact that Rasmussen sought to include
perspectival deformation during movement into architectural experience, indicates
an awareness of the structure of experience. A scene is experienced from a perspective, with the prior scenery in retention, and the immediate impression actively
constructed. An enduring temporal continuity is essential for any particular experience (Gallagher and Zahavi 2012, Chap. 4). However, the cadence of succession in

temporal experience is not fixed (Rasmussen 1959, p. 137). Architecture itself contains no rhythm nor frequency, yet the architecture is experienced depending on how it may afford fast or slow movements e.g. in a curvy or perpendicular corner, a steep staircase or soft ramp. Here, architecture induces a certain velocity and complexity of movement, which comes at different energetic costs. Besides including active perception, movement, and time into architecture, his descriptions also approach cognitive science ideas on a neural level. Predictive-errors as neural currency, in terms of invested energy of the body, seems mysteriously obscure in Rasmussen's account of experience, as he states that impressions are best made by unpredictable features in architecture *"demanding an energetic effort of the viewer, in a continuously shifting perception"* (own translation) (2012, p. 61). The surprising resonance between phenomenology of Rasmussen, amongst other phenomenologists, and predictive processing presupposes a common acknowledgeable approach, appropriate for mutual discourse on perceptual experience. As illustrated, movement plays a central role in phenomenology and in predictive processing.

Phenomenology gives us a structure for understanding architectural experience, which links with cognitive science in terms of an active perceptual experience. The designed movement becomes a form of artistic communication between the body of an experienced architect and the body of the experiencer. The poetry and art of architecture thus evidently describes how an architect propels the experiencer's body through space and spatial events. The non-radical claim is that investigations in action-perception are able to provide a better understanding of how architecture affects and shapes our experiences. We thus set out to understand the advancement of movements, through active inferences and affordances.

9.3 Active Inferences and Affordances

The concept of affordances was introduced by Gibson (1979) as action possibilities. Considering the traditional cognitive conception of action, as nicely summarized by Cisek (2007), the perceptual system builds a representation of the external world by collecting sensory information (Marr 1982), which then is used, together with current needs and prior experience, to judge a course of action (Johnson-Laird 1988). From this, a plan is generated and realized through action (Keele 1968), which means the brain operates sequentially by first building knowledge about the world, utilizing that knowledge to make a decision and finally compute an action plan, realized through action. Clark (2015) argues that such a model cannot account for the fluent and rapidly changing situations, which the human nature ecologically must be inclined towards. Instead, the rolling cycle suggested by active inferences, seems much stronger, more thorough and steady.

A Bayes-optimal perspective of the brain (Friston 2003; Friston et al. 2006), infers mainly two properties; the current state of the world and the uncertainty of that state. These active inferences are constructed through an embodied (enacted) form of a continuous action-perception cycle, where perception minimizes exteroceptive

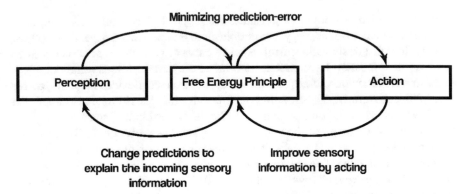

Fig. 9.1 Active inference conceptually illustrated as a self-sufficient principle. The free energy principle reduces entropy by either changing the prediction to explain the sensory signal, or by improving the signal due to action. In this sense, prediction-error are minimized in both action and perception. Figure inspired from Friston (2013)

prediction-errors and action minimizes proprioceptive prediction-errors (Fig. 9.1). This link between the world and a systematic change of sensory stimuli, positions action in a decisive seat that has direct implications for the outcome of the future state of the world. Being proactive in nature, the brain thus computes multiple action possibilities, each with different cost of energy, and partially projects information to the motor system to prepare for a probable action (Cisek 2007). Such an account finally means the brain seeks to be as ready as possible to unfold the appropriate response as the evidence of the world evolves and improves. These affordances compete against each other while minimizing prediction-error to bias the competition towards a single action possibility (Cisek 2007, p. 1585). Note that *movement* is not necessarily intended, however, *action* is. For instance being pushed, riding the train or driving a car moves the body, but one does not act (Gallagher and Zahavi 2012, pp. 171–174). We are only concerned here with intentional action, which, in this sense, is selected through a process of minimizing prediction-error of the systematically changing patterns of sensory input (or optimizing uncertainty of the state).

Affordances, the possible actions, compete against each other beyond the initiation of an action, as the action-perception cycle is an ongoing continuous flow of prediction-error minimizing. According to Cisek's (2007, p. 1586) affordance competition hypothesis, one can thus expect the affordance competition to be an ongoing process even as certain actions are continuously selected while unfolding. The selected action arguably depends on the affordances of the environment and the bodily proficiency of a given agent. In other words, action selection depends on affordances. Perhaps due to the debate on free will (Bergson and Pogson 1960; Libet et al. 1983), affordances came to play a major role in brain science, leading to investigations of movement and brain components measured with EEG (Brunia 2003; van Boxtel and Böcker 2004; Walter et al. 1967). It lead to an understanding of actions as an embodied decision-making process. According to active inference, what is perceived is conditioned by what is done, and what is done depends upon what is perceived (Clark 2015, p. 176), which ultimately places affordances in the centre of the continuous perceptual and motor-related processes in the brain. Decisions on body trajectory are complex processes, specified by spatio-temporal information dependent on the geometry of the environment that will overwhelm the brain if not attended dynamically. In this sense, relevant variables such as outcome values, success probability, action costs, must be in constant bidirectional flux (Cisek and Pastor-Bernier 2014). The suggested temporally organized hierarchical model of decision and action offers that various brain areas are processing in parallel various aspects of decisions (Pezzulo and Cisek 2016, p. 421).

The main argument here is that decisions regarding action possibilities are much faster than hitherto thought. In fact, affordances, in terms of action selection, are an intrinsic part of cognition, necessarily positioned in relation to perception, as suggested by active inferences. The implications are then, firstly, that the brain processes (cognitive and sensory) and body trajectories (actions) are much closer related than previously thought. Secondly, that affordances, being attributes of a cue (Friston et al. 2012), are actively inferred by way of an action-perception cycle, meaning affordances play a key role in a continuous perceptual experience. Thirdly, that one must be able to measure different cortical responses associated with perceptual stages by systematically varying action possibilities (affordances) in the environment. According to active inferences, action potentials are attended to as they occur as fast sensorimotor inference, thus being hierarchically low level in the generative model (Kiebel et al. 2008). Finally, that prior to movement onset, one must be able to measure a difference in cortical activity due to immediate affordance competition—if affordances are a rolling cycle. Essentially, affordances emerges as a pragmatic and natural outcome of action-oriented predictions (Clark 2015, p. 184). To return to the initial departure, experiencing art, we argue that perceptual experience is constructed through a rolling cycle of action and perception, ranging from eye saccades and subtle head movements to full body actions.

9.4 An Obstacle of Heterogeneous Environment

Investigating architecture statically (e.g., Ma et al. 2015; Vartanian et al. 2015, 2013) has been criticized for not sufficiently corresponding to natural architectural experiences. Precisely because during action, experience is shaped by a heterogeneous environment, as a perspectival deformation that is continuously synthesized in temporal transitions—the fundament of architectural experience is that of a continuously heterogeneous environment. This is a central empirical obstacle of investigating perceptual experience: linking body trajectory with meaningful events in time. Such an insight would arguably lead to an improved understanding of active inferences, cortical responses and perceptual experiences. Inferences about the state of the world are continuous over time, constantly struggling to improve the uncertainty of that state. In this respect, an art investigation must necessarily be reduced to the absolute minimum, to faithfully understand the influence and relation of action, perception and cognition. The necessary requirement of any reductionism is that the reduced entity is properly understood (Gallagher and Zahavi 2012, p. 10, following Nagel 1974). With this idea, one can limit actions and events by systematically varying the art, which in the case of architecture, as argued, is the landscape of affordances inducing a narration of actions. For dance, the velocity and configuration of body parts during the movement perhaps characterize meaningful events. For paintings, it is perhaps eye saccades or subtle head movements, while for sculptures it is full body movements. Actions interacts with perception, which in turn forms perceptual experience, and are therefore exceptionally important to control.

Active inference illustrates a sense of effective use of time, by allowing multiple processes to predict already at the level of partial sensory information, embedding the agent acutely. Walking, talking, listening, and watching etc. are dynamic processes continuously active in the body and brain, handled at the same time. Technologically, this requires a high and continuous temporal resolution of brain activity. The emerging obstacle, rightfully highlighted, relates to the investigation of experience due to technological limitations. In terms of a predictive and embodied theory, the architectural investigation shows how the world we shape shapes our thoughts and experiences through affordances. Perhaps the most radical notion presented is that action modulates perceptual processes throughout the brain, and vice versa. To investigate this, human brain activity has to be recorded in synchrony with action while human observers perceive and act according to the affordance of a given environment. This however, is not the established approach in human neuroscience that aims at restricting active behaviour as far as possible as it is considered a source of artefact for recordings of brain activity (Makeig et al. 2009). Using a Mobile Brain/Body Imaging (MoBI) approach, however, allows for investigating differences in cortical responses when modulating apparent affordances and thus allows for illustrating how the continuous action selection similarly differs according to affordances.

9.5 Mobile Brain/Body Imaging Approach

The MoBI approach (Gramann et al. 2014, 2011; Makeig et al. 2009) allows to record activity of the human brain in actively moving participants using mobile brain imaging devices like electroencephalography (EEG) or functional near-infrared spectroscopy (fNIRS) synchronized to motion capture and other data streams. Optionally, head mounted virtual reality (VR) systems can be coupled to the setup to allow full control over visual and auditory stimulation while human participants move through and interact with virtual worlds. The method was developed to allow investigation of the relationship of action, cognition, and brain activity and aims at overcoming the limitations of traditional brain imaging modalities that restrict active movement of participants to avoid artefacts originating from movement. By synchronizing recordings of brain activity with motion capture, MoBI allows us to investigate the interplay of sensation, perceptual experiences, and action, while recording the accompanying brain dynamics. This contrasts MoBI studies from mobile EEG studies that do not record specific aspects of participants' behaviour but rather compare brain activity in different movement conditions like sitting as compared to walking (Jungnickel and Gramann 2018). Using data driven analyses approaches with the help of information from movement recordings to advance the signal decomposition, MoBI allows separation of brain and non-brain activity for further analyses. This way, MoBI is the method of choice to investigate the brain dynamics underlying the impact of architecture on perception and action in freely behaving humans.

Early MoBI studies mainly focused on demonstrating the feasibility of the approach using treadmills that allowed movement of participants without necessitating large physical spaces. These studies demonstrated that it is possible to investigate human brain dynamics accompanying cognitive processes including attention to relevant rare stimuli during active behaviours like walking (Gramann et al. 2010). In the study by Gramann and colleagues (2010), participants were standing or walking with different speed on a treadmill while, at the same time, responding to rare target stimuli in a visual oddball task presented on a screen in front of them. Using independent component analysis (ICA) and subsequent clustering of independent components (ICs), the authors demonstrated that the P300 component, a positive deflection in the event-related potential (ERP), could be reconstructed for target stimuli irrespective of the behavioural state (standing or walking; see Fig. 9.2).

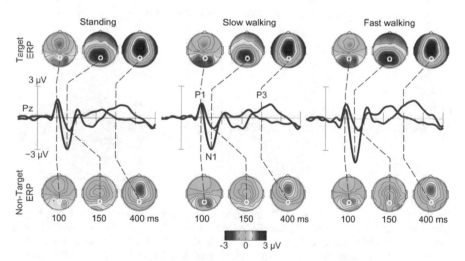

Fig. 9.2 Grand-average ERPs following ICA-based artifact removal in a standing, slow walking, and fast walking condition. Middle-row traces show ERP time courses at electrode Pz (red, target ERPs; blue, non-target ERPs). Scalp maps show the grand-average ERP scalp distributions at 100, 150, and 400 ms after onsets of target stimuli (upper row) and non-target stimuli (lower row). White dots indicate the location of electrode Pz. Note the scalp map similarities across movement conditions. Figure from Gramann et al. (2010) Frontiers in Human Neuroscience

Later studies using MoBI in participants walking on a treadmill further shed light on the supraspinal control of locomotion and providing new insights into the modulation of different frequency bands involved in gait (Gwin et al. 2010, 2011).

Recent studies started to investigate different kinds of movement and explored human brain dynamics during upper and full body actions. The first MoBI study investigating human brain dynamics during pointing responses in a dynamically changing environment (Jungnickel and Gramann 2016), demonstrating significantly increased activity in brain regions underlying the integration of multi-modal information supporting active pointing behaviours as compared to simple button press responses reflecting established minimal experimental behaviours (Fig. 9.3).

In this experiment, participants stood in front of a large projection screen and followed the movement of a sphere bouncing of the borders of the screen. Unpredictable

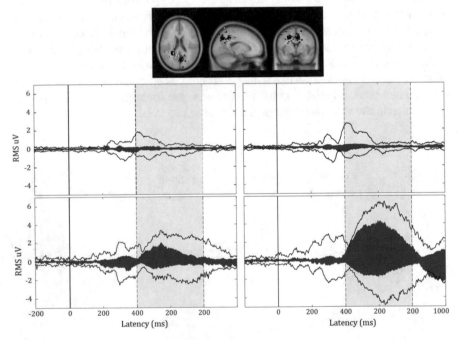

Fig. 9.3 Contribution of two parietal IC-clusters back-projected to the sensor level ERPs. Bigger spheres represent cluster centroids and smaller spheres individual ICs with cluster centroids located in or near the parietal cortex. One cluster was located to the left parietal cortex (Talairach coordinates: x = −19, y = −42, z = 39, corresponding to Brodmann area 31) including eleven ICs and one cluster located to the posterior parietal cortex (Talairach coordinates: x = 11, y = −66, z = 38, corresponding to Brodmann area 7) comprising seven ICs. Middle and lower row present ERP contributions of the clusters located in or near parietal cortex to the back-projected sensor ERP computed by back-projecting all clusters located in the grey matter of the brain model. The dark grey area displays the latency range of the P3 component from 400 to 800 ms after a color change. The left and right columns display envelopes for the button press condition and the physical pointing condition, respectively, with the upper row displaying standard stimuli and the lower row target stimuli. Adapted from Jungnickel and Gramann (2016) Frontiers in Human Neuroscience

on a trial, the sphere could change its colour to either a target, a distractor, or a standard colour and participants were instructed to respond to a target colour change only. In a first condition, participants simply responded by pressing a button on a remote mouse they held in their hand. This response requirement replicated traditional brain imaging setups with a simple button press. In a second condition, participants had to point at the moving target sphere and motion capture was used online to track the position of the participants' finger and to stop the trial when the target sphere was touched. Response times in this experiment revealed faster responses in the pointing condition as compared to the button press condition, revealing faster motivated actions. Importantly, the parietal cortex revealed increased activity for target trials as compared to standard trials reflecting higher computational demands when humans have to continuously update action-relevant information to allow time-sensitive interaction with the environment. In the experiment Jungnickel and Gramann (2016), this included the continuous computation of a motor command for a complex pointing movement that had to be adapted to the dynamically changing position of the stimulus. Thus, the affordance of the environment significantly influenced brain dynamics.

More recent MoBI studies provided new insights into multi-modal sensory integration during full body rotations (Gramann et al. 2018) and conflict processing during active reaching. The latter studies demonstrating that active behaviour significantly modulates brain dynamics and that proprioception resulting from movement execution plays a central role in the subjective experience of the environment and conflict resolution to solve ongoing tasks (Singh et al. 2018; Töllner et al. 2017). Importantly, all results from the above reported studies demonstrate an impact of the behavioural state on the brain dynamic state of the participant and an impact of the affordance of the environment.

Only very recently, studies started directly addressing architectural questions with the use of MoBI. The first study by Banaei and colleagues (2017) investigated the impact of different architectural forms on human brain activity during exploration of different virtual spaces. The results replicated previous findings from fMRI studies revealing increased anterior cingulate cortex activity for architectural spaces with curvature geometries. More specifically, theta band activity in anterior cingulate cortex (ACC) co-varied with specific features and the geometry immediately upon entering the environment. While these results were promising in showing the advantage of MoBI to better understand the brain dynamics underlying perception of and action in architectural spaces, the study did not however, address different affordances of architecture.

This was done in a study by Djebbara and colleagues (under review), specifically testing the assumption that the affordance of a given environment affects perceptual and motor processes. In this experiment, participants explored virtual environments with the task to move between different rooms passing through doors of different width, reflecting different affordances of the environment. Assuming the presented enactive account on the nature of action, perception and cognition is correct; the affordances of a given environment should directly relate to perceptual processes. In fact, as depicted by active inferences, action and perception become a looped cycle, substantiating one another. The study by Djebbara and colleagues aimed at investigating whether brain activity changes according to the affordances of the environment. The authors hypothesized the emotional evaluation to correlate with reaction times as well as early cortical potentials to co-vary with the environmental affordances. Furthermore, motor-related cortical potentials were also expected to co-vary with environmental affordances, with onset of an imperative stimulus (Go/NoGo) indicating to move through the environment. EEG data was acquired continuously using a 64 channels EEG system (eegoSports, ANT Neuro, Enschede, Netherlands) and data were analysed using adaptive mixture independent component analysis (AMICA; Palmer et al. 2011) and subsequent computation of time-domain activity of sensor data (Fig. 9.4a).

The experiment used a Go/No-Go paradigm that required the participants to walk from one room to a second room to find a goal object (red circle) in the second room to receive a monetary reward. To transit into the second room, participants had to pass through doors of different widths, ranging from unpassable (*Narrow*) to passable (*Mid*) to easily passable (*Wide*) (Fig. 9.4b). This was the critical manipulation of the

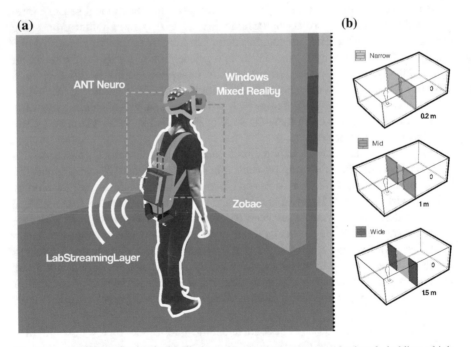

Fig. 9.4 a An illustration of the MoBI setup. The participants wore a backpack, holding a high-performance gaming computer, powering the VR head-mounted displays. The computer was attached to two batteries, making it completely wireless and mobile. Additionally, we attached a wireless EEG amplifier using 64-channels cap. All necessary events and time synchronization were assembled through LabStreamingLayer (Kothe 2014). **b** Three conceptual diagrams of the three possible doors. The Narrow door (yellow) measured 20 cm and was impossible to pass. The Mid door (cyan) measured 1 m and was difficult but passable. The Wide door (red) measured 1.5 m and was easily passable

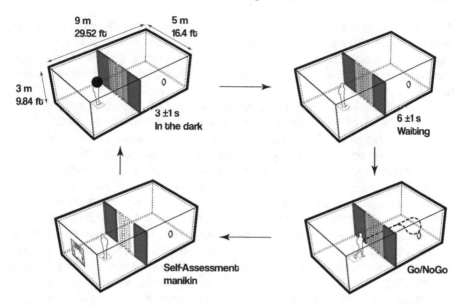

Fig. 9.5 The participants were masked with a black sphere around their head, restricting their visual perception to pitch black. After 3 s, the sphere would disappear ("lights on"), and the participant would be able to perceive the pseudo-randomly selected door. The participant was instructed to wait until the door would change colour. If the door changed to green (Go-trial), the participant was instructed to transit and virtually touch the red circle, return to start and answer the self-assessment questionnaire. If the door turned red (NoGo-trial), the participants were instructed to directly fill in the self-assessment questionnaire and restart

affordance of the environment reflecting different potentials to walk from one into the next room. One trial consisted of a participant starting in a dark environment on a predefined starting square (see Fig. 9.5). The participants would then face a room with a closed door and were instructed to wait for a colour change of the door. If the door changed to green (Go-trial), participants were instructed to walk towards the door, which would slide aside. Upon entering the second space, participants walked towards and picked up the red rotating circle to receive their monetary reward. Afterwards, they went back to the starting square, and filled in the virtual Self-Assessment Manikin (SAM) questionnaire. If the door changed to red (NoGo-trial), participants were instructed to directly answer the SAM questionnaire (Bradley and Lang 1994).

In the Go-condition with the unpassable door, participants were instructed to walk towards the door and into the second room even in the case that the door was too narrow to pass. This was done to control for motor execution in the Go-condition and to allow movement towards the goal irrespective of the affordance (passable vs. unpassable). If the participants touched the surrounding walls, the walls would turn red and inform the participants they have failed to pass, and thus must return to the start square. Participants would quickly notice that the narrow door (20 cm) was impossible to pass without producing the warning feedback that they have failed to pass. All subjects had a training phase to get accustomed to the VR environment and the different conditions. After each trial, participants had to give an emotional rating for the environment irrespective of whether they transitioned through the door (Go condition) or whether they remained in the same room (NoGo condition) without transition.

The acquired results demonstrate how affordances modulate perceptual processes as early as 200 ms after perception of the environment. Specifically, the event-related data showed significant differences in amplitudes of the visual evoked P100 component over the occipital sites dependent on the affordance of the door. The P100 component is a positive peak in the ERP approximately 100 ms post stimulus, which reflects early perceptual processes. In line with the affordance competition hypothesis, difference over fronto-central sites were observed starting approximately 50 ms until 200 ms after onset of the doors display. The findings indicate fast, lower sensorimotor active inferences, as explained in hierarchical and dynamic models of the world. The differences found after 200 ms indicate that action modulated perception, and therefore, similar to Cisek (2007) the results point towards active inference (Friston 2013; Kiebel et al. 2008). The findings support the assumption that action is processed in parallel to processes of sensory information, which ultimately situates action in an intimate position with perception. As these early processes may be involved in the impression of the environment, one may speculate whether the impression of an environment compose the immediate experience. Such an account for instance fits with the developed architectural concept *atmospheres* as defined by Zumthor (2006) and Böhme (2017).

Recall that affordances are a rolling cycle, meaning that prior to movement onset, one must be able to measure a difference in cortical activity due to immediate affordance competition. The results of Go-trials showed that the post-imperative negative variation (PINV) was modulated depending on the affordances starting 600 ms after Go-display until 800 ms (see Fig. 9.6). No such differences could be found for NoGo-trials. *Narrow* was significantly different from *Mid* and *Wide*, while the passable conditions did not differ from one another. Further, there were no significant differences in the PINV component in cases of NoGo instructions, emphasizing the importance of the motor execution itself to evoke the PINV component. The results indicate that the PINV component is an expression of the embodied willingness to execute an act that is restricted by affordances, and there might be a reflection of the continuous action selection. Such an account fits with the affordance competition hypothesis and active inferences.

Summarizing the finding by Djebbara et al. (under review), the early perceptual cortical responses varied according to affordances of the environment, relating the perceptual processes to action potentials as suggested through active inferences. Moreover, prior to approaching the door opening, a difference was also found varying according to the affordances, corresponding to the immediate rolling processing of the affordances, as described in Cisek's affordance competition hypothesis (Cisek 2007). The findings in general indicate that brain and body responds differently depending on the offered affordances in the environment, which in turn mean, in light of embodied theories, the spaces were experienced differently.

(a)   **(b)**

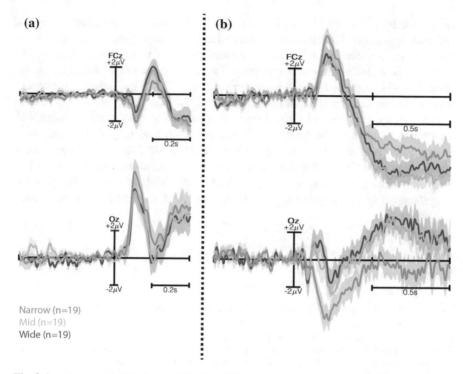

Narrow (n=19)
Mid (n=19)
Wide (n=19)

Fig. 9.6 a Two event-related potential plots (FCz and Oz) depicting the cortical response at the "lights on" event described in Fig. 9.5. **b** Two event-related potential plots (FCz and Oz) depicting the cortical response after the Go-display. Note the difference between 600 and 800 ms

9.6 Conclusion

This chapter set out to highlight the importance of action in the experience of art by investigating the art of architecture, namely action. We have given an embodied account of cognition, action and perception, which form the perceptual experience. By investigating architecture using a MoBI approach, it is argued that MoBI serves as a suitable approach for artists evaluating their specific art, and for cognitive neuroscientists to investigate the nature of an enacted brain. We conclude that action alters perceptual processes in the brain, and vice versa, calling attention to the importance of movement and affordances for perceptual experiences. Even small informed movement of the head and eye saccades while perceiving a painting or digital media, might have an impact of the total experience. Such an account of the relation between action and perception also fits with an ecological perspective of the perceptual system, which is to keep the organism in constant contact with its current affordances, and thus, actions (Rietveld and Kiverstein 2014).

In terms of active inferences, our results indicate that the brain is not solely addressing what the world state is, but it seeks to decrease the uncertainty of that world state by acting upon it. In other words, the brain and body does not merely ask *"what do I perceive"*, they also ask *"how can I act"*. It can be argued that affordances are omnipresent, because affordances are attributes of a cue (Friston et al. 2012), they must be actively inferred everywhere. Thus, the experience of art might not be solely rooted in perception itself, but in action as well. As an outlook of this chapter, we ask whether art, in any given form, is at all experienced if not actively explored.

Embodied cognition and architecture, as a field of research, offers insights to the dynamic nature of the human body-brain-environment interaction, creating a novel fundament for architectural experience. Architectural experience has been a matter of interest for philosophers and architects for centuries and we here offer a novel approach, using cutting-edge methods and models of embodied cognition to systematically investigate the underlying impact of architecture on the human experience.

References

Alberti LB, Bartoli C, Leoni G (1986) The ten books of architecture: the 1755 Leoni edition. Dover Publications

Bachelard G (1969) The poetics of space. Beacon Press

Banaei M, Hatami J, Yazdanfar A, Gramann K (2017) Walking through architectural spaces: the impact of interior forms on human brain dynamics. Front Hum Neurosci 11. https://doi.org/10.3389/fnhum.2017.00477

Bergson H, Pogson FL (1960) Time and free will, an essay on the immediate data of consciousness. Harper

Böhme G (1937–), Engels-Schwarzpaul A-C (2017) Atmospheric architectures the aesthetics of felt spaces. Bloomsbury Academic, an imprint of Bloomsbury Publishing Plc

Bradley MM, Lang PJ (1994) Measuring emotion: the self-assessment manikin and the semantic differential. J Behav Ther Exp Psychiatry 25:49–59

Brown H, Adams RA, Parees I, Edwards M, Friston K (2013) Active inference, sensory attenuation and illusions. Cogn Process 14:411–427. https://doi.org/10.1007/s10339-013-0571-3

Brunia CHM (2003) CNV and SPN: indices of anticipatory behavior. The Bereitschaftspotential. Springer US, Boston, MA, pp 207–227. https://doi.org/10.1007/978-1-4615-0189-3_13

Cisek P (2007) Cortical mechanisms of action selection: the affordance competition hypothesis. Philos Trans R Soc B Biol Sci 362:1585–1599. https://doi.org/10.1098/rstb.2007.2054

Cisek P, Pastor-Bernier A (2014) On the challenges and mechanisms of embodied decisions. Philos Trans R Soc Lond B Biol Sci 369. https://doi.org/10.1098/rstb.2013.0479

Clark A (2015) Surfing uncertainty: prediction, action and the embodied mind. Oxford University Press, New York

Dickie G (1997) Introduction to aesthetics: an analytic approach. Oxford University Press

Fazio MW, Moffett M, Wodehouse L (2008) A world history of architecture, 2nd edn. Laurence King

Friston K (2003) Learning and inference in the brain. Neural Netw 16:1325–1352. https://doi.org/10.1016/j.neunet.2003.06.005

Friston K (2005) A theory of cortical responses. Philos Trans R Soc Lond B Biol Sci 360:815–836. https://doi.org/10.1098/rstb.2005.1622

Friston KJ (2013) Active inference and free energy. Behav Brain Sci 36:212–213. https://doi.org/10.1017/S0140525X12002142

Friston KJ, Kilner J, Harrison L (2006) A free energy principle for the brain. J Physiol 100:70–87. https://doi.org/10.1016/J.JPHYSPARIS.2006.10.001

Friston KJ, Shiner T, FitzGerald T, Galea JM, Adams R, Brown H, Dolan RJ, Moran R, Stephan KE, Bestmann S (2012) Dopamine, affordance and active inference. PLoS Comput Biol 8:e1002327. https://doi.org/10.1371/journal.pcbi.1002327

Gallagher S, Zahavi D (2012) The phenomenological mind, 2nd edn. Routledge

Gibson J (1979) The ecological approach to visual perception. Hought Mifflin, Boston

Gramann K, Gwin JT, Bigdely-Shamlo N, Ferris DP, Makeig S (2010) Visual evoked responses during standing and walking. Front Hum Neurosci 4:202. https://doi.org/10.3389/fnhum.2010.00202

Gramann K, Gwin JT, Ferris DP, Oie K, Jung T-P, Lin C-T, Liao L-D, Makeig S (2011) Cognition in action: imaging brain/body dynamics in mobile humans. Rev Neurosci 22:593–608. https://doi.org/10.1515/RNS.2011.047

Gramann K, Jung T-P, Ferris DP, Lin C-T, Makeig S (2014) Toward a new cognitive neuroscience: modeling natural brain dynamics. Front Hum Neurosci 8:444. https://doi.org/10.3389/fnhum.2014.00444

Gramann K, Hohlefeld FU, Gehrke L, Klug M (2018) Heading computation in the human retrosplenial complex during full-body rotation. biorxiv 417972. https://doi.org/10.1101/417972

Gwin JT, Gramann K, Makeig S, Ferris DP (2010) Removal of movement artifact from high-density EEG recorded during walking and running. J Neurophysiol 103:3526–3534. https://doi.org/10.1152/jn.00105.2010

Gwin JT, Gramann K, Makeig S, Ferris DP (2011) Electrocortical activity is coupled to gait cycle phase during treadmill walking. Neuroimage 54:1289–1296. https://doi.org/10.1016/J.NEUROIMAGE.2010.08.066

Hohwy J (2013) The predictive mind. Oxford University Press. https://doi.org/10.1093/acprof:oso/9780199682737.001.0001

Holl S, Pallasmaa J, Pérez-Gómez A (2006) Questions of perception: phenomenology of architecture. William Stout Publishers, San Fransisco

Johnson-Laird PN, Philip N (1988) The computer and the mind: an introduction to cognitive science. Harvard University Press

Jungnickel E, Gramann K (2016) Mobile brain/body imaging (MoBI) of physical interaction with dynamically moving objects. Front Hum Neurosci 10:306. https://doi.org/10.3389/fnhum.2016.00306

Jungnickel E, Gramann K (2018) MoBI—mobile brain body imaging. In: Ayaz H, Dehais F (eds) Neuroergonomics: the brain at work and in everyday life. Academic Press, p 374

Kawabata H, Zeki S (2004) Neural correlates of beauty. J Neurophysiol 91:1699–1705

Keele SW (1968) Movement control in skilled motor performance. Psychol Bull 70:387–403. https://doi.org/10.1037/h0026739

Kiebel SJ, Daunizeau J, Friston KJ (2008) A hierarchy of time-scales and the brain. PLoS Comput Biol 4:e1000209. https://doi.org/10.1371/journal.pcbi.1000209

Kothe C (2014) LabStreamingLayer. https://github.com/sccn/labstreaminglayer

Libet B, Gleason CA, Wright EW, Pearl DK (1983) Time of conscious intention to act in relation to onset of cerebral activity (Readiness-potential). Brain 106:623–642. https://doi.org/10.1093/brain/106.3.623

Luck SJ, Kappenman ES (2011) Oxford handbook of event-related potential components. Oxford University Press, New York, USA

Ma Q, Hu L, Wang X (2015) Emotion and novelty processing in an implicit aesthetic experience of architectures: evidence from an event-related potential study. NeuroReport 26:279–284. https://doi.org/10.1097/WNR.0000000000000344

Macpherson F (2014) Is the sense-data theory a representationalist theory? Ratio 27:369–392. https://doi.org/10.1111/rati.12085

Makeig S, Gramann K, Jung T-P, Sejnowski TJ, Poizner H (2009) Linking brain, mind and behavior. Int J Psychophysiol 73:95–100. https://doi.org/10.1016/J.IJPSYCHO.2008.11.008

282   Z. Djebbara et al.

Marr D (1982) Vision: a computational investigation into the human representation and processing
of visual information. W.H. Freeman
Merleau-Ponty M, Edie JM (1964) The primacy of perception: and other essays on phenomenolog-
ical psychology, the philosophy of art, history, and politics. Northwestern University Press
Merleau-Ponty M, Lingis A, Lefort C (1968) The visible and the invisible: followed by working
notes, 1st edn. Northwestern University Press
Miller GA (2003) The cognitive revolution: a historical perspective. TRENDS Cogn Sci 7:141–144.
https://doi.org/10.1016/S1364-6613(03)00029-9
Nagel T (1974) Philosophical review what is it like to be a bat? Philos Rev (Duke University Press)
Noë A (2004) Action in perception. MIT Press
Norberg-Schulz C (1997) Nightlands. MIT Press, Cambridge, Mass.; London
Palmer JA, Kreutz-Delgado K, Makeig S (2011) AMICA: an adaptive mixture of independent
component analyzers with shared components
Pezzulo G, Cisek P (2016) Navigating the affordance landscape: feedback control as a process
model of behavior and cognition. Trends Cogn Sci 20:414–424. https://doi.org/10.1016/j.tics.
2016.03.013
Rasmussen SE (1959) Experiencing architecture. Chapman & Hall, London
Rasmussen SE (2012) Om at opleve arkitektur. Linde Tryk, Aarhus
Rietveld E, Kiverstein J (2014) A rich landscape of affordances. Ecol Psychol 26:325–352. https://
doi.org/10.1080/10407413.2014.958035
Robinson S (2011) Nesting: body, dwelling, mind. William Stout Publishers, US
Singh AK, Chen H-T, Cheng Y-F, King J-T, Ko L-W, Gramann K, Lin C-T (2018) Visual appearance
modulates prediction error in virtual reality. IEEE Access 6:24617–24624. https://doi.org/10.
1109/ACCESS.2018.2832089
Thagard P (2009) Why cognitive science needs philosophy and vice versa. Top Cogn Sci 1:237–254.
https://doi.org/10.1111/j.1756-8765.2009.01016.x
Thompson E (2007) Mind in life: biology, phenomenology, and the sciences of mind. Belknap Press
of Harvard University Press
Töllner T, Wang Y, Makeig S, Müller HJ, Jung T-P, Gramann K (2017) Two independent frontal midline theta oscillations during conflict detection and adaptation in a Simon-type manual reaching
task. J Neurosci 37:2504–2515. https://doi.org/10.1523/JNEUROSCI.1752-16.2017
van Boxtel GJM, Böcker KBE (2004) Cortical measures of anticipation. J Psychophysiol 18:61–76.
https://doi.org/10.1027/0269-8803.18.23.61
Varela FJ, Thompson E, Rosch E (1991) The embodied mind: cognitive science and human experience
Vartanian O, Navarrete G, Chatterjee A, Fich LB, Leder H, Modroño C, Nadal M, Rostrup N, Skov
M (2013) Impact of contour on aesthetic judgments and approach-avoidance decisions in architecture. Proc Natl Acad Sci USA 110:10446–10453. https://doi.org/10.1073/pnas.1301227110
Vartanian O, Navarrete G, Chatterjee A, Fich LB, Gonzalez-Mora JL, Leder H, Modroño C, Nadal
M, Rostrup N, Skov M (2015) Architectural design and the brain: effects of ceiling height and
perceived enclosure on beauty judgments and approach-avoidance decisions. J Environ Psychol
41:10–18. https://doi.org/10.1016/j.jenvp.2014.11.006
Vesely D (2004) Architecture in the age of divided representation: the question of creativity in the
shadow of production. MIT Press
Vitruvius, Morgan MH, Morris H (1960) Vitruvius: the ten books on architecture. Dover Publications
Walter WG, Cooper R, Crow HJ, McCallum WC, Warren WJ, Aldridge VJ, van Leeuwen WS,
Kamp A (1967) Contingent negative variation and evoked responses recorded by radio-telemetry
in free-ranging subjects. Electroencephalogr Clin Neurophysiol 23:197–206. https://doi.org/10.
1016/0013-4694(67)90116-2
Zeki S (1998) Art and the brain. Dædalus 127:71–103
Zeki S (2002) Trying to make sense of art. Nature 418:918–919. https://doi.org/10.1038/418918a
Zumthor P (2006) Atmospheres. Birkhäuser, Basel

Chapter 10
Your Brain on Art: A New Paradigm to Study Artistic Creativity Based on the 'Exquisite Corpse' Using Mobile Brain-Body Imaging

Jesus G. Cruz-Garza⬛, Girija Chatufale, Dario Robleto and Jose L. Contreras-Vidal⬛

Abstract We propose a novel experimental paradigm to investigate the human creative process in artistic expression using mobile brain-body imaging (MoBI) technology, which allows the study of brain dynamics in freely behaving individuals performing in natural settings that promote authentic artistic experiences. Our proposed multimodal experimental protocol is based on the '*Exquisite Corpse*'—a collaborative, chance-based game created by the Surrealists in the 1920s. In this protocol, three artists collaborate to create the start, middle, and end of an improvisational piece of artwork, which can be implemented across artistic domains, including the visual arts, dance, music, creative writing, acting and even gastronomic art. Performers are instrumented with wireless scalp electroencephalography (EEG) to record brain activity and inertial measurement units (IMUs) to capture body movement, while video cameras capture the evolving gestures of the participants and the art pieces. Sample adaptive denoising algorithms, computer vision, visualization, sonification and machine learning methods allow for the pre-processing, tagging, parsing, storing, aggregating, analyzing, and sharing of complex containerized multimodal data. These MoBI data and associated behavioral, cultural, demographic, and situational data collected under the Exquisite Corpse paradigm holds the promise of a better understanding of functional (affective, cognitive and motor) and dynamic brain processes, the study of the neuroscience of individuality and group behavior, and the

J. G. Cruz-Garza (✉) · D. Robleto · J. L. Contreras-Vidal
Laboratory for Non-Invasive Brain-Machine Interface Systems, NSF IUCRC BRAIN, University of Houston, Houston, USA
e-mail: jgcruz@uh.edu

D. Robleto
e-mail: robletodario@gmail.com

J. L. Contreras-Vidal
e-mail: jlcontreras-vidal@uh.edu

G. Chatufale
Department of Psychology, University of California Los Angeles, Los Angeles, USA
e-mail: gchatufale@g.ucla.edu

© Springer Nature Switzerland AG 2019
A. Nijholt (ed.), *Brain Art*,
https://doi.org/10.1007/978-3-030-14323-7_10

design of robust affective and artistic brain-computer interfaces (BCI) and other diagnostic and therapeutical devices.

Keywords Creativity · EEG · MoBI · Creative process · Neuroscience · Neural interfaces · BCI

10.1 Introduction

The nature of the human creative process, both in the production and contemplation of art has been extensively debated among philosophers, historians, anthropologists, artists, and more recently neuroscientists. The inclusion of the latter has been not without controversy and skepticism from established schools of thought (Noë 2011; Holt 2013), but nevertheless, neuroscience studies have provided alternative and often competing approaches and tools for understanding the neural underpinnings of the human creative process with empirical neuroscience data and methods. More recently, computational neuroscience and advanced mobile brain-body imaging (MoBI) technology to record the brain and the body "in action and in context" have allowed researchers to study the dynamic brain of freely behaving individuals in complex natural and creative settings (Kontson et al. 2015; Contreras-Vidal et al. 2017b). The underlying framework is that by engaging in meaningful collaborations at the nexus of the arts, science and engineering, emergent bottom-up (data-driven) and top-down (e.g., from first principles) analyses, complemented by input from artists and philosophers, can lead to reconciliation of high-level personal perspectives, and a balanced body of fundamental knowledge from which to build models and hypotheses for further study.

The development of MoBI technology, typically comprised of mobile scalp electroencephalography (EEG) and/or functional near infrared spectroscopy (fNIRS) and motion sensors in its simple technical instantiation, has made it possible to study directly human brain activity (or indirectly via measurement of blood oxygenation profiles from the surface of the scalp with fNIRS) in unconstrained and freely behaving individuals acting in real world settings (Cruz-Garza et al. 2017a, b). MoBI experiments require the integration of synchronized mobile bio-sensor technology for brain and body data collection, and context monitoring devices such as video and event tagging.

Along with the capability of studying freely behaving participants in complex settings over short or long periods of time, MoBI technology provides the means to study brain responses in a wide range of subject populations encompassing healthy participants, people with a history of neurological disease, children, older adults, and it allows for the participation of spontaneous volunteers in public spaces (Kontson et al. 2015; Cruz-Garza et al. 2017a, b; Herrera-Arcos et al. 2017). IMUs on the headset itself and on the participants' bodies enable for acceleration, magnetometer, and gyroscope data to be collected to understand both how users move through and navigate space, and to help identify potential motion induced artifacts on the

EEG signals (Kline et al. 2015). Additionally, electrodermal activity, electromyography, electrooculography (EOG), heart rate monitoring, and virtually any biosensor that can be synchronized to the brain-monitoring device enables the measurement of embodied physiological contextual components of behavior. Finally, video cameras, motion tracking sensors, machine vision, and human annotators provide the context-awareness mechanism that enables the systematic study of neural and body dynamics in complex natural settings. As such, the sensors in concert with computer vision algorithms provide valuable contextual information for labeling the brain-body data according to environmental cues, movement type, or tasks to mention a few possibilities.

10.1.1 Chapter Organization

This chapter provides an overview of neuroscience research in the human creative process and recent developments in MoBI data collection that allow for its study in freely moving, real world settings. First, we highlight neuroimaging studies that provide evidence for the human creative process as emerging from the interaction of affective, cognitive, and movement-related processes, and brain areas associated to them. Second, we propose an integrative experimental protocol that allows the study of the production of an artistic composition implemented across artistic domains, where the artists create in a freely moving environment. Third, we provide an example of a data analysis technique to extract important features in an artist's individual creative process. Then we discuss how such an experimental protocol addresses the question of authenticity in the study of creative production. Finally, we consider how neuroscience knowledge gathered in authentic creative experiences can enhance artistic BCIs.

10.2 In Search for a Universal Model of the Human Creative Process

10.2.1 From the Mystical to the Neural

Initially regarded as the product of a "mystical" mental state, or of an unexplainable "divine intervention," creativity during and before the early nineteenth century was largely understood as a spiritual process—one that was untouchable by the grasp of scientific reasoning or study, and only experienced by those who were able to use an otherworldly introspection to create product from inspiration (Sternberg and Lubart 1999). From viewing creativity as an inaccessible, ethereal state, the early twentieth century paved way to understanding creative thinking by means of a theoretical lens—a movement that heavily relied on a psychodynamic approach of study.

This approach was not only headed by Freud, who popularized the psychoanalytic theory and pointed to the importance of the emergence of unmodulated thoughts in consciousness, but also highlighted the idea that creative thought arose from the tension between reality and unconscious motivations. While this approach could be regarded as successful in pulling creativity out of its mystical background and into a more scientific realm, this method of study relied largely on tightly-controlled laboratory settings keeping this progress in creativity research somewhat isolated. Consequently, some of the first truly objective, measurable, and widely-applicable research on creativity was incited by the 1950 American Psychological Association (APA) Presidential address (Guilford 1950) delivered by J. P. Guilford, who not only emphasized the prevalence of creativity in "everyday subjects" and proposed that this phenomenon could be studied through simple paper-and-pencil tasks, but also propagated the distinction between convergent and divergent streams of thinking (Sternberg and Lubart 1999; Kozbelt et al. 2010). Utilizing methodologies such as the *Unusual Uses* or *Alternate Uses* tests (i.e., *how many uses are there for a brick?*), Guilford jumpstarted creativity research, proposing ways in which individuals' creative abilities could be measured and placed on a standard scale. This approach, however, was only meant as a starting point for the field. While some of these psychometric measures are still being used within creativity research today, and allow for everyday individuals' creative abilities to be measured, researchers have continued to question its application to real-world settings (Contreras-Vidal et al. 2017a, b).

Research on human creativity today draws not only from an acknowledgement of creativity as a deeply personal, introspective process but also as one experienced by all. Further, progress in research concerning the human creative process is evidenced in the increasingly creative methods researchers are relying on to study its origin by going beyond simple paper-and-pencil tasks or measures, and instead focusing on more context-relevant settings. For example, studies on creative performance have been conducted in dance through MoBI technology; while functional magnetic resonance imaging (fMRI) has been deployed to investigate creative writing in poetry composition and revision (Liu et al. 2015), action planning while imitating chord progressions comparing classical and jazz-trained pianists (Bianco et al. 2018), musical improvisation using pitch sets or cue words in pianists (Pinho et al. 2015), or semi-professional visual artists sketching drawing ideas for a book cover based on sets of descriptions (Ellamil et al. 2012).

Taken together, these studies make an important suggestion: creativity is likely to emerge from the interaction of multiple affective, cognitive and movement processes, and therefore the study of creativity should not be reduced to one measure or task. These studies, as reviewed in Sect. 10.2.2, are consistent with a model proposed in Liu et al. (2015), showing inhibition of the dorsolateral prefrontal cortex (DLPFC) in the production of the creative product, increased cooperation between the DLPFC and ventromedial prefrontal cortex (VMPFC) during revision and evaluation of the work, and increased coupling between these two regions during the planning component of the activity.

10.2.2   Neuroscience of Creativity

We postulate that creativity lies in an individual's ability to produce a composition, object, artifact, sensory experience, act or thought that is novel, timely, with reward-eliciting attributes (valued), and relevant within a socio-cultural context.

Although the exact neuroanatomical network that underlies creativity still remains unknown, recent neuroimaging studies have consistently implicated the prefrontal cortex (PFC) as an essential, fundamental structure involved in creative cognition, e.g., expressive movement execution and imagery as well as in many cognitive abilities such as processing complex information, abstract thinking, conceptual expansion and cognitive flexibility (Beaty et al. 2016). Thus, research suggests fundamental cognitive functions (integrating highly processed information, abstract thinking, cognitive flexibility, etc.) of the prefrontal cortex as central in forming the foundation for original thoughts from which a moment of creative insight can emerge. Further, these prefrontal functions can be understood as originating mainly from two regions within the prefrontal cortex: the VMPFC and DLPFC (Liu et al. 2015).

The VMPFC and DLPFC each represent one of two broader neural systems within the brain—the emotional (i.e., instinctive, visceral) system, and the computational (or cognitive) system, respectively. More specifically, the VMPFC, or the emotional system, is thought to draw from life events and assesses the emotional, personal content contained within them (Dietrich 2004; Junghofer et al. 2017; Winker et al. 2018). This emotional system attaches value to an experience by evaluating its relevance to an individual's life experience, memories, and training. This follows from the finding that the VMPFC is strongly connected to the limbic system, which regulates important functions such as emotion, motivation, the internalization of values/rewards, and the evaluation of the consequences of one's actions (Motzkin et al. 2015). Moreover, research has shown that the DLPFC, or the computational system, receives sensory input from the TOP (temporal, occipital, and parietal lobes) as well as is involved in working memory and, consequently, cognitive flexibility—thought to be important components of the creative process.

Working memory not only produces temporary representations of the immediate, real-time events occurring around an individual, but also creates a buffer, which allows one to momentarily hold these representations, integrate incoming and past knowledge and stimuli that is relevant to solve a particular problem, and manipulate those stimuli to generate creative work. A review and meta-analysis performed by (Brunoni and Vanderhasselt 2014) examined the effects of two non-invasive brain stimulation techniques: repetitive transcranial magnetic stimulation (rTMS) and transcranial direct current stimulation (tDCS) on the DLPFC as well as working memory performance, specifically through an n-back task[1]—a widely-used measure of working memory. Stimulation of the DLPFC resulted in faster and more accurate

---

[1] The n-back task is a common measure of working memory capacity. In order to complete this task, subjects are presented with a series of stimuli (such as numbers or letters), and are asked to identify when a given stimulus corresponds to one seen n number of steps earlier.

responses on this n-back task, suggesting that the DLPFC is heavily connected to working memory.

In addition to the significance of the prefrontal cortex for creative thinking, studies have also implicated the parietal lobe as heavily connected to creative activity—both spatially and emotionally. Overall, parietal regions have been recognized as significant for body-environment interactions (specifically for "visual exploration," motor use of the hands, and tool use). Recent research also supports the importance of the parietal region in higher-order processes such as multisensory and sensorimotor assimilation, spatial orientation, motivation and intention, and the representation of the external environment's relationship to the body (Fogassi et al. 2005; Rathelot et al. 2017). Further, research has also cited the contributions of the parietal lobe as extending to cognitive functions such as episodic memory retrieval—consciously accessible memory for specific events that allow humans to retrieve past experiences and employ them for future goals. A literature survey performed by Wagner et al. (2005) revealed that fMRI as well as EEG studies on episodic retrieval have highlighted significant activity in the temporal and lateral posterior parietal cortex. These tools, including visual exploration, motor capabilities, tool use, spatial orientation, motivation, and memory retrieval, amongst others are central to the creative process of generating art.

10.2.3   Uncovering a Neural Signature for Creativity

Within the highly interconnected functional brain networks, and based in the consistent findings summarized in Sect. 10.2.1, we hypothesize that there is a cortical neural signature that emerges in the brain during aesthetic experiences, both during production and contemplation of a work of art. To study this potential electrophysiological neural signature, we propose an innovative experimental protocol to study the human creative process in authentic experiences.

The investigation of this hypothesis has the potential to provide a unifying view that informs traditional art theory and art practice. The neural signature associated with creative output would be likely expressed in distinct, distributed, and temporally evolving cortical activation patterns that can be measured with MoBI technology and characterized with functional connectivity and neural decoding analyses (see, for example, Kontson et al. 2015). We also expect that such brain patterns tagged to creative output may show neural individuality and variance across participants and art forms modulated by situational context, skill level, demographics and other factors yet unknown.

Uncovering a neural signature for creativity would likely lead to new metrics or biomarkers associated with the creative process, which could guide potential interventions for acquiring and tracking the development of new creative skills, and evaluating art therapies (King 2016). Critically, such a model ought to integrate links to existing art theory, art practice, and art therapy. From the detailed understanding

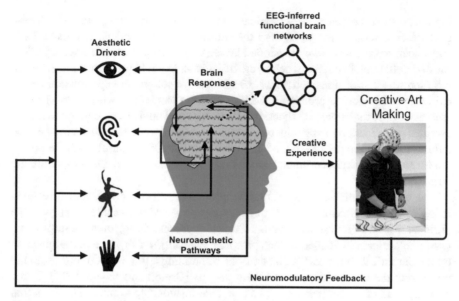

Fig. 10.1 MoBI technology enables the study of the human creative process in freely behaving individuals performing in complex, natural, and authentic settings allowing us to (a) understand the brain in action and in context, (b) study the neuromodulatory effects of the arts on brain activity and (c) develop robust artistic BCIs

of the neural mechanisms of human creative expression, we can develop BCIs for artistic or therapeutic purposes that interact adequately with the user input (Fig. 10.1).

10.3 The Exquisite Corpse as an Experimental Protocol to Study Creativity in Action and in Context

We propose a transdisciplinary and multimodal experimental approach to study the human creative process using MoBI technology. This approach is based on four principles set forth for an effective transdisciplinary collaboration. First, transdisciplinarity between fields requires the convergence and synthesis of different research methods. This convergent research requires equal input from scientists and artists on experimental design to the interpretation and applicability of the data. In this case, bridging a data-driven bottom-up approach with top-down analysis from the artist's perspective and first principles will be crucial to investigate the creative process. Second, we considered an experimental protocol that would allow for the inquiry into common and unique neural patterns of brain activity across artistic domains and individuals. We therefore need an experimental protocol that can be implemented across different creative categories (e.g. visual, dance, writing, etc.), people of different skill level (e.g. novices, experts, children, adults), and demographic factors including age, gender, language, geographical location, etc. Third, to create an authentic creative experience—and to explore the meaning of "authenticity" across artistic and scientific domains—we envisioned an experimental protocol that would allow for data collection from freely behaving individuals in a real world setting. A fourth criterion was that of practicality and scalability. We sought an experimental protocol that would allow to produce a work of creative expression within a reasonable amount of time that would be accessible to experts and novices, be enjoyable for the eventual participation of spontaneous in situ participant volunteers from the general public (e.g. children patrons at the Children's Museum of Houston), with the potential for scalability, and a common framework from which to extend into other artistic domains.

The effort to define a protocol that fit into the criteria described above resulted in the recontextualization of the *Exquisite Corpse* as a MoBI-enabled neuroscience protocol from which to study the human creative process during creative improvisation. The protocol is defined in the spirit of the *Exquisite Corpse,* a game invented by the Surrealists in the 1920s that consists of building a three-part improvisational piece from the contributions of different players (Breton and Yaylor 1972). In the growing field of neuroaesthetics, it has become fashionable to make the claim that artists were our first neuroscientists. Studying painters of the past, for example, offers insight into how artists illuminated brain structure and the mechanisms of perception through inventive techniques of luminosity, rendering of shadows, and an understanding of the visual illusions our brain plays on perception (Grossberg and Zajac 2017). Less explored is an analogous argument: the rich tradition of artist's inventive performances, games, "actions," or "prompts" holds similar insights for the brain sciences today. By adapting the *Exquisite Corpse*, which incorporates improvisation, collaboration, and novel problem solving as experiment design, we can merge the long tradition of the arts exploring the inner workings of the mind with a replicable scientific protocol.

10.3.1 History of the Exquisite Corpse

First gaining popularity in the 1920s, *Cadavre Exquis*, or *Exquisite Corpse*, was originally conceived as a word-based parlor game relying on collaboration, chance, and unexpected juxtaposition. The game typically involved three to four players who would each secretly write a word or phrase on a shared piece of paper, then fold and pass the sheet to the next player. When opened to reveal all sections, this process often produced nonsensical phrases like "*Le cadavre exquis boira le vin nouveau*" ("*The exquisite corpse will drink the new wine*"), wherein the game obtained its name. The game was soon expanded to visual imagery through drawing and collage, where the players would attempt to create a "body" consisting of head and shoulders, torso and arms, legs and feet. In this version, players are allowed to see the edge of the previous composition to begin their own. Other art forms such as dance, music, and poetry have also adapted the game for their respective genres.

Around 1925, members of the artistic movement known as Surrealism began to explore the game's possibilities within the arts. Seeking ways to break freely of what they considered the limitations of the rational mind, and rejecting the 19th c. approach to purely representational and observational painting, the Surrealists were deeply invested in exploring ways to disrupt the conscious mind's need for order. They were drawn to the elements of chance, randomness, and unpredictability that the game produced and believed that this revealed a more authentic view into the creative subconscious mind. As the founder of the Surrealist movement, André Breton, stated, "With the Exquisite Corpse we had at our disposal—at last—an infallible means of sending the mind's critical mechanism away on vacation and fully releasing its metaphorical potentialities." (Breton and Yaylor 1972).

10.4 Recording MoBI Data in the Exquisite Corpse Protocol

The human creative process is a multi-dimensional and multi-stage process that does not happen in isolation; rather, it is fueled by environmental stimuli (Slepian and Ambady 2012; Kandler et al. 2016). The protocols outlined below attempt to capture the creative production process as it happens in freely behaving participants, involving elements of social interaction and environmental and other contextual factors occurring in a real-life scenario.

10.4.1 Instrumentation

In this protocol, brain activity is typically collected with 64 active-electrode wireless EEG sampled at 1000 Hz (e.g., BrainAmpDC with actiCAP, Brain Products GmbH; see (Cruz-Garza et al. 2017a) for examples of MoBI headsets); eventually down-sampled to 200 Hz. Four electrodes are used for EOG recordings. IMUs are used to track head and body motion data from the artists that capture the creative gestures of the performers, while providing useful information for identifying potential motion artifacts. Typically, for the visual artists, musicians and writers, data are collected from the head and forearms. In the case of the dancers, six IMUs are placed on the head, both wrists, torso, and both ankles of the dancers. Video cameras capture the creation of each work of art and the group dynamics. After the experiment, the artists are asked to annotate the video recordings to mark significant behavioral and cognitive events they recall. Annotators during the performance also provide event tagging, which is complemented by regions of interest identified from other sensor data (e.g., arousal from electrodermal activity). An example of a typical experimental setup with sample EEG, acceleration, and video data is shown in Fig. 10.2.

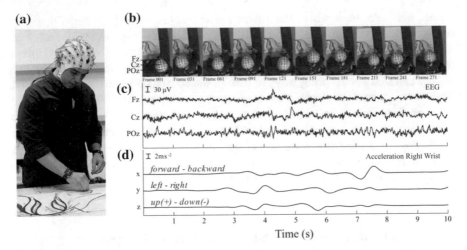

Fig. 10.2 Recording of behavioral, neural, and motion data from an artist constructing an improvisational artistic piece in an unconstrained environment. **a** Data recording setup: EEG, IMUs, video recording. **b** Sample frames from the video recording of the artist. **c** EEG traces shown for three spatially representative electrodes. **d** Acceleration data from the IMU located at the right wrist

10.4.2 The Exquisite Corpse as an Experimental Protocol

The Exquisite Corpse protocol includes baseline and experimental conditions, with the baseline conditions, with the Baseline conditions introduced before and after the experimental session and consisting of closing eyes for at least 60 s, and looking at a blank sheet of paper for at least 60 s. The experimental conditions are detailed below.

10.4.2.1 Visual Arts

In the visual arts modality of the Exquisite Corpse, three artists typically work on a "body" consisting of three sections: head, torso, and tail/legs. The artists are provided with a foldable triboard (32″ × 40″ four-ply chipboard), a 2-layered panel comprised of three sections that can be folded or 'blinded'. At the end of each section, the staff covered the art piece with a strip of cardboard, leaving approximately 3 cm uncovered at the bottom, and then transported the piece for the next artist to view before beginning the next stage. The artists worked on the three art pieces simultaneously, on three different triboards. The artists are separated from each other by opaque curtains to prevent interactions during the experiment.

The artists were asked to provide or identify basic art materials such as pencils, pastels, chalk, charcoal, water-based painting materials, glue, and scissors for use during the performance. Artists are also requested to bring "surprise" materials for one another as a way to bring an element of surprise as well as personalize—and construct meaning through—the process. Examples of materials brought by the artists include insects, stickers, ink, film, stencils, and printed color paper.

Figure 10.3 shows the experimental setup and timed protocol. The artists (labeled S1, S2, S3) work on separate boards (A, B, C) on the head (Section 1) of the figure for 15 min. The boards are rotated, and the artists continue to work on the body for 15 min (Section 2), and subsequently the tail/legs for the last 15 min (Section 3). Versions of this protocol for children typically limit the duration for each session to 5 min given time limitations and attention span of the children (Fig. 10.4).

10.4.2.2   Creative Writing

In this instantiation of the Exquisite Corpse, three creative writers work simultaneously on three compositions (which can include poetry and/or prose). The writers start by writing on a blank notebook, and for each consecutive session, they continued from where their collaborators finished their writing at the end of each session. The writers are able to see the last two lines of the previous text. The sections are 15 min long with 1-min vocal warnings before the end of each. Three Exquisite Corpse texts (A, B, C) are produced at the end of the 45 min experiment (Fig. 10.5).

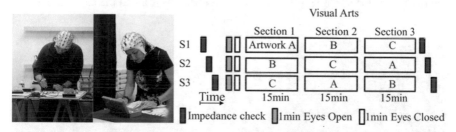

Fig. 10.3 (Left) Three artists (two pictured) worked simultaneously creating the head, torso, and tail/legs of a figure in the spirit of the Exquisite Corpse. (Right) Experimental protocol designed for data collection on the improvisational creative process in visual artists. The subject ID is denoted by S1, S2, S3 respectively

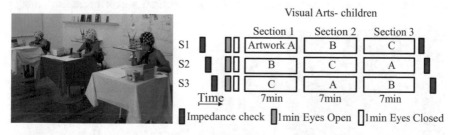

Fig. 10.4 Three children at a time participated in the Exquisite Corpse protocol, with a 32-electrode EEG headset. The children followed a similar procedure as the visual artists

10.4.2.3 Music

In the musical adaptation of the Exquisite Corpse, three musicians work on three improvisational jazz pieces, each divided into three sections. In the first section, one musician plays while the others listen. For the second section, a second musician joins in for a duet, while the other listen. The last musician joins the others for the last section. Each of the sections was 5 min long. A timer is placed visible to the musicians so that the subsequent musician joined at the 5 min mark. The process is repeated three times, rotating the order for the musicians. The musicians' performance sequence is represented in Fig. 10.6.

10.4.2.4 Dance

The dance adaption of the Exquisite Corpse involves three dancers separated by curtains so that they could not see each other during their performance. In the first section, the dancers performed in silence, dancing with external cues or music. The second section of the Exquisite Corpse features a 144 bpm *Alegría* (with cajón and palmas) flamenco metronome (Fitzgerald 2016). The third section features an instrumental musical piece: *Raff's Ode au printemps in G major Op. 76 200*. The songs were edited to the length of the section (10 min) prior to the experiment.

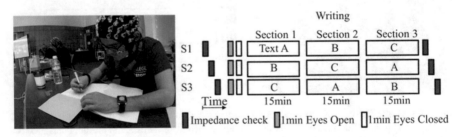

Fig. 10.5 (Left) Three artists worked simultaneously in parallel, creating the beginning, middle, and end or a creative writing piece in the spirit of the Exquisite Corpse. (Right) Experimental protocol designed for data collection on the improvisational creative process for creative writing

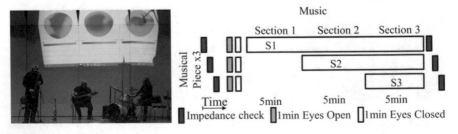

Fig. 10.6 (Left) Three musicians participate in the study, playing a five-piece drum-set, a saxophone X/Y, and a saxophone W/Z. (Right) Experimental protocol for improvisational music performance

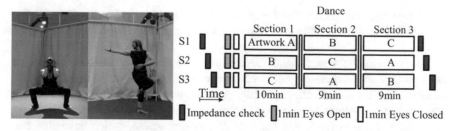

Fig. 10.7 (Left) Three dancers (two pictured) participated in the study. (Right) Experimental protocol for the dance version of the Exquisite Corpse

Each dancer performed improvisational movement for 10 min in isolated stages. The first section was followed by a 1 min collaborative performance where they stepped into view of each other and shared movements among them. They then returned to their isolated stages for 9 min, and repeated this procedure for the third section of the experiment. Figure 10.7 summarizes the protocol followed. The sections were 10 min long with a 1-min vocal warning.

10.5   MoBI Data Analysis Through Machine Learning

We present, as an example of an analytic methodology, the data processing for one participant in the Exquisite Corpse for visual artists. The machine learning methodology proposed requires label actions from the artists, with labels relevant to the artistic modality, and a classification approach with automatic feature extraction and visualization.

Data driven neuroscience studies have found great success in applying supervised and unsupervised machine learning techniques to find relationships between the data collected and a behavioral response observed. Classical machine learning requires the researcher to identify, obtain, and select features of the data to analyze. In EEG, these features usually take the form of power in specific frequency bands or commonly used frequency bands: e.g. delta 1–4 Hz, theta 4–8 Hz, alpha 8–12 Hz, beta 12–30 Hz, gamma 30–50 Hz; or time domain features involving temporal and spatial relationships in the data. These features are used decode movement intent in mobile settings (Kilicarslan et al. 2013; Bulea et al. 2014; Cruz-Garza et al. 2014; Luu et al. 2017; Zhang et al. 2017). Coherence metrics, which measure the functional connectivity between electrodes, have also shown to be promising features for EEG analysis (Gaxiola-Tirado et al. 2018). Quantitative neuroscience based on EEG has developed through a combination of spectral, temporal, and spatial features, with which researchers are able to build a set of descriptors to feed into machine learning algorithms to learn about the data and to build models for intentionality prediction (Lotte et al. 2018). Classical machine learning techniques in neuroscience involve a combination of features selected by the researcher, based on previous neuroscience or a promising new metric. The features are tracked and averaged over hundreds of trials to find an overall pattern of brain activity that can be associated to a specific task.

In order to study the neural basis of a complex cognitive task such as the human creative process across demographics and artistic domains, we find that automatic feature extraction algorithms offer a promising new approach to find new data descriptors and predictors. Automatic feature selection algorithms have shown rapid progress in recent years, in particular in the field of machine vision, which have also been applied to EEG data (Schirrmeister et al. 2017). Promising automatic feature extraction algorithms include those based on deep neural network architectures such as convolutional neural networks (CNNs), long-short term memory networks, Boltzmann Machines, or a tactful combination of these.

Feature visualization remains a key aspect of automated feature extraction methods. Hypotheses and feature visualization techniques based on previous neuroscience (e.g. we expect alpha power changes in prefrontal cortex; is that what the computer finds?) help the researcher understand if the algorithm is learning useful and relevant information. Therefore, it is necessary to have a top-down, artist-informed framework from which to base the feature visualization methods and overall data analysis when using automatic feature extraction methods. Data mining techniques, however sophisticated, will fall blind to the task and rendered ineffective, if not outright counterproductive, to the field if they are not accompanied by appropriate feature visualization methods.

The proposed machine learning method described below requires labeled datasets. We annotated the data by having human annotators watch the video recording of the artists as they worked on their composition. Because the experiment is unconstrained by design, there are two critical aspects to consider in this approach: (1) what classes to label the artists' actions into, and (2) inter-annotator consistency. Relevant labels were discussed and analyzed with the professional artists that participated as subjects in our study through interviews.

10.5.1 Labeling Creative Tasks

The Exquisite Corpse protocol in the visual arts consisted of elements from drawing and collage. The video recordings were visually segmented by annotating the behaviors and tasks done by the artists, relevant to drawing and collage. A second person validated the annotations.

The MoBI data were segmented in terms of the artistic action each artist displayed: planning/observing, cutting, placing/pasting, correction, outlining, tracing, coloring, spreading, drawing, and writing. In addition, the baseline eyes open and baseline eyes closed were also segmented. In this example, four classes were selected for illustration purposes: baseline eyes closed, baseline eyes open, planning, and coloring.

10.5.2   *Automatic Feature Extraction and Classification*

In an unconstrained behavioral task, where artists work with elements of chance and improvisation to create a composition, we consider that a machine learning approach with automatic feature extraction would enable us to capture neural dynamics and processes that are hard to predict a priori (e.g. by having the researcher select what features to analyze).

CNNs have shown impressive results in the field of machine vision due to their capacity to learn local patterns in data through convolutions. With the proper architecture, CNNs can find important features of the data automatically, potentially opening the possibility for discovery of previously unknown relevant features. These networks are built by adding convolutional layers that map local patterns in the data. CNNs make good candidates for end-to-end decoding: from raw EEG data to a prediction about behavioral intent. However, they require a large number of hyper-parameters, so they also require a large amount of training data and representative variations in that data. They also take a long time to train compared to simpler models often used in neuroscience studies.

We used a CNN for automatic feature extraction and classification of the creative tasks. Figure 10.8 shows the CNN architecture selected for the study. Our parameters were selected based on the discussion in Schirrmeister et al. (2017), fine tuning them to our data. Deep learning approaches require a large amount of data to iterate over, in which by means of backpropagation, the weights of the computation units in each layer are updated such that the metric of interest (mean-squared error before the Softmax layer) is minimized. The EEG data was augmented by taking 1 s time windows with 99% overlap. The first temporal 80% of the data was used for training and validation, while the latest temporal 20% of the data per class was used as the test set. This partition enables the learned model to be tested in pseudo real-time data: the test set. To build the classification model, each of the four classes were set to contain 5000 samples using random sub-sampling without repetition for the training and validation sets. From the 20,000 samples, 13,000 were selected for the training set and 7000 for the validation set. The network ran 10 times to compute a distribution of the classification accuracies, with randomized selection of the samples to be used for the training and the validation sets. 4000 samples were selected for the test set, with 1000 samples per class.

To illustrate the performance of the CNN on our 4-classes problem, the CNN was tested on artist one (S1). The accuracy for the training and validation sets reached near 80–90% in both cases, with classification accuracy dropping to near 66.5% in the test set (Fig. 10.9). The classification accuracy in the CNN improved after utilizing the temporal properties of EEG: there is a higher probability that the classification for sample x is similar to the classification of the temporally adjacent sample. In this application, our tied-weights consisted of averaging the classification output (before Softmax) of the immediately previous 5 samples before running the network through the Softmax layer and finally selecting a class label for the sample.

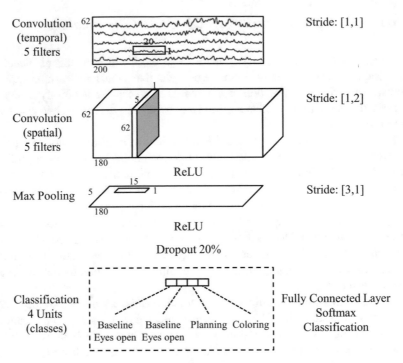

Fig. 10.8 CNN architecture proposed. The EEG inputs are windows of 62 channels by 200 time samples (1 s, at 200 Hz)

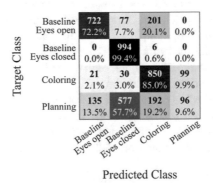

Fig. 10.9 Confusion matrix for EEG data classification of artist 1. Each row contains 1000 test set samples for each of the four classes: Baseline Eyes open, Baseline Eyes closed, Coloring, Planning. The whole numbers are number of samples classified into a particular class. The average classification accuracy is 66.5%

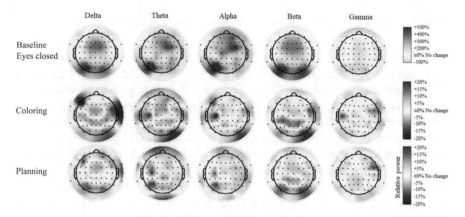

Fig. 10.10 EEG feature visualization, as learned by the CNN: spectral differences in the best examples from each class. The 200 highest-activation samples in their correct class were selected. The color maps show the percentage spectral power change with respect to 'baseline with eyes open'

10.5.3 Feature Visualization

In neuroscience, we are interested in understanding the neural features that contribute to the classification of tasks. In this unconstrained experimental setting with multiple and varied actions performed by the artists, these features may be a combination of several different cognitive processes acting together. Therefore, visualizing the features learned automatically is critical for understanding the performance of the classifier, and thereby the relevant feature spaces associated with the task.

A method used to identify the most relevant features for the network was to find the best examples (highest activation in the last layer before Softmax) for each class and compare the spectral differences between them. Figure 10.10 displays the results of the spectral power in the 200 best examples from each class: those which yielded the highest activation in the last layer before the Softmax for each class and therefore those which the network found to be most representative of each class. The spectral power in each class was compared to "Baseline Eyes open". In this visualization method, there is an increase in power in the occipital area expected for "Baseline Eyes closed" (Fig. 10.10). There is a decrease in power in the theta and alpha power in left central scalp areas for the "Coloring" and "Planning" tasks: the artist worked with their right hand. An increase in delta, theta, and alpha bands in left-parietal regions is found for the "Planning" task. Although these observations are for one subject at the sensor level and they not necessarily reflect the cortical sources of brain activity, the method shows promise for understanding the neural features and channel locations that the network found to be most relevant for classification.

10.5.4 Top-Down Analysis of the Creative Process

A top-down analysis, using insights from the experts in the creative compositions
—the artists themselves, was used to interpret the feature visualization and feature
relevance results. The corresponding interviews of the artists were conducted the day
after the experiment. The video recording of the experiment was shown to each artist
and their recollection of their process was recorded.

The feature visualization techniques showed importance of scalp areas over the
frontal and left motor regions during the execution task in the delta and alpha bands.
Parietal and frontal scalp areas were relevant in the planning tasks, in the theta and
beta frequency bands (Fig. 10.10). Artist one (S1) not only utilized many different
colored pastels, but also incorporated small film strips, pieces of paper, felt, and
carefully rolled strips of tape and stickers into the artwork (Fig. 10.11). Each of
these tasks—coloring, aligning strips of film, cutting and placing paper and felt,
and rolling and positioning tape—are largely spatially dependent as well as involve
careful planning and attention to detail, and thus, involve the parietal and frontal
areas.

With further source analysis, we hypothesize that we would find involvement of the
VMPFC in the artists. Research identifies the VMPFC as heavily connected with the
limbic system, which regulates emotion, instinct, motivation, and the internalization
of values, and these personal and meaningful emotions, reflections, and beliefs of the
artists are clearly manifested through their expressive and telling work. Both artists
two (Fig. 10.12) and three (Fig. 10.13) reported to have felt a "real connection to each
other and the space" around them, which they described, "allowed them to give into
someone else's sensibilities." Additionally, each of the two artists reflected on their
work and mentality during their moments of creation, citing that they each thought
more about themselves rather than the state of others. Artist three created a powerful
message— "How Can I Resist?"—that was central to her artwork and influenced
by thoughts she had earlier that day, reported to have felt a sense of "authenticity,
familiarity, and relief" while creating her work as well as remembered that she had
"less moments of reflection" during her creative process itself—intimating that the
process was more intuition-driven, an important feature of the VMPFC. Moreover,
these artists incorporated additional materials in their artwork, such as paper, dead
butterflies, plastic eyes, and tape, as well as utilized coloring, and placing paper,
amongst others, pointing to the parietal activity that was seen in the feature selection
data. This raw, unfiltered integration of external and internal stimuli present in the
works of each of the artists not only motivated the production of novel arrangements
of ideas, experiences, and sensory inputs, but also facilitated the transition of these
arrangements into a meaningful, creative work (Figs. 10.11, 10.12 and 10.13).

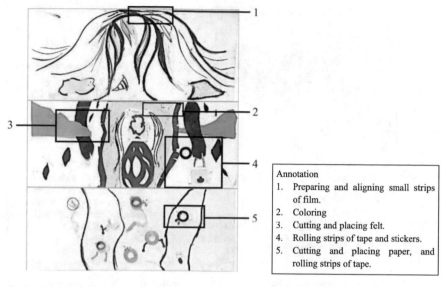

Annotation
1. Preparing and aligning small strips of film.
2. Coloring
3. Cutting and placing felt.
4. Rolling strips of tape and stickers.
5. Cutting and placing paper, and rolling strips of tape.

Fig. 10.11 Artwork created by artist one (S1), with annotations. *Inset* Examples of annotated tasks performed by the artists

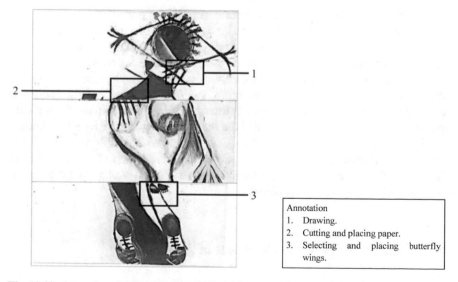

Annotation
1. Drawing.
2. Cutting and placing paper.
3. Selecting and placing butterfly wings.

Fig. 10.12 Artwork created by artist two (S2), with annotations

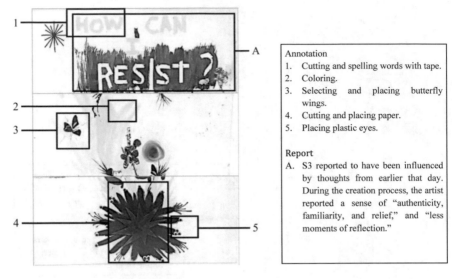

Annotation
1. Cutting and spelling words with tape.
2. Coloring.
3. Selecting and placing butterfly wings.
4. Cutting and placing paper.
5. Placing plastic eyes.

Report
A. S3 reported to have been influenced by thoughts from earlier that day. During the creation process, the artist reported a sense of "authenticity, familiarity, and relief," and "less moments of reflection."

Fig. 10.13 Artwork created by artist three (S3), with annotations and an example of a report provided by one of the artists

10.5.5 MoBI Data Analysis Across Artistic Modalities

A similar data analysis pipeline, as described in this section, can be applied for other artistic modalities. The data can be labeled from a discipline-specific annotation framework. For example, in dance, where research typically involves studying expressive movements, a labeling system based on Laban Movement Analysis provides the appropriate tags for the MoBI data; see, for example, Cruz-Garza et al. (2014). A CNN, with parameter fine-tuning, could be implemented for automatic feature extraction for the set of labels defined, and a similar feature visualization approach would be useful to understand the features being learned by the computer.

Classical machine learning approaches with predefined and well-known EEG features offer performance baseline and comparisons for the automatic feature extraction algorithms. See Lotte et al. (2018) for a review of machine learning algorithms often used for EEG-based BCI applications.

10.6 Discussion: On the Question of Authenticity

The burgeoning field of neuroaesthetics, of which our proposed protocol falls under, is an excellent example of the necessity of transdisciplinary problem-solving. A mystery as complex as human creativity cannot only be understood through a single approach and requires the synthesis of expertise from multiple disciplines. While creativity and aesthetic experience undoubtedly have a physical, neurological underpinning, this should not be misunderstood as an "explanation" of art, but rather a characterization of the creative process. *A rigorous neuroaesthetics needs to account for the lived, emotional and experiential aspects of art, as well as its ability to construct and represent values and meaning for the individual and society.* This is why neuroaesthetics represents a rare instance wherein the advancement of a scientific field hinges on meaningful interactions with the arts. This interaction should not only be with art of the past, but with living artists and contemporary institutions of art such as museums, galleries and artist's studios. The Exquisite Corpse experiment was designed to address this issue of collaboration with contemporary artists while producing valuable data for both the scientist and artist.

From the artists' point of view, not only is it an unusual experience to be in the role of test subject, but it offers a novel lens through which to reflect on their creative process. In our proposed approach to study creativity as detailed in the next sections, research required each artist to examine their own creative practice in order to better articulate processes and parse specific moments, while learning about how creativity is perceived within the parameters of neuroscience. The artists found that their self-reflections offered them a more nuanced understanding of their own creative process and how it was in tension with the scientific assumptions of it, either through working definitions of creativity and aesthetics, experiment design, expectations about the end results, or even the post-experiment evaluative process. This intersection of artistic reflection and neuroscientific discovery is of great importance as we build a common language with the hope of advancing each of our respective fields in unexpected ways.

Although many questions were provoked, a recurring theme appeared to anchor them, which can add valuable insight and inform the development of future experiments: What does *authenticity* mean in relation to creative processes, and how do we measure it? Like "aesthetics" or "creativity," the concept of "authenticity" from both the creator and observer's points of view, has a complex meaning that is usually understood as highly personal and subjective. But, in the context of these studies, there is an expectation between artist and scientist for a common definition and, perhaps most problematic for the artist, a quantifiable categorization of authentic creativity. The question of authenticity is particularly relevant in light of major advancements in MoBI technology: the ability to record real-time data from a diverse group of freely behaving individuals makes it possible to study creativity outside of highly controlled and artificial laboratory settings. The assumption is that a typical site of artist production, such as a studio or museum, will facilitate a more authentic experience and, hence, the resultant scientific reading will be more accurate than data gathered in a traditional laboratory setting.

But if the innovation of this technology partly hinges on more accurate, i.e. "authentic," recordings, then the artist's understanding of authenticity as it relates to creativity must be given an equal consideration in the experiment design and the evaluative process. Because even though we have moved this experimental procedure away from the laboratory setting, the situation presents a new set of highly artificial variables that could disrupt the artist's sense of an authentic experience. Additionally, breaking down the constituent physical and measurable aspects of the creative act (e.g. stroke, cut, pasting, coloring, drawing, planning, etc.) has been an enlightening process for both artist and scientist. In the process, preconceived definitions of creativity (at least on a process level) must be challenged from the viewpoint of each discipline. Through the continued development of language and systems with which to articulate and report the experiences recorded in collected data, we hope to contribute to this technology's potential therapeutic goals, as well as investigate the rich artistic and philosophical questions posed by neurological understandings of creativity and aesthetics.

10.7 Applications

Creativity is not only integral to the actions and decisions of many individuals throughout their lives, but has also served as the foundation for bringing about substantial change and advancement within a wide variety of fields, including those of education, politics, economics, science, medicine, technology, and art. The human quality of creative abstraction has been championed by politicians, leaders, and educators alike as the answer to many of a nation's pressing issues (Moran 2010); as a method of teaching as well as a quality to cultivate within the education system (Shaheen 2010) as a path to improving the products and services offered by corporations and institutions (Bobirca and Draghici 2011); and as a means to aid individuals on their journey of personal growth and healing (Belkofer et al. 2014). In the next two sections, we describe applications and potential impact of studies on the neural basis of creativity.

10.7.1 Creative Art Therapy for Neuro-rehabilitation

Creative art therapy allows an individual to articulate personal sensory experiences through the various visual and tactile properties of tools such as paints, pencils, stickers, charcoal, and stamps—for example—and the muscle pressure an individual must exert in order to manipulate these raw materials to form something meaningful (Lusebrink 2004; Sarid and Huss 2010). Further, it has been commonly found to be associated with numerous positive outcomes such as decreased stress (Martin et al. 2018), depression (Bar-Sela et al. 2007), fatigue (Bar-Sela et al. 2007), anxiety (Morris 2014), PTSD (Walker et al. 2016), improvements in behavioral functioning, mood (De Petrillo and Winner 2005), speech (Pachalska et al. 2001), self-image, self-esteem (Hartz and Thick 2005), communication, responsiveness, and sociability (Rusted et al. 2006), amongst others. As a result, art therapy has improved the quality of life of many individuals from various walks of life and backgrounds—including not only those inflicted by Alzheimer's and other forms of dementia (Wang and Li 2016), but also of those facing the daily stresses of life.

Fig. 10.14 An interactive artistic BCI that uses selective neural features to control the sculpture's position, color, and sound in "Self-conscious/Physical Memory", by Eric Todd. A dancer, Shu Kinouchi, interacts with the space in real time. Photo by Ronald L. Jones

These studies highlight the effectiveness and potential of art therapy and provoke further questions that can only be answered through the neuroscientific study of the human brain in artistic production in real world settings: How can medical professionals, therapists and neuroscientists collaborate more effectively with artists to personalize creative art therapies as a form of precision medicine? Empirical neuroscientific data from collected in mobile settings during the process of creating a work of art offers the possibility to create better, more effective, personalized therapeutic interventions. By analyzing the neural dynamics associated to the human creative process, art therapy methods can be personalized for optimal performance.

10.7.2   Artistic Brain-Computer Interfaces (BCIs)

Understanding the neural basis of creativity has the potential to develop artistic BCIs that can promote creativity in art making and also provide alternative ways of visualizing brain data. The chapter by Todd et al. (Chap. 11 of this book) is an example of how EEG activity can be used to represent and visualize multiple aspects of brain activity through motion, lights and sound (Fig. 10.14). Closed loop artistic BCIs can also be deployed as powerful neuromodulators of brain activity to augment the repertoire of the artist by allowing brain control of the environment or stage.

Acknowledgements The authors thank Jo Ann Fleischhauer and Lily Cox-Richard, artists in residence at the University of Houston, for assisting in creating and participating in the experimental design. We also thank James Rosengren, Claudia Schmuckli, and Colleen Maynard from the Blaffer Art Museum for facilitating the space for experiments and scientific outreach. This research is funded in part by NSF award #BCS1533691, NSF IUCRC BRAIN Award CNS1650536, a Seed Grant from the Cullen College of Engineering at the University of Houston, and the SeFAC grant from the Center for Advanced Computing and Data Science (CACDS) at the University of Houston.

References

Bar-Sela G, Atid L, Danos S et al (2007) Art therapy improved depression and influenced fatigue levels in cancer patients on chemotherapy. Psycho-. https://doi.org/10.1002/pon

Beaty RE, Benedek M, Silvia PJ, Schacter DL (2016) Creative cognition and brain network dynamics. Trends Cogn Sci

Belkofer CM, Van Hecke AV, Konopka LM (2014) Effects of drawing on alpha activity: a quantitative EEG study with implications for art therapy. Art Ther. https://doi.org/10.1080/07421656.2014.903821

Bianco R, Novembre G, Keller PE et al (2018) Musical genre-dependent behavioural and EEG signatures of action planning. A comparison between classical and jazz pianists. Neuroimage 169:383–394

Bobirca A, Draghici A (2011) Creativity and economic development. Int J Soc Behav Educ Econ Bus Ind Eng 5:887–892

Breton A, Yaylor SW (1972) Surrealism and painting, 1st U.S. edn. Harper & Row, New York

Brunoni AR, Vanderhasselt M-A (2014) Working memory improvement with non-invasive brain stimulation of the dorsolateral prefrontal cortex: a systematic review and meta-analysis. Brain Cogn 86:1–9

Bulea TC, Prasad S, Kilicarslan A, Contreras-Vidal JL (2014) Sitting and standing intention can be decoded from scalp EEG recorded prior to movement execution. Front Neurosci 8:376. https://doi.org/10.3389/fnins.2014.00376

Contreras-Vidal JL, Cruz-Garza J, Kopteva A (2017a) Towards a whole body brain-machine interface system for decoding expressive movement intent challenges and opportunities. In: 2017 5th international winter conference on brain-computer interface (BCI). IEEE, pp 1–4

Contreras-Vidal JL, Kever J, Robleto D, Rosengren J (2017b) At the crossroads of art and science: neuroaesthetics begins to come into its own. Leonardo

Cruz-Garza JG, Hernandez ZR, Nepaul S et al (2014) Neural decoding of expressive human movement from scalp electroencephalography (EEG). Front Hum Neurosci. https://doi.org/10.3389/fnhum.2014.00188

Cruz-Garza JG, Brantley JA, Nakagome S et al (2017a) Deployment of mobile EEG technology in an art museum setting: evaluation of signal quality and usability. Front Hum Neurosci 11. https://doi.org/10.3389/fnhum.2017.00527

Cruz-Garza JG, Brantley JA, Nakagome S et al (2017b) Mobile EEG recordings in an art museum setting. IEEE Dataport. https://doi.org/10.21227/h2tm00

De Petrillo L, Winner E (2005) Does art improve mood? A test of a key assumption underlying art therapy. Art Ther. https://doi.org/10.1080/07421656.2005.10129521

Dietrich A (2004) Neurocognitive mechanisms underlying the experience of flow. Conscious Cogn 13:746–761

Ellamil M, Dobson C, Beeman M, Christoff K (2012) Evaluative and generative modes of thought during the creative process. Neuroimage 59:1783–1794

Fitzgerald A (2016) Alegría. In: Ravenna Flamenco. https://ravennaflamenco.com/metronomes/. Accessed 1 Feb 2019

Fogassi L, Ferrari PF, Gesierich B et al (2005) Neuroscience: parietal lobe: from action organization to intention understanding. Science (80–). https://doi.org/10.1126/science.1106138

Gaxiola-Tirado JA, Salazar-Varas R, Gutierrez D (2018) Using the partial directed coherence to assess functional connectivity in electroencephalography data for brain-computer interfaces. IEEE Trans Cogn Dev Syst. https://doi.org/10.1109/tcds.2017.2777180

Grossberg S, Zajac L (2017) How humans consciously see paintings and paintings illuminate how humans see. Art Percept. https://doi.org/10.1163/22134913-00002059

Guilford JP (1950) Creativity. Am Psychol 5:444–454. https://doi.org/10.1037/h0063487

Hartz L, Thick L (2005) Art therapy strategies to raise self-esteem in female juvenile offenders: a comparison of art psychotherapy and art as therapy approaches. Art Ther. https://doi.org/10.1080/07421656.2005.10129440

Herrera-Arcos G, Tamez-Duque J, Acosta-De-Anda EY et al (2017) Modulation of neural activity during guided viewing of visual art. Front Hum Neurosci 11:581

Holt J (2013) Neuroaesthetics and philosophy. SAGE Open. https://doi.org/10.1177/2158244013500677

Junghofer M, Winker C, Rehbein MA, Sabatinelli D (2017) Noninvasive stimulation of the ventromedial prefrontal cortex enhances pleasant scene processing. Cereb Cortex 27:3449–3456

Kandler C, Riemann R, Angleitner A et al (2016) The nature of creativity: the roles of genetic factors, personality traits, cognitive abilities, and environmental sources. J Pers Soc Psychol. https://doi.org/10.1037/pspp0000087

Kilicarslan A, Prasad S, Grossman RG, Contreras-Vidal JL (2013) High accuracy decoding of user intentions using EEG to control a lower-body exoskeleton. In: Proceedings of the annual international conference of the IEEE Engineering in Medicine and Biology Society, EMBS, pp 5606–5609. https://doi.org/10.1109/embc.2013.6610821

King JL (2016) A review of "Art therapy and the neuroscience of relationships, creativity and resiliency: skills and practices." Art Ther. https://doi.org/10.1080/07421656.2016.1126691

Kline JE, Huang HJ, Snyder KL, Ferris DP (2015) Isolating gait-related movement artifacts in electroencephalography during human walking. J Neural Eng 12:46022

Kontson KLKL, Megjhani M, Brantley JAJA et al (2015) Your brain on art: emergent cortical dynamics during aesthetic experiences. Front Hum Neurosci 9:626. https://doi.org/10.3389/fnhum.2015.00626

Kozbelt A, Beghetto RA, Runco MA (2010) Theories of creativity. Camb Handb Creat 2:20–47

Liu S, Erkkinen MG, Healey ML et al (2015) Brain activity and connectivity during poetry composition: toward a multidimensional model of the creative process. Hum Brain Mapp 36:3351–3372. https://doi.org/10.1002/hbm.22849

Lotte F, Bougrain L, Cichocki A et al (2018) A review of classification algorithms for EEG-based brain-computer interfaces: a 10 year update. J Neural Eng

Lusebrink VB (2004) Art therapy and the brain: an attempt to understand the underlying processes of art expression in therapy. Art Ther. https://doi.org/10.1080/07421656.2004.10129496

Luu TP, Nakagome S, He Y, Contreras-Vidal JL (2017) Real-time EEG-based brain-computer interface to a virtual avatar enhances cortical involvement in human treadmill walking. Sci Rep. https://doi.org/10.1038/s41598-017-09187-0

Martin L, Oepen R, Bauer K et al (2018) Creative arts interventions for stress management and prevention—a systematic review. Behav Sci (Basel). https://doi.org/10.3390/bs8020028

Moran S (2010) 4: The roles of creativity in society. In: The Cambridge handbook of creativity

Morris FJ (2014) Should art be integrated into cognitive behavioral therapy for anxiety disorders? Arts Psychother. https://doi.org/10.1016/j.aip.2014.07.002

Motzkin JC, Philippi CL, Wolf RC et al (2015) Ventromedial prefrontal cortex is critical for the regulation of amygdala activity in humans. Biol Psychiatry 77:276–284

Noë A (2011) Art and the limits of neuroscience. New York Times

Pachalska M, Frańczuk B, Macqueen BD et al (2001) The impact of art therapy on the intelligibility of speech in children with cerebral palsy. Ortop Traumatol Rehabil 3:508–518

Pinho AL, Ullén F, Castelo-Branco M et al (2015) Addressing a paradox: dual strategies for creative performance in introspective and extrospective networks. Cereb Cortex 26:3052–3063

Rathelot J-A, Dum RP, Strick PL (2017) Posterior parietal cortex contains a command apparatus for hand movements. Proc Natl Acad Sci. https://doi.org/10.1073/pnas.1608132114

Rusted J, Sheppard L, Waller D (2006) A multi-centre randomized control group trial on the use of art therapy for older people with dementia. Gr Anal. https://doi.org/10.1177/0533316406071447

Sarid O, Huss E (2010) Trauma and acute stress disorder: a comparison between cognitive behavioral intervention and art therapy. Arts Psychother. https://doi.org/10.1016/j.aip.2009.11.004

Schirrmeister RT, Springenberg JT, Fiederer LDJ et al (2017) Deep learning with convolutional neural networks for EEG decoding and visualization. Hum Brain Mapp 38:5391–5420. https://doi.org/10.1002/hbm.23730

Shaheen R (2010) Creativity and education. Educ Theory. https://doi.org/10.1111/j.1741-5446.1961.tb00062.x

Slepian ML, Ambady N (2012) Fluid movement and creativity. J Exp Psychol Gen. https://doi.org/10.1037/a0027395

Sternberg RJ, Lubart TI (1999) The concept of creativity: prospects and paradigms. Handb Creat 1:3–15

Wagner AD, Shannon BJ, Kahn I, Buckner RL (2005) Parietal lobe contributions to episodic memory retrieval. Trends Cogn Sci

Walker MS, Kaimal G, Koffman R, DeGraba TJ (2016) Art therapy for PTSD and TBI: a senior active duty military service member's therapeutic journey. Arts Psychother. https://doi.org/10.1016/j.aip.2016.05.015

Wang Q-Y, Li D-M (2016) Advances in art therapy for patients with dementia. Chin Nurs Res. https://doi.org/10.1016/j.cnre.2016.06.011

Winker C, Rehbein MA, Sabatinelli D et al (2018) Noninvasive stimulation of the ventromedial prefrontal cortex modulates emotional face processing. Neuroimage 175:388–401

Zhang Y, Prasad S, Kilicarslan A, Contreras-Vidal JL (2017) Multiple kernel based region importance learning for neural classification of gait states from EEG signals. Front Neurosci. https://doi.org/10.3389/fnins.2017.00170

Chapter 11
Self-conscience/Physical Memory: An Immersive, Kinetic Art Installation Driven by Real-Time and Archival EEG Signals

Eric Todd, Jesus G. Cruz-Garza, Austin Moreau, James Templeton and Jose Luis Contreras-Vidal

Abstract The intermingling of art and science has often been seen as equivocal, as noncommittal, the art world dubious of the certainty of science and science seeking function in art, but both disciplines very often are in search of the same thing, something we can generalize, something common among us. Neuroscience in particular seeks to give definition to questions that art and poetry, mysticism, religion, and psychology have been asking since the beginning of recorded history and presumably longer. These are fundamental questions about the nature of identity, ideas we've named and redefined many times over. *Self-conscience/Physical Memory* is a brain-controlled robotic sculpture in the University of Houston's Noninvasive Brain Machine Interface Laboratory. Motorized and illuminated acrylic ceiling tiles shift the architecture of the space itself in response to EEG data. The height of the panels is driven by alpha power suppression in the central cortical areas, and the tiles' color shifts with alpha power changes in the occipital and frontal lobes. The EEG data can be input in real-time by a single participant, or, in the absence of user input, the work also serves as a playback device for archival EEG recordings, a physical manifestation of a past experience, of a moment in someone's life, a person both absent and present in that new moment.

E. Todd (✉) · J. G. Cruz-Garza · A. Moreau · J. Templeton · J. L. Contreras-Vidal
Noninvasive Brain-Machine Interface Systems Lab, Department of Electrical & Computer
Engineering, Cullen Engineering Building II, University of Houston, Houston, TX 77204-4005,
USA
e-mail: eamtodd@gmail.com

J. G. Cruz-Garza
e-mail: jgcruz@central.uh.edu

A. Moreau
e-mail: amoreau876@gmail.com

J. Templeton
e-mail: realteeeth@gmail.com

J. L. Contreras-Vidal
e-mail: jlcontreras-vidal@uh.edu

© Springer Nature Switzerland AG 2019
A. Nijholt (ed.), *Brain Art*,
https://doi.org/10.1007/978-3-030-14323-7_11

Keywords Installation · EEG · Neuroscience · Kinetic sculpture ·
Brain-computer interface

11.1 Background

The University of Houston's Non-invasive Brain Machine Interface Laboratory at the
National Science Foundation Industry-University Cooperative Research Center for
Building Reliable Advances and Innovations in Neurotechnology (IUCRC BRAIN)
is the home of a multidisciplinary team of neuroscientists, engineers, health professionals, artists, and students developing neurotechnologies for the restoration and
rehabilitation of cognitive-motor function in people with mental and movement disabilities. The lab works at the interface of the arts, science and engineering, uncovering the relations between perception, cognition and action. Our work is funded
by the National Institutes of Health, the National Science Foundation, the Cullen
Foundation and others, and the Center for Advanced Computing and Data Systems
at the University of Houston.

The Brain on Arts Program, funded by the National Science Foundation Award
#BCS 1533691, is led by neuroengineering professor Jose Luis Contreras-Vidal—to
study connections between the brain and creativity, expression, and the perception
of art.

Eric Todd is a visual artist and co-founder of {exurb}, an art collective that
explored the conjunction of science, art, and technology often through large scale,
kinetic and interactive sculpture. {exurb} had the solo exhibitions *waveForms*, University of Indianapolis (2013) *Topologies*, Lawndale Art Center, Houston (2015)
and *Array*, Art League Houston (2016), among others. Todd was featured artist at *Day
for Night* festival in 2016. His work is held in the State of New Mexico's Art Collection. Originally from West Tennessee, he received his BFA from the University of
Houston in Creative Writing and Theatre and is currently pursuing post-baccalaureate
work in Electrical Engineering.

Jesus G. Cruz-Garza develops strategies for decoding of brain activity signals
through mobile brain-body imaging in real world settings. His work includes decoding and characterization of EEG data in expressive movement, freely moving aesthetic production and contemplation. Jesus graduated from BS Engineering Physics
at Tecnologico de Monterrey, Mexico. He is currently a PhD candidate in Electrical Engineering at the University of Houston and Graduate Fellow on Translational
Research with Houston Methodist Research Institute and University of Houston.

Austin Moreau is currently studying Computer Science as a post baccalaureate
student in The University of Houston undergraduate Computer Science program,
and previously received a Bachelor's of Fine Arts, Studio Art—Painting, from the
University of Houston as well as a minor in mathematics. At present, he is applying
to programs to pursue a Ph.D. in the field of Human Computer Interaction with the
intention of creating autonomous and responsive environments.

Fig. 11.1   Scale rendering of completed installation in situ

James Templeton is a Houston artist, musician, animator, and founder of the creative production company, Dinolion. He is also the man behind the experimental electronic music of LIMB. In 2013 James performed at TEDx Houston with his own custom instrumentation. In 2016 LIMB was the creative lead on the interactive audio-visual installation OCTA at the Day for Night festival. In the spring of 2018 James composed a piece entitled "What We Keep" for the Houston Ballet's mixed rep program, "Play." Figure 11.1 below shows a scale rendering of the proposed final installation. As of this writing, a four-panel prototype, which will be revised and expanded later this year, has been completed.

11.2   Introduction

Our understanding of ourselves is a fluid and nervous thing. We create entire paradigms, competing bodies of knowledge, shifting historical hegemonies to help our brains better apprehend themselves. We have longed for an accurate mirror, are motivated by a desire to see ourselves clearly, and continuously design new lenses through which to do so.

The language we use to describe achievements in art, science, and engineering might be considered variations on a theme. We say that scientists *discover* while engineers *invent*, and artists *create*. The notions of *creation* and *invention* can be easily aligned through control and choice, while we might consider a *discovery* distinct, something borne from observation of existing, if yet undefined, facts (Gowers 2012). But in practice the lines separating these ideas, and disciplines, can blur. In Biology, species distinctions, especially those separated by only a few traits, can seem arbitrary, but the same might be said of how we define Calculus. We might consider Calculus a collection of incontrovertible facts, but its shape and how we describe those facts, what we include or don't, is a matter of choice, of design.

What our investigations into scalp electroencephalography (EEG) offer is the quantifiable, voltages that we can measure, locate, and specify. Mobile EEG technology provides a real glimpse into the electrical activity of the brain in action and in context, making tangible what has long been vague and ethereal, locating our identities in physical space and in time. Our aim is not strict robotic control, a one-to-one button press, but a sort of reflection pool, a reification of lived experience.

In this work, the architecture of the space is given primacy. It becomes a metaphor for the body, the room itself a vessel for consciousness, a mirror for the real-time user into her unseen biology or an intimate glimpse of a person's past, a projection of a stranger's inner life.

This piece lives in context and dialogue with many works of art that precede it and many contributions to science that make it possible. Artists such as Raphael Lozano Hemmer, Roxy Paine, Tim Hawkinson, Ryoji Ikeda, and Kevin Beasley have experimented with motion, sound, data and direct user-input to make active sculpture, sculptures and installations that shift and move and respond directly to the viewer, creating art experiences, not simply passive aesthetic objects.

BCIs specifically have been employed in a variety of artworks: audio/visual solo performances such as Dmitry Mororzov's *eeg_deer* (Morozov 2014), multi-person performances such as Marina Abramović's *Measuring the Magic of Mutual Gaze* (Abramović 2011), single-user responsive installations like Lisa Park's *Eunoia* series (Park 2014), among others.

We believe a major strength of this work is its versatility. We are excited by its ability to function as a single-user closed-loop neurofeedback experience, as a playback device for visualization and sonification of archival EEG recordings, and as a tool for the University's continued neuroengineering research efforts on neuroaesthetics, neural individuality, and BCIs. Its multi-use functionality is essential to its design and, we hope, central to its success. In Fig. 11.2 below, we can see our initial design prototype, completed and installed.

Fig. 11.2   First completed panel prototype. *Photo credit* James Templeton

11.3   Space/Sculpture

Self-conscious/Physical Memory lives on the University of Houston's campus, on the 4th floor of an Engineering building home to classrooms and offices and research labs and student organizations and lecture halls. The space itself is multi-use, a part of three rooms afforded the lab to conduct its research, with experiments and meetings and presentations planned out like a beachfront time-share nearly every day. The work was designed with this in mind, to not just live in this environment but to be at home here.

The room is a 16′ by 18′ rectangle with standard foam 2′ by 2′ square tiled ceiling ten feet overhead. The sculpture is built into the suspended ceiling itself, the foam tiles removed and replaced by acrylic panels, lit on edge by 160 individually addressable RGB LEDs, 80 per side on two sides with two sides left exposed. The choice to do so was an effort to both feature and hide the space. The effect is both an empty and inundated room.

Each panel is hung from the suspension ceiling frame by 0.5 mm wire rope affixed to its four corners with one of the four cables carrying the signal for the LEDs. The cables run through a simple pulley system that gathers them in the center above each tile and winds them around a four-channeled winch spool on a shaft connected to a stepper motor. The full height of the ceiling is 10 ft, and the panels travel 3 ft in the space to a nadir of 7 ft.

The space above the drop ceiling frame was a design limitation itself. A large HVAC duct runs the length of the room, centered and hanging just 10″ above the frame, forcing the support mount and motor system to be compact and lightweight enough to be secured directly to the suspension frame. The installation can be viewed from inside the room or from a window looking in from a hallway just outside. Figure 11.3 details the mechanical components and how they fit into the architecture of the room.

The sculpture is accompanied by sonification of EEG activity, by applying 4th order Butterworth filters to one frontal EEG channel, AF4, in the delta (1–4 Hz), theta (4–8 Hz), alpha (8–12 Hz), beta (12–30 Hz), and gamma (30–50 Hz) frequency bands. The band-power in each of these frequency bands is projected to an electronically generated sine wave sound at the C minor 9th cord pentatonic scale.

Several parts of the cabling system were designed and 3D printed to custom fit the suspension ceiling and mount: the above-mentioned spool, a corner segment that fits onto the cross-joints of the ceiling frame and guides each cable down, and another cable guide mounted below the spool to steer the cables up (Fig. 11.4).

The motors each include a power-off brake (not pictured) that holds the panel in position in case of a power outage and when the system is powered down. A limit

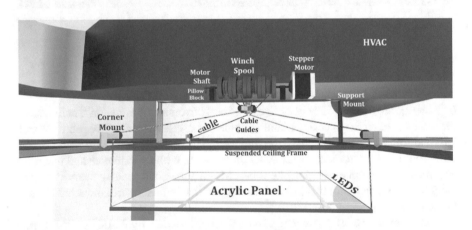

Fig. 11.3 Prototype model

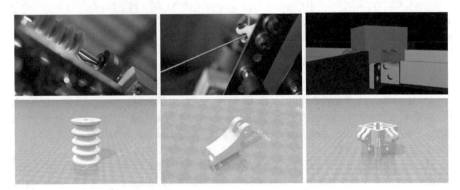

Fig. 11.4 (From left to right) Design and implementation of winch spool, cable guides, and corner mount

Fig. 11.5 Completed mechanical system design. *Photo credit* James Templeton

switch (not pictured) is bolted to the bottom of each support mount that serves as both a safety mechanism and a tool to calibrate each panel's location.

The mechanics are left exposed. Lit from behind, they are featured as a part of the sculpture, not hidden like components of a magic trick. There is no magic here. The mechanics are essential to a work about the mechanics of the mind and serve as a reminder that machines are collections of discrete parts imagined and combined by actual living people and a call-back to the body itself (Fig. 11.5).

11.4   Signal Processing and Application

Among tools to measure brain activity, EEG is unique in that it can capture data at a high sampling rate, short setup time in case of dry-electrode systems, low operating cost, safety, and, most importantly, it is mobile. This allows researchers to record small moments of cognition in real-time and in (something close to) their natural settings. Given all of these characteristics, mobile EEG also makes for an ideal input device for the development of Brain-computer interfaces (BCI).

Neural oscillations large and synchronous enough to be measured by EEG have been consistently linked to a variety of brain processes including cognition and focus, creativity, language processing, motor control, and visual processing (Jeon et al. 2011; Levy et al. 2013; Fink and Benedek 2014; Lobier et al. 2018). A key facet to our understanding of the connection of these signals to assumed causes is event-related desynchronization (ERD), a phenomenon in which a known cognitive function or stimulus creates a signal that results in a power decrease or "blocking" of a certain frequency range in a particular region of the brain (Pfurtscheller and Lopes da Silva 1999).

For two core elements of this project, we looked at ERD in the alpha band (8–12 Hz). The panel LED animations are generated with alpha power input from occipital lobe channel Oz, and the height of each panel is driven by alpha power values in central cortical channels. Research has tied momentary decreases in local alpha power to movement related events in the motor cortex and to visual input and shifts in focus and attention in the occipital lobe (Jeon et al. 2011; Woertz et al. 2004; Lobier et al. 2018).

A third element, sonification, includes signal power from all five of the major EEG frequency bands: delta (1–4 Hz), theta (4–8 Hz), alpha (8–12 Hz), beta (12–30 Hz), and gamma (30–50 Hz). Here we have chosen channel AF4 reading the pre-frontal cortex and are mapping its various frequency power to harmonics in C minor 9th in a 432 A4 tuning.

To filter the raw EEG signal we are utilizing the SciPy Python library's 4th order bandpass Butterworth filter with 8–12 Hz in the passband. As an indicator feature, we chose the relative alpha power, calculated by obtaining the average power over the alpha band and diving by the total power from 1 to 50 Hz.

The treatment of these signals is hopefully one that provokes the feeling of sitting directly inside a human moment, a moment of wonder and meditation at the strangeness of something so intimately familiar.

11.4.1 Hardware/ERD Motor Cortex Test

To calibrate the changes in the alpha power for hand-movement induced ERD, we demonstrated and collected EEG data from two volunteers who were asked to move their hands. In five 25 min trials of gripping with the right hand, gripping with the left, with both, and at rest, we ran a simple FFT on one second of data prior to each event onset.

We did this to both verify that we could observe motor correlated ERD in the central region and to verify that we could reliably observe this data using the Cognionics Quick 20 (San Diego, CA) wireless, dry EEG system that we employed as the main input device for the real-time functionality of the installation (Yuan et al. 2011; Abbasi et al. 2018).

The results in Fig. 11.6 show unfiltered frequency domain data for the C1 channel of one subject transformed over the 1 s before each event and averaged across all 25 trials. The graph clearly shows a decrease in alpha (as well as beta) power for all non-rest trials. These results give us confidence that our choice to focus on alpha band ERD in the central region should reliably correlate to sensorimotor action.

Fig. 11.6 Changes in alpha power in channel C1 during left and right-hand movements compared to a resting state

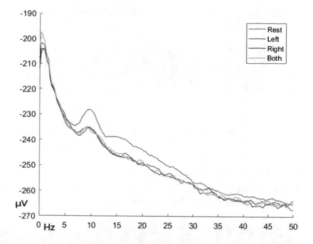

11.4.2 Panel Movement

While our calibration test did show that we should reliably see alpha ERD in the central channels just before a physical movement, in real-time applications, it will not be practical to take the average of multiple trials. Additionally, the viewer's experience with the installation is intended to be open-ended, not a prescriptive, controlled outcome-oriented interaction, but rather a moment of exploration, an invitation to survey his or her inner life and help to create the work itself. So while we do tie the EEG signals most likely to be associated with a person's intention to move to the motion of his/her environment, these signals will not serve as perfect triggers but rather a direct representation of the signals themselves, which as noted above, while located in the region most associated with intention to move, are very likely to also include a variety of other signals from adjacent regions with other purposes as well.

Alpha power in central area channels are used to drive the height of the acrylic tiles. The higher the relative alpha power in a channel, the closer to the ceiling its corresponding tile will move. The layout of the tiles and their related channels mirrors the flattened layout of the EEG cap as it is placed on the wearer's scalp. The effect is that the tiles move together but not necessarily in unison as adjacent electrodes share portions of their signal composition and change gradually as their distance from each other increases. The result is an undulating wave that reflects the changing and constant nature of brain processes (Fig. 11.7).

However, because we are sampling an oscillating EEG signal at 100 Hz, the output changes very rapidly in time, a change not ideal for our purposes. To compensate for this we apply a left half Gaussian filter over a 2 s time window to smooth the changes in alpha power in time.

Further smoothing occurs at the hardware level where motor positions are accelerated and decelerated to each new position. Taken all together, the result, despite being driven by rapidly changing oscillations, is a gentle, undulating affect created by

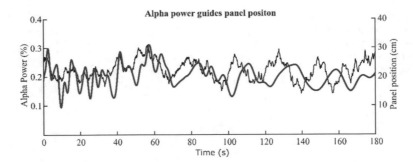

Fig. 11.7 A representation of how EEG signal in channel CP1 is transformed into the movement of a single panel in the sculpture over the course of 180 s. The alpha power is shown as a moving average from the last 4 s of data. Panel height is scaled from 0 to 100 cm from the top of the ceiling

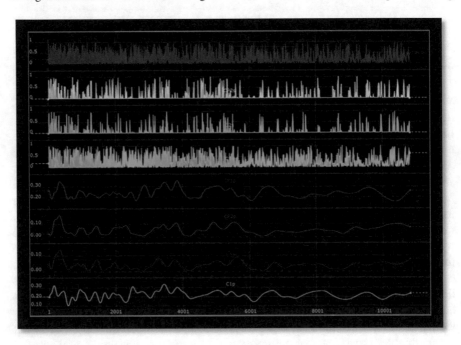

Fig. 11.8 How relative alpha power in EEG channels CP1, CP2, C1, and C2 are transformed into the physical movement of the ceiling tiles over 3 min

the motion of each panel. Figure 11.8 illustrates this idea in practice, displaying the raw EEG signal on four channels and their corresponding outputs in the sculpture's movement.

Fig. 11.9 (left) Animated image mapped across all ceiling tiles (right) panel LED mapping in 3D space representing 30 ceiling tiles—the LED strips are mapped in a grid onto the square color palate image

11.4.3   LED Animations

The LEDs that illuminate each panel are all animated together, mapped across a single image. The image is a shifting blend of 5 vertically aligned complimentary colors, with each raw EEG channel subtly modulating the white balance along the horizontal.

The center occipital lobe Oz channel modulates the LED animations. The LED strips placed edgewise along two sides of each panel are mapped to a slowly-wiping five-part square 2D color gradient seen in Fig. 11.9. The alpha PSD percentage in the Oz channel causes the color gradient to rotate through 8 different color palates depending on the input value and one solid black color set that corresponds to peaks in relative alpha power corresponding to the cap wearer's eyes being closed. Additionally, frontal lobe channel Fp2 shifts the relationship of the colors within each palate and the speed of the wipe through the five colors.

The choice of occipital lobe data driving changes in color followed the region's known connection to vision processing. ERD in the alpha band has been seen in response to visual stimulus as well as changes in visual attention. Frontal, parietal, and occipital lobe alpha band suppression has also been shown to relate to attention and visual focus thus providing a natural feedback loop to drive the changing look of this immersive environment (Woertz et al. 2004; Quaedflieg 2016; Belyusar et al. 2013; Lobier et al. 2018).

11.4.4   Sonification

In addition to the subtle sounds of the motors, sheaves spinning, and cables winding, we've included audio generated by EEG signal. The power in the delta, theta, alpha, beta, and gamma frequency bands is mapped to the amplitude of sine waves in the C minor 9th chord (C_3, $D_2^{\#}$, G_3, B_3, D_4) pentatonic scale, with frequencies at 128.43,

152.74, 192.43, 242.45, 288.33 Hz respective to each frequency band. Figure 11.10 shows the original EEG frequency activations in a spectrogram, and the mapping to the signifying sine waves corresponding to channel AF4.

Here we see the fundamental C_3 and $D_2^{\#}$ mapped to the theta band predictably producing the most common incidences and greatest volumes. The result of this, and thus of our choice to use these signals to drive a single chord, creates a harmonic drone, the minor key slightly dissonant, the alpha, beta, and gamma adding complexity as they synchronize. The choice to do simplify here is also intended to be unifying, the low drone calming amidst the shifting visual stimulus, a reminder of how continuously our brains are processing information (Fig. 11.11).

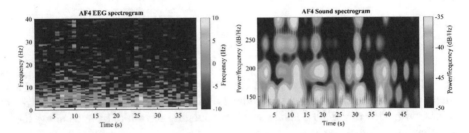

Fig. 11.10 Spectrogram of the EEG data in AF4 signified as a mixture of sine waves. Left: EEG activation of the AF4 channel across the whole frequency spectrum analyzed. Right: Spectrogram of the sine waves used to generate sound from the EEG channel

Fig. 11.11 Photo of installation progress in situ. *Photo credit* Ronald L. Jones

11.5   Function/Operational Modes

11.5.1   Playback

When no viewer-input is present, the installation will serve as a playback device for selected EEG recordings. Under normal circumstances, this is how a viewer will most likely find the work. The subject's gender, age, date and circumstances of recording will be displayed. In this way, viewers can be intimately present with a person who is absent, a person who may have never set foot in the room where she is standing or may no longer even be alive. These recordings will take a variety of forms, from recently recorded as a part of university research to novel early experiments with EEG hardware and many things in between.

11.5.2   Real-Time

During select times, viewers will be able to interact with the sculpture directly in real-time via a 20-channel dry EEG system. Like Raphael Lozono-Hemmer's *Pulse Room*, a single patron will for a time control the entire installation, her brainwaves mirrored in the sensory shifts (movement, light, sound, color, shadow) of the room and its architecture. Others can watch from the viewing window outside or be present in the space and wait to participate themselves.

11.5.3   Research

In addition to the work's function as art, we hope it will have some use in closed-loop BMI research as a tool to explore non-anthropomorphic control systems or experiments involving response to various stimuli. We also imagine it as a therapeutic tool to explore maladies like claustrophobia or balance disorders. These applications are not currently in development but could augment the existing hardware and software framework in the future.

11.6   Collaboration

The real-time interactive functionality of this work makes it well-suited to further artistic collaboration. In Fig. 11.12, we see Shu Kinouchi, a dancer with the Houston Ballet, giving an improvised closed-loop performance with the sculpture. In the future, we have plans to explore similar experiments with writers, musicians, and multi-media artists as well.

Fig. 11.12 Photo of dancer, Shu Kinouchi, giving a closed-loop performance with the installation, 2019. Photo by Ronald L. Jones

11.7 Conclusion

My practice as an artist is most often a highly collaborative one in which I conceptualize and design work with in conjunction with others—as a group or with our names listed individually—but nearly always as a collective effort, and this is a departure from that. While the final outcome of this project was certainly the cause of a collaborative process, and the people whose names are listed here as co-authors were essential to its realization, Dr. Contreras-Vidal trusted me with a space, funds, and mentoring to drive the project forward, to take the lead in defining its aesthetic and conceptual choices and to be solely responsible for its result, and for that I am very grateful.

When we started this work, I proposed an idea that I knew was possible but did not have the theoretical foundations to make the space represent brain activity dynamics. I spent the first six months on this project just learning and planning, taking meetings with people in the lab and others outside learning the practical and tangible ways to make this idea real. The design here went through many iterations, some failed outright, some revised, but all edifying and all essential to the finished sculpture.
—Eric Todd

Acknowledgements This work was partly funded by the University of Houston Cullen College of Engineering, NSF Awards 1219321 and 1302339, NSF NCS-FO 1533691, NSF Award #1650566 I/UCRC BRAIN Center, NSF I/UCRC BRAIN Award CNS1650536, NSF REU Site Award #EEC-1757949, and the University of Houston's 2017 SURF program.
We would also like to personally thank Atilla Killacarslan, Yongtian He, Andrew Paek, and Daniel Schaeffer for their patient consultation and unwavering support.

References

Abbasi O et al (2018) Unilateral deep brain stimulation suppresses alpha and beta oscillations in sensorimotor cortices. NeuroImage 174:201–207

Abramović M (2011) Measuring the magic of mutual gaze. https://www.ericforman.com/marina-abramovic-mutual-gaze#1

Belyusar D et al (2013) Oscillatory alpha-band suppression mechanisms during the rapid attentional shifts required to perform an anti-saccade task. NeuroImage 65:395–407

Fink A, Benedek M (2014) EEG alpha power and creative ideation. Neurosci Biobehav Rev 44:111–123

Gowers T (2012) Is mathematics discovered or invented? In: Pitici M (ed) The best writing on mathematics. Princeton University Press, pp 8–21

Jeon Y, Nam CS, Kim Y-J, Whang MC (2011) Event-related (de)synchronization (ERD/ERS) during motor imagery tasks: implications for brain–computer interfaces. Int J Ind Ergon 41(5):428–436

Levy J, Vidal J, Oostenveld R, FitzPatrick I, Démonet J-F, Fries P (2013) Alpha-band suppression in the visual word form area as a functional bottleneck to consciousness. NeuroImage 78:33–45

Lobier M, Matias Palva J, Palva S (2018) High-alpha band synchronization across frontal, parietal and visual cortex mediates behavioral and neuronal effects of visuospatial attention. NeuroImage 165:222–237

Morozov D (2014) eeg_deer. http://vtol.cc/filter/works/eeg_deer

Park L (2014) Eunoia. http://www.thelisapark.com/eunoia-ii/

Pfurtscheller G, Lopes da Silva FH (1999) Event-related EEG/MEG synchronization and desynchronization: basic principles. Clin Neurophysiol 110(11):1842–1857

Quaedflieg CWEM et al (2016) The validity of individual frontal alpha asymmetry EEG neurofeedback. Soc Cogn Affect Neurosci 11(1):33–43

Woertz M et al (2004) Alpha power dependent light stimulation: dynamics of event-related (de)synchronization in human electroencephalogram. Cogn Brain Res 20(2):256–260

Yuan H et al (2011) Differential electrophysiological coupling for positive and negative BOLD responses during unilateral hand movements. J Neurosci 31(26):9585–9593

Part IV
Using Brain Art in Therapy

Chapter 12
Advancing the Rehabilitative and Therapeutic Potential of BCI and Noninvasive Sensing Systems

Stephanie M. Scott, Chris Raftery and Charles Anderson

Abstract Brain-Computer Interface (BCI) technologies have the capacity to go beyond serving as a basic communication medium for the motor-impaired, as they can also extend creative expression for application within therapeutic processes. By examining the ways in which BCI technologies can mediate communication, provide a venue for embodied interaction, and act as a medium for sharing both individual and collective experiences through the creative process, more can be understood about the ways these technologies can advance and improve the user-system interactions that occur within both digital and physical spaces. This study articulates the differences between neurofeedback and biofeedback, and explores a proof of concept application of both types of interventions. In doing so, it explores conceptual frameworks from a transdisciplinary approach, and demonstrates the potential new interface design strategies have for enabling original forms of creative expression through technology. It also examines the potential of noninvasive tools for expanding boundaries of digital spaces to offer more inclusive means for self-expression and identity formation for users. Using newly designed TCRE electrodes, these efforts explore how more precise scalp recordings may help uncover new methods for discovering patterns of brain activity while engaged with artistic creation for therapeutic processes.

Keywords Tripolar EEG · Biofeedback · Neuroadaptive applications · Art therapy · Digital communication

S. M. Scott (✉) · C. Raftery
Department of Media Communications, Colorado State University, Fort Collins, CO, USA
e-mail: SMS.Scott@colostate.edu

C. Raftery
e-mail: CM.Raftery@colostate.edu

C. Anderson
Department of Computer Sciences, Colorado State University, Fort Collins, CO, USA
e-mail: Chuck.Anderson@colostate.edu

© Springer Nature Switzerland AG 2019
A. Nijholt (ed.), *Brain Art*,
https://doi.org/10.1007/978-3-030-14323-7_12

12.1   Introduction

Evolving technologies often influence the way newer tools are received and integrated within social environments. As a technology develops, changes that alter the function or form of a device can invite new customs and behaviors that surround use, and can present challenges that impact user experience. Media research suggests that technological shifts can not only vary adoption and retention rates among users, but can also lead to interface modifications that directly affect the quality and character of communication afforded to users (Schulz 2004). Nijholt et al. (2018) note that while traditional BCI research prioritizes robustness and efficiency as primary characteristics of concern, those who design artistic BCIs are more likely to focus on their use within multimodal or multiparty contexts. While efficiency and usability remain integral components for evaluating technological progress (Kübler et al. 2015), these interactive processes that occur between devices and users often go overlooked. This may consequently impede efforts directed towards discovering the communicative-use potentials that these devices support. In a focus group conducted amongst individuals with ALS and their caregivers, Blain-Moraes et al. (2012) identified both personal and relational issues as being factors towards acceptance of BCI system use. Participants identified that personal factors such as physical, physiological and psychological issues as being less of a concern than those which were deemed relational, including corporeal, technological and social factors, when considering the acceptance of Augmented and Alternative Communications (AAC) (Daly and Huggins 2015). Although post-positivist methodologies have been instrumental for guiding technological improvement for AAC and Assistive Technologies (AT), combining a qualitative approach (such as participatory questionnaires and open-ended surveys) with quantitative methods can provide valuable information for understanding user priorities and opinions towards developing systems that feature effective user-end based designs (Blain-Moraes et al. 2012; Huggins et al. 2011; Morone et al. 2015) (Figs. 12.1, 12.2, 12.3 and 12.4).

Research within the field of Computer-Mediated Communications (CMC) research was initially guided through two facets: efforts directed towards understanding new media alters informative capabilities of a medium, and examination of how interactive characteristics of a new medium may impact communication (Papacharissi 2005). Initial definitions of CMC focused on synchronous and asynchronous message delivery and encoding processes from senders to receivers (Walther 1992), but have since expanded to include people and technology communicating within certain contexts in which media is appropriated for specific, personal needs. Progressive technologies that enable interpersonal methods of communication have encouraged more contemporary definitions of CMC, which include "processes of communication facilitated by information technologies, involving people in online and offline contexts" (Papacharissi 2005, p. 218). Using a modern CMC lens to examine accessibility and usability issues, theoretical approaches towards the acceptance and adoption of new technologies argue that the question of whether a technology will be acceptable depends on whether the changes new technologies

Fig. 12.1 Jewel Crasta (left), Elliott Forney (top center), Katie Bruegger (right), members of the Colorado State University Brain-Computer Interface Laboratory team, work to ensure that the sensors of an EEG cap are properly placed on a research participant (bottom center). The team's research largely concerns improving the efficiency and usability of in-home systems for users, and requires assessing how this technology functions outside of a controlled setting (Photograph by Stephanie Scott, Fort Collins, Colorado, 2017)

present can be translated within existing productions and routines in ways that make sense for the user (Webster 2002). This concept helps illustrate the need for critical assessment on behalf of stakeholders within the technology community to be able to recognize how BCIs can impact individual and social communication processes, as well as influence interactions that occur between the user and interface (Cox and Depoe 2015). Mediated interactions between the user and a technology are shaped not just by the inherent properties of the device, but also the individual characteristics of the users themselves, in conjunction with the behaviors used to approach and interact with the interface within a specific environment. As each user requires a specific and unique result from these mediated interactions, we must consider the need for individualized variance within the design and implementation process to ensure that technological development maximizes communicative potential for all types of users. Simply improving the function of a device does not guarantee user engagement, as it may prompt users to opt out of using various devices if technical solutions fall short of addressing individual needs (Huggins et al. 2011).

Fig. 12.2 Participant is shown wearing an EEG cap while painting. The purpose of this intervention was to enable creative experience for the BCI system user and to demonstrate how visualization of EEG data could help to enable artistic self-expression. Active EEG analysis was combined with visual neurofeedback output to enable a process that replicated methods of traditional analog painting. The finished product consisted of visual overlays projected upon a transparent canvas (Photograph by Lukas Gehrke, Valencia, Spain, 2017)

As new communication technologies continue to be integrated within established societal practices, populations inevitably face new barriers, questions and configurations emerging from interactions between user and interface. Each new technology placed within contextual elements of a certain place and time brings with it a direct or indirect influence over behaviors and attitudes concerning both the "how" and "why" of communicative processes. As such, it is especially important to assess how these tools are both designed and applied. When the technical properties of the tools themselves change, so too do our responses to them (Webster 2002). When these principles are applied to BCI technologies, we see that there is room for improvement though including target users within the processes of design, research and application (Huggins et al. 2011), as well as within projects directed towards integrated art therapy-based interventions.

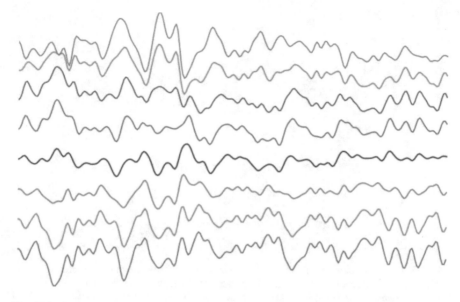

Fig. 12.3 A still capture from an mp4 animation showing 8 channels of EEG recorded from one of Colorado State University's Brain-Computer Interface Laboratory research participants in a relaxed state. The participant was not engaged in any BCI paradigm at the time of the capture. From this example time sample, and for a full feedback visualization using a Brain-Computer Interface system, the team subtracted the average of the 8 channels using the common average reference (CAR). The order of channels pictured, from bottom to top are F3, F4, C3, C4, P3, P4, O1, O2 (mp4 animation created by Chuck Anderson, May 2018)

12.2 BCI Technology Past and Present

BCI technologies enable communication unaided by physical movement by providing a direct link between a functional brain and the outside world. BCI systems operate by translating central nervous system (CNS) activity into a generated output to replace and restore functions for individual users (Brunner et al. 2015). This research traditionally focuses on efforts to establish a channel of communication for paralyzed or locked-in patients who would not be able to communicate effectively without these systems. However, "recent scientific, technological, and societal events have changed this situation," prompting additional objectives to be applied to BCI research to improve, enhance, supplement, and allow these technologies and subsequent interactions to serve as a research tools (e.g., Wolpaw et al. 2002, p. 768; Brunner et al. 2015). Despite the intent to further explore and identify the potential BCI devices have to serve as communicative tools for users within and among target populations that stand to benefit from these technologies (e.g., patients who rely on them for communicating and maintaining external connections and motor rehabilitation purposes) there are still challenges and limitations that serve as barriers towards widespread implementation. Though studies that have addressed in-home

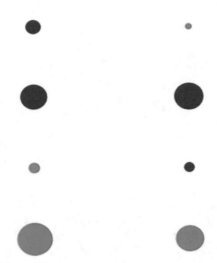

Fig. 12.4 Instead of controlling BCI through EEG features, this project explores the effect of simple and immediate feedback of EEG characteristics translated directly into a dynamic visual display. To start, 8 circles (varying in both size and color) are digitally "drawn" upon the interface screen. At each sample in time, the EEG voltage from across the 8 electrodes is subtracted from the 8 voltages as a common average reference. The EEG signals greater than average are shown in green, and those lower than average in red. The radius of each circle is proportional to the measured absolute value of the difference from the average. By filtering the EEG signals in order to preserve the slower calculated frequencies, we were able to differentiate the greater visual changes from those with a lesser variance (mp4 animation created by Chuck Anderson, May 2018)

independent use for motor-impaired users (e.g., Wolpaw et al. 2002; Wolpaw and Wolpaw 2012; Pfurscheller et al. 2006; Kuebler and Mueller-Putz 2007; Nam et al. 2018) more efforts need to be directed towards active analysis of target populations in order to improve reliability and contribute towards long-term and translational research (Kübler et al. 2013, 2015). For example, researchers have demonstrated than fewer than 10% of published studies address this group and further involve people with severe disabilities (Zickler et al. 2013). These statistics highlight the need for research that aims to bridge the translational gap between contemporary BCI research and the potential application of these systems for users within their home or healthcare environments (Kübler et al. 2015).

By translating brain signals into new kinds of multimodal outputs that can be accessed by various groups of users, BCI systems possess tremendous potential as transformative, rehabilitative and communicative tools (Huggins et al. 2011). Currently, there are two ways that we are able to utilize BCI technology: The first allows a user to have "control of the external environment (light switch, temperature control) or communication devices," while the second "involves using the system as a motor learning-assist device" (Daly and Huggins 2015, p. 1). These systems offer new modes of communication for those who otherwise would remain limited or unable to communicate independently due to a neurological condition or physical injury.

BCI devices can serve as a bridge for those who experience limited mobility and the means for external communication while affording users a sense of autonomous expression that would otherwise not be feasible without the assistance of a care giver or translator.

Although an original objective of BCI research and corresponding technological development was to enable basic forms of social interaction for patients, the potential exists to not only restore and enhance communication for the motor-impaired but to extend them to include opportunities for creative expression and therapeutic care (Zickler et al. 2013; Wolpaw and Wolpaw 2012). By using a multidisciplinary approach to examine the ways in which technology can mediate communication, enable a furthered sense of embodiment and self-identification, and permit users to express themselves in a manner that is unique to their own abilities through the creative process, we realize the need for a shift towards improving the quality of interactions that occur between user and device while incorporating these findings into a user-centric design process to take into account the needs of a highly-diverse user population.

12.3   Noninvasive Considerations and Applications

Despite decades of research within the field, the accuracy and reliability of BCI systems remains limited for several reasons. First, current EEG scalp electrodes lack the necessary spatial and temporal precision. Activity from a relatively large area of the cortex of the brain affects the signal that is recorded by one electrode, so differences in brain activity in neighboring regions of the brain cannot be detected. Second, current BCI research is focused on detecting known characteristics in EEG. Given that variation is present in these recordings, and that that most research relies on patterns in brain activity that can only be revealed by averaging the signals over multiple trials, these patterns are not sufficient for accurately discerning differences in brain activity as a user's mental state varies. The BCI field requires more precise scalp recordings and new methods for discovering the patterns within.

This first of these limitations is currently being addressed by Dr. Walter Besio at The University of Rhode Island. Besio is developing and researching a new EEG sensor; the Tripolar Concentric Ring Electrode (TCRE) with a signal-to-noise ratio 4 times more accurate than that of conventional EEG electrodes (Besio et al. 2014). Dr. Besio has demonstrated the ability to find high-frequency oscillations preceding seizures with TCRE electrodes; oscillations that are completely absent from recordings taken with conventional EEG electrodes. His research with TCRE electrodes has also discovered that movements of different fingers can result in different EEG signals, as opposed to recordings from conventional EEG electrodes that show no difference (Anderson et al. 2018). These findings illustrate how TCRE electrodes can record brain activity that is occurring directly beneath the electrodes themselves, as opposed to conventional electrodes that record activity from a wider area of the brain.

To overcome the second limitation, new approaches to signal analysis are needed. For example, it is not yet understood exactly which patterns in brain activity correspond to feelings of heightened creativity, engagement, stress, or depression, due to the involvement of many parts of the brain and biological variations inherent within individual users. Recent developments from the machine learning community in the use of multilayered neural networks promise increased capabilities in discovering and recognizing such patterns (Forney et al. 2018).

Electrocorticography electrodes (ECoG), which are placed directly on the surface of the brain, have 20–100 times better signal-to-noise ratio than conventional EEG placed on the scalp (Ball et al. 2009). An example of the potential for ECoG in neurorehabilitation was recently demonstrated by Gharabaghi et al. (2014), in which a patient with a lesion in the cortex was able to reliably perform neurofeedback training to alter brain oscillations for developing better control of impaired limbs. When using conventional scalp EEG electrodes, researchers found that the patient was unable to alter brain oscillations, suggesting that more accurate sensing would help for recording more accurate neuro and biofeedback responses.

Likewise, the impracticality of the surgical procedure for implanting ECoG electrodes restricts the availability of the traditional approach. It is in this context that Besio's new TCRE scalp electrodes may lead the way for a more practical approach to neurofeedback therapy and rehabilitation, and that the combination of the increased spatial and temporal resolution and signal-to-noise ratio provided by TCRE electrodes and the advances in spatial and temporal pattern discovery afforded by deep learning algorithms will lead to a better understanding of how brain processing is affected by disease, injury, and psychological trauma. This in turn can lead to more precise neurofeedback training and therapy applications with the technology.

12.4   Integrating Neurofeedback and Biofeedback

Integrating neurofeedback applications within BCI potentials can present the opportunity to offer clinical benefits to users through novel interventions grounded in psychological and neurosciences practices. Effective tools may potentially enhance neural functions by assisting users through the process of regulating their neural functions, as well as contribute to efforts aimed towards developing therapies for those with neural disabilities (Ordikhani-Seyedlar et al. 2016). Research conducted using controlled studies to evaluate cognitive and affective outcomes from both EEG and neurofeedback assessments has favored "sustained attention, orienting and executive attention, memory, spatial rotation, RT, complex psychomotor skills, implicit procedural memory, recognition memory, perceptual binding, intelligence and wide ranging aspects of mood and well-being" (Gruzelier 2013, p. 155). A re-emergence of EEG-neurofeedback research has followed controlled evidence of clinical benefits and validation of cognitive/affective gains in healthy participants including correlations in support of feedback learning mediating outcomes in order to provide sufficient evidence validating the role of EEG-neurofeedback in enhancing function

(Gruzelier 2013). This offers initial results for proof of concept studies, and high-lights important questions and considerations for moving forward in order to support the need for further exploration into EEG and different applications.

Orndorff-Plunkett et al. (2017) highlight the benefits of manipulating neuronal activity via neurofeedback, defined by the authors as "an operant conditioning-based technique in which an individual can sense, interact with, and manage their own physiological and mental states" (1). When the intention is to normalize the system, (e.g. via neurofeedback), we describe it as self-directed neuroplasticity, with an outcome that is persistent in functional, structural, and behavioral changes (Orndorff-Plunkett et al. 2017). EEG frequency bands reflect information processing, such as concentration and attention, as well as aspects of arousal including tension, wakeful-ness, relaxation, or sleep, and neurofeedback technique. This type of analyzation also makes "individuals aware of these processes by *feeding back* a representation of their own electrical brain activity and allowing them to change it" (Gruzelier and Egner 2004, p. 1; Folgieri et al. 2017). Furthermore, research within this field suggests control of brain oscillations could be harnessed intentionally to mediate plasticity (Orndorff-Plunkett et al. 2017). The authors suggest that continued research in this area may "collect additional measures of community functioning, including the direct impact on the quality of relationships and occupational functioning" (11).

EEG feedback typically combines two challenges. The first concerns the frequency-based organization of brain communication, while the second targets inap-propriate state transitions. Biofeedback systems, on the other hand, involve two dif-ferent categories: intrinsic and extrinsic feedback. Biofeedback modalities use human senses to receive feedback information generated in the biofeedback loop. The most commonly used modalities in biofeedback processes are visual, auditory, and tactile (Sigrist et al. 2013). Biofeedback data collection is most successful when the user is appropriately reacting to the given feedback information (e.g. able to correct errors in movement or abandon its execution given the appropriate amount of feedback information), as well as when the biofeedback loop is closed.

Umek et al. (2016) suggest that concurrent biofeedback "can reduce the frequency of improper movement executions and speed up the process of learning the proper movement pattern" (327). These movement learning methods may be adapted to users of various skill levels for the first stages of the learning process (Liebermann et al. 2002). Likewise, it has been shown that "modern technical equipment can help both the learner and the instructor by providing additional, parallel feedback infor-mation that is not obtainable by traditional observation methods." (Kos et al. 2018, p. 7). Given these findings, it is possible to suggest that design and implementa-tion of interfaces that provide concurrent feedback loops within BCI systems could improve the learning potentials for both the process of the user becoming familiar with the system and the process the system uses to calibrate with the user. Engaging multimodal systems through these types of feedback loops could extend the lens of discovery to self-reflection and community participation.

Through exploring newly designed applications of BCI systems as motivating and engaging communicative tools, the integration of neurofeedback with biofeedback systems for creative expression could encourage new methods of discovery. In biofeedback systems, a user's biological data is read, measured and processed before being sent back to the user. At this point, the biofeedback is translated into visible data, and the user attempts to act on this received information in order to produce a desired effect (Sigrist et al. 2013; Kos et al. 2018). Neurofeedback, which records the brainwave activity measured through BCI, provides researchers and users with an alternative yet complimentary set of data. This intrinsic data is combined and augmented by its respective interface, allowing us to learn more about the user through the mediation process.

12.5 Mediations of Digital Environments

Though biofeedback and neurofeedback each offer certain inherent values through individual application, the product and potential use-value of the accumulated data can be far greater than the sum of its parts. However, concerns remain about whether these tools can be designed to create a more meaningful whole without minimizing the importance of their respective individual pieces. As such, this approach asks; how can we utilize the data (technical) in order to enable and maximize user communication on both an interactive (individual) and greater (social) scale? Orndorff et al. (2017) note that these neurotherapeutic interventions provide numerous benefits for users, as they "give individuals a more active role in their own health care, utilize a holistic approach to body, mind, and spirit, are non-invasive, and elicit the body's own healing response" (14). When considering the potential of neurofeedback training for the purposes and goals of social neuroscience, it appears that experimental neurofeedback designs with specific behavioral outcomes could inform both the social neuroscience and clinical communities alike.

In order to better understand the phenomenon of how BCI systems continuously adapt to evolving technologies, assuming a perspective that emphasizes the role of media ecology towards understanding the total environment in which human/technology interactions take place can be constructive. This particular metadiscipline investigates how the media of communication can affect our perception, understanding, feeling, and value (Postman 1970) while recognizing that "any understanding of social and cultural change is impossible without a knowledge of the way media work as environments" (McLuhan and Fiore 1967, p. 26). Contemporary discourse regarding the ecology of media involves the analysis and interpretation of new media technologies; specifically, an identification of the various structural and environmental components that make up both the communicative device as well as the environment in which the communication occurs (Star 1999). This may include examination of social norms, formal rules/laws/codes, and language in order to influence participation or behavior within a given space.

Individual creative expression may further subjectivity of meaning in brain-computer/human-computer interactions. When applied to the development of BCI systems, this stresses that cultural processes have come to bear equal weight with those that are economic or political in nature (Schulz 2004). Further, due to substantive and epistemological reasons, an understanding of cultural elements has also become vital towards how we make sense and develop collective meaning within the communities in which we live (Du Gay et al. 1997). In recognizing media ecology as it pertains to user environments, it is important to discuss the role of visual materials in cultural process and message design (in this case, the creation of art). In order to maximize the benefits of precise data obtained from TCRE electrodes, these proof of concept projects propose protocols that provide BCI users with opportunities to engage in creative expression through artistic-creative processes. Prioritizing visual content creation allows users to gain understanding through observation of, and inter-action with, visual representations of their biological information in order to provide communicative and therapeutic benefits both technical and social in nature. In order to understand how these designs may affect the neuroscience and clinical communities, researchers should be willing to look beyond intrinsic individual user experiences to externalize the BCI process in order to promote optimal self-development practices through socially communicative and self-expressive means.

12.6   Art and Visual Design in Health Communication

Traditional definitions of literacy are no longer adequate for describing user comprehension abilities across a multimodal landscape, as both visual materials and corresponding layout techniques are common components of communication mediated by digital screen-based technologies. There is a need for a new form of visual literacy in a digital environment; one in which users actively negotiate with images in order to derive meaning. Researchers argue that health professionals can improve communication about new technologies by incorporating visual materials in their message designs. This type of information can effectively demonstrate to users that they possess agency concerning both how and when to use new technologies, which in turn may encourage further understanding and adoption of new health technologies into everyday health practices and behaviors (Harrison 2002). Incorporation of visuals into the characteristics of a complicated health message can prompt patients to respond more favorably to the information presented, due to increased levels of comprehension. With the increasing progress of mediated forms of communication, looking towards visual-based media can help us better understand the relationship of an image to the text that accompanies it.

The power of art in health communication may be related to its capacity for emotional engagement and stimulation of critical thought. Incorporation of visuals has been shown to be more persuasive and helpful for stimulating an emotional response within the viewer (Houts et al. 2006). Schweizer et al. (2009) highlight key principles for effective communication including the uniting of a message through cultural values, experiences and traditions to cause the audience to react in a more positive manner. Additionally, crafting a piece which communicates meaningful and relevant ideas can help cultivate critical social thinking about an issue. The creative process also allows individuals to understand something about themselves that may otherwise be hidden or obscured by providing a forum for self-understanding and reflective distance (Potash 2010). Allen (1995) reminds readers through her own creative experience that the creation of art is a way of knowing, while Potash (2010) demonstrates that engaging with imagery allows for an alternative way in which to make sense of one's own unique experiences. Critical theory and sociocultural studies examine discourse and forms of representation as key areas of inquiry and, within this, seek to demonstrate the importance of images as forms of communication which are culturally and socially situated within and mediated by a user's own ideology and subjectivity. By positioning the artistic creative process as one which is intrinsically relevant as well as socially valuable, these systems can be perceived as a valuable therapeutic tool for BCI users that face constrained means of communication or individual forms of self-expression.

12.7   Benefits of Art Therapy

Art therapy has been shown to reduce anxiety through the interactive process that occurs between the art, the creative process, and the artist/patient-therapist interaction (Czamanski-Cohen and Weihs 2016). The creation of art allows for therapy patients to communicate that which might otherwise go unexplored or uninterpreted, as it provides a medium for conveyance of subconscious emotions through the use of metaphor (Angheluta and Lee 2011). It also has the ability to stimulate user self-expression through stimulation of the senses, enablement of social interaction, and provision of the opportunity to be creative (Stewart 2004; Ehresman 2014; National Organization for Arts in Health 2017). Art therapy may be a safe and cost-effective intervention as an adjunct to traditional medical management; however, it still faces challenges pertaining to the mediated process which occurs between user and interface. In order to achieve the most successful therapeutic benefits through art therapy, it is important to recognize the potential for BCI to act not only as an interface with which to develop a creative product, but as a communicative and social technology centered around the enablement of discussion and understanding of the art which is created. This prompts reconsideration of not just the interface design, but the entire environment in which the user/technology interaction occurs, as well as the social interactions that take place once the creative process is complete. It is only when the user is allowed expression through a process that is specifically tailored to their individual capabilities and needs that full therapeutic benefits of identity-formation and self-expression can be realized.

12.8 BCI as a Communication Technology

Communication technologies, by their very nature, are social technologies. In claiming that "sociocultural factors are subject to technical analysis just as technical factors are subject to social and cultural analysis," Ito (2008, p. 4) prompts us to recognize the inseparability of studying the characteristics of tools we use and the questions surrounding our own use habits. Though some may view communicating health messages as creating a reality molded by social, cultural, ethnic and economic forces taking shape over time, others believe that the reality of this discipline is dependent upon individual and group constructions (Guba and Lincoln 1994). When discussing the relationship between a specific environment and self-expression as it relates to BCI systems, it may be helpful to examine interactions between user and interface as reciprocal exchanges of information. In doing so, it becomes difficult to make general assumptions about the personal, social and cultural influences inherent within the processes (Lowery and DeFleur 1983), thus lending credence to the necessity for examining each user/technology interaction as a unique and special exchange based on both user-specific needs as well as goals for enablement of a communal-based program.

In using social representation theory to assess exposure impact of science and health efforts and effects outside of a laboratory setting, Hwang and Southwell (2009) acknowledge that attitude formation and adoption are a fluid process dependent on these user/technology interactions. Nisbet and Scheufele (2009) also suggest that emergent technologies are more likely to advance change through resonation with discourse of the individual and community. These ideas imply that development, design and implementation efforts for these enhanced BCI systems should not be conceived with a "one-size-fits-all" approach in mind, but instead based on a systematic empirical understanding of a user's own values, knowledge and attitudes, their interpersonal and social contexts, and their preferred media sources and communication channels. In this, it may be beneficial to utilize both a critical and a constructivist approach to analyzing the composite product of interface design, mediated interaction with the technology, and the communication that occurs between the user and researcher/therapist.

These separate yet arguably complimentary paradigms outline two separate frameworks upon which these issues can be approached. First, from a critical standpoint, the basis of inquiry into this discipline requires a dialogue between an investigator and the subjects, and supposes that the subsequent findings are value mediated. This means that taking a critical approach towards communicating health issues requires the researcher to acknowledge that the communication processes will be both transactional and subjective (Guba and Lincoln 1994). The second framework is derived from a constructivist point of view, as it approaches outcomes as continuously created and alterable processes. This differs from critical in that the nature of inquiry and knowledge is dependent upon individual user constructions and reconstructions of information, instead of a critique into the structural and historical insights of a situation. Constructivism also facilitates change through reconstructions formed through engagement while stimulating participants to act on those changes (Guba and Lincoln 1994). Taking into consideration the value that each individual method has to offer, it is not surprising that many health communication scholars have taken a transdisciplinary approach towards solving multi-faceted problems using these paradigms in conjunction with one another, as they maintain similar epistemological assumptions.

12.9 BCI as a Digital Environment

Likewise, when discussing user engagement, researchers should not ignore all characteristics of the specific environment in which the communicative act takes place. The "environment" in this case includes but is not limited to: the physical location in which the user is situated, the objects and other people within the space, and the mediating digital interface that enables the communicative/creative act. Research pertaining to the study of Social Network Structures (SNS) suggests that the technology's interface (or, the architecture of digital communication media) may be largely responsible for dictating the type of behavior that is seen as socially "normal" within the context of each space. Stromer-Galley and Martey (2009) echo this sentiment in suggesting that consideration must be given to the role the environment plays in the interaction process, while Eco (1986) posits architecture as consisting of spatially-embodied forms that communicate their function as a result of the social and cultural forces that have brought them into being. Social interactions that occur within digital spaces naturally encounter boundaries and limitations similar to those found within the built environment, and these structural components are elements that can both afford liberties to and place constraints upon users within a space, in turn affecting how users behave within a given space. Similar to physical spaces, the digital environment influences the individual's choices about engaging with other individuals and communities within and through use of these environments. In considering the positioning of true self within a digital context, the formation of identity in online SNSs is often based on the fundamental claim that one's identity is a complex product; one that is socially constructed from both inherent individual characteristics as well as various social contexts experienced through exposure to different environments (Postmes et al. 2005). This suggests that identity is formed not solely as a result of who an individual *is*, per se, but also derived from the negotiation process of whom others perceive an individual to be (Walther et al. 2009). This is important to consider when attempting to evaluate the ethical nature of identity formation through digitally-enabled BCI interactions, as these spaces provide the potential for expanded agency concerning presentation of one's identity to both individuals and communities alike. The developmental and design considerations that follow then become the framework for how an individual will practice this behavior, as well as how they will interact with others within these spaces (Papacharissi 2002).

12.10 Identity and Self-expression Through Art

Furthermore, online and digital identity formation can affect the development of boundaries for thoughtful behavior amongst individuals while simultaneously constraining a user's ability to develop and nurture an individuals' sense of self (Davis 2011). Research into the affordances and constraints presented by digital interfaces has demonstrated that the design of digital environments can present users with physical, psychological and structural limits that lead to repercussions for self-expression. In other words, although digital spaces may offer users a sense of freedom from the social norms and values that exist in everyday life, many recognize that the boundaries of these virtual environments are similar to those found in physical environments. Rather than providing an escape from social constraints of the offline world, that these arenas are more likely to mirror the social norms established outside of the digital spaces (Papacharissi 2005). However, BCI technologies cannot be expected to immediately enable a utopia of an unrestrained universal language, as users will likely face difficulties at the onset concerning the process of identity formation. For those that rely on and engage with BCI systems for communication and self-expressive means, this suggests that interface architecture plays a role in identity formation through the process of social interaction, and that evaluating the architecture of these communication systems is a crucial step towards understanding how to fulfill a user's need for self-expression and provide a sense of connectivity; both of which are social processes considered to be crucial to the formation and maintenance of human identity.

12.11 Information Through Visual Representation

The study of visuals and the creation of art involves the role that aesthetic design plays within the mediated process between user and technological interface. Visual images, as Midalia (1999) explains, "are never innocent or neutral reflections of reality... they re-present for us: that is, they offer not a mirror of the world but an interpretation of it" (131). Aesthetic design affects how we as social beings experience meaning, as our brains and bodies are continuously interacting with the environments that surround us. As such, embodiment as a factor in self-identification posits itself as an important theoretical construct in discussing experiences for users of BCI.

Different disciplines approach embodiment from different starting points, ranging from theoretical grounding and analysis (Philosophy) to evidence-based investigation (Artificial Intelligence) (Farr et al. 2012). Phenomenologist Maurice Merleau-Ponty's account of embodiment distinguishes between two different types of bodies; for one there is the objective body, which is regarded as our actual physiological body, Alternatively, there exists the phenomenal body, which transcends the corporeal physical body and instead acts as the body which is "experienced." Embodiment in the latter sense extends beyond recognition of the tenets and abilities of the corporeal body itself to recognize the importance of interaction as fostering the creation of the bodily experience. Ryan (2010) describes the primary notion of embodiment as "the ideas that mind and body cannot be separated, and that thinking is profoundly affected by the mind's embodiment" (485). Dourish (2001) writes that embodiment "is not a property of systems, technologies, or artifacts; it is a property of interaction" (189) and Wood (2007) notes that it is important to analyze the different embodied encounters that different technologies can offer. When applied to the context of BCI for a therapy or rehabilitation-based environment, the limitations of the physical body alter the traditional interaction between mind and body, in turn inducing a reconsideration of the idea of embodiment and how it may be considered as a vital component of identity formation for users. Farr et al. (2012) note that "this reunification of body, action, and mind is a key consideration in contemporary debates around embodiment" (3), and researchers across a transdisciplinary spectrum must be cognizant of the various approaches for reunifying mind and body across differing user populations.

Embodied experience is approached not only as a multimodal field to be explored, but also as an ability to refine. A study conducted by Banakou et al. (2008) demonstrated that virtual embodiment placing the user within the body of Albert Einstein lead to changes in implicit attitudes and biases, as well as alterations of cognitive perception. This illusion of ownership over the body resulted in improvements of participant mood and happiness, as the self-aligned with characteristics of the "new" body and allowed the user to access mental resources that differ from traditional ways of thinking about themselves and their own abilities. Dix and Gill (2018) suggest that a philosophical approach to embodiment is one that encourages us "to see perception, cognition and action not as separated stages in a pipeline, but as a single process, with a rich intertwining of self and world" (138). This dynamic view of embodiment has implications for design both as a means of acting as well as an end product. It can be a source of potential insights and methodologies for design, and as such, should be considered a high priority in future development of systems for supporting enhanced cultivation of the self.

12.12 Interventions

Communication research has shown that interaction in digital spaces is perceived among users to be equally as important as interaction in real life social engagements (Papacharissi 2005). Recognizing this, there should be efforts to integrate these considerations when designing digital spaces that offer more affordances to users. Through more accurate EEG signal acquisition and a responsive closed neurofeedback loop with BCI interactions, this experimental concept has the potential to expand boundaries for existing language potentials for BCIs by blending the physical with the digital (Papacharissi 2002). Integrating this type of application during the process of creative expression could construct new interpretations for technological spaces by engaging explanation and understanding of the interactions that occur between the user and their environment.

These proof of concept efforts explore how applied communicative theoretical frameworks and Human-Computer Interface (HCI) concepts can advance potential therapeutic opportunities for neurofeedback and biofeedback applications. Through integrating a transdisciplinary framework derived art, technology and therapy disciplines, these applications look to enhance communication within existing channels of language behaviors for users. The overarching goals of these efforts were varied, yet each was designed to contribute to the neurophysiological creative experience for users. Potential objectives for the interactive applications were identified as efforts that:

- Encourage new methods of internal and external discovery for users
- Provide a unique entry point to BCI technologies for neurologically-impaired individuals
- Offer a tool to improve QOL among disabled individuals
- Promote rehabilitation and recovery through artistic self-awareness and reflection
- Position the technology as an innovative assessment tool for therapists
- Offer an interactive outlet and provide an emotional connection with medical caregivers
- Offer alternative perspectives towards mobility perception
- Adaption of how users can learn for efficacy
- Encourage transdisciplinary discourse designed for problem solving within these fields (Scott and Gehrke 2019)

Beyond this, these efforts aim to bring together leading intellectuals and innovators from the art, engineering and scientific communities for the purpose of cultivating and promoting strategic efforts for encouraging artistic and creative developmental initiatives within emerging neuroscience and new medical technologies research. Ideally, this work may help to highlight the need for scientists and artists alike to reimagine and modernize their own boundaries to develop shared meanings and embrace supportive spaces designed to foster these important types of multi-modal and multidisciplinary discourses for future development and possible future integration of their fields.

12.12.1   Neurofeedback and Artmaking

The purpose of the intervention was to enable creative experience for the BCI system user, and to demonstrate how visualization of EEG data could help to enable artistic self-expression (Fig. 12.2). To do so, an active EEG analysis was combined with visual neurofeedback output to enable a process that replicated methods of traditional analog painting. The finished visual product consisted of visual overlays projected upon a transparent canvas.

The main objectives this application hoped to address through this type of setup included promotion of an increased sense of self-awareness through artistic self-expression and increased capability for users to identify with, and to tell their own "stories" through individual biological information integrated within an interface. In assisting users to subjectively create their own narratives through this interface design, the project aimed to extend the social sense of comprehension towards the technology itself, as well as towards the complex issues inherent within users' creative expressions.

As the primary objective was to enable self-awareness through the creative artistic process, challenges for future art and therapeutic applications of this intervention method include examining potential use-cases. First, these efforts identified that patient populations may stand to benefit from exposure to visualizations of their own EEG data as a component of an art-therapy program. Likewise, this project recognized the potential these applications for promoting both awareness and understanding of complex medical issues (such as brain trauma) towards a variety of audiences through visualization of the data. Lastly, it is possible that situating this technology within the art-therapy field could disrupt existing schemas pertaining to BCI processes, and position itself as a viable alternative tool within existing options for interaction modalities.

EEG data was collected from brain cap[1] with 32 dry electrodes using an amplifier[2] sampled at 250Hz before being streamed from a recording PC to the presentation computer. Once the raw data was collected[3] in Python[4] format, data of 2s was buffered before applying a bandpass filter (1–125 Hz) on the 2s data window with a subsequent time frequency decomposition using fast fourier transform to estimate power spectral density. We then extracted power values for 5 common EEG bands (delta, theta, alpha, beta, and gamma)[5] to feed into a visualization scheme (see Figs. 12.2 and 12.5). At this point, the visualization output from the collected data was projected onto transparent podium paper, which completed the closed-loop neurofeedback setup. By incorporating a watercolor painting setup into the experiment (paint type, brushes, white drawing sheet) researchers were able to achieve the desired transparency effect. Participants were not given any instructions or parameters concerning the creation of their art (Scott and Gehrke 2019).

Moving forward, there are additional considerations to be considered when evaluating parameters of both user circumstance as well as goals of each individualized therapy program. For one, we can examine the possibility of source-level instead of sensor-level EEG dynamics with the according spatial filtering techniques (Cohen 2017). Strong comprehension of affected EEG signatures is necessary in order to

---

[1] actiCAP Xpress Twist, Brain Products Available at: https://pressrelease.brainproducts.com/twist/.

[2] LiveAmp compact wireless amplifier, Brain Products Available at: https://www.brainproducts.com/productdetails.php?id=63&tab=1.

[3] Data received by LabStreamingLayer inlet available at: https://github.com/sccn/labstreaminglayer.

[4] Python Software Foundation. Python Language Reference, version 2.7. Available at http://www.python.org.

[5] https://en.wikipedia.org/wiki/Electroencephalography.

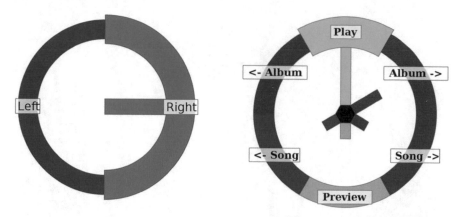

Fig. 12.5 Featured images of a motor imagery user-interface design (left) and a multiplayer module (right) for the CEBL3 Brain-Computer Interfaces platform. (CEBL-software program designed by Colorado Electroencephalography and BCIL laboratories). For the user-interface on the left, a user simply imagines moving their left or right arm for item selection. To do so, the machine learning algorithm identifies alterations of power in the mu band within the contralateral motor region of the brain. This interface may also incorporate an Event-Related Potential (Error-Related Negativity) in order to identify incorrect selections made by the BCI system. The multiplayer module on the right is used to elicit a P300 event-related potential for item selection, allowing a user to focus on their selection. Through advanced machine learning algorithms, users can perform actions by concentrating on more simple visualizations that interfaces that use flashing or flickering components. It also allows users to use a variety of imagined mental tasks that may evoke changes in activity over various regions of the brain (see Forney et al. 2013a, b, 2015; Forney and Anderson 2015)

efficiently target the EEG features and source locations specific to user conditions, as well as to optimize results from intervention processes (Lizio et al. 2011; Monge-Pereira et al. 2017). Also, by acknowledging the challenges relevant to a specific user population, we stand to achieve a better understanding of how a technology will or won't be accepted by that population. We also propose the use of digital drawing tablets to serve as canvases tailored to the needs of each individual user (Scott and Gehrke 2019).

12.12.2   Biofeedback and Artistic Expression and Engagement

Instead of controlling BCI through EEG features, this project explores the effect of simple and immediate feedback of EEG characteristics translated directly into a dynamic visual display by using the following methodology: To start, 8 circles (varying in both size and color) are digitally "drawn" upon the interface screen. At each sample in time, the EEG voltage from across the 8 electrodes is subtracted from the 8 voltages as a Common Average Reference (CAR). The EEG signals that

Fig. 12.6 A conceptual design of the application menu of a hybrid program that uses the CEBL3 platform and real-time neurofeedback to generate a full feedback loop for users. The first step of the application involves designing a visual overlay for the CEBL3 software platform (see Fig. 12.5) that complements the menu navigation screen. After making "tool selections" the user would be able to control the on-screen art creation tools through neurofeedback. As part of a rewards-based program, more tools would become available once the user meets certain predefined performance standards developed in advance. Art therapists would evaluate responses from user questionnaires in order to design customized multi-stage art therapy programs to suit each user's individual needs (e.g. see Jones et al. 2018 for patient-led treatment model, see National Organization for Arts in Health 2017 for patient-centered care, p. 17). In this, this gradual learning process would be likely to reduce frustration, increase motivation and engagement, and strengthen the reciprocal learning process between user and interface. This type of multi-step learning process would serve not only for calibration between human/computer, but also provide therapists an opportunity for expanded conversation pertaining to these experiences. This idea is a proof of concept, and we anticipate further research in examining these hypotheses

are greater than this average are drawn in green, and those lower than average in red. The radius of each circle is proportionate to the measured absolute value of the difference from the average. By filtering the EEG signals in order to preserve the slower calculated frequencies, we are able to differentiate the greater visual changes from those with a lesser variance (Figs. 12.3, 12.4 and 12.6).

BCI systems typically follow two types of paradigms: synchronous and asynchronous. Synchronous paradigms tend to align with Event Related Potentials (EPR) in which the dependency relies on a user's individual response with regards to a directed stimulus within a given time frame. However, asynchronous paradigms differ in that they do not depend upon external stimuli, are individually-paced, and deliver frequencies know and Event-Related (de)Synchronization (ERD/s) (Forney et al. 2015). In pairing the user-interface module software offered through Colorado Electroencephalography and Brain-Computer Interfaces Laboratory (CEBL currently CEBL3) with the above visual feedback prototype, we stand to learn more about the potential for both online and offline types of BCI hybrid applications (Forney and Anderson 2015) (Fig. 12.5).

Müller-Putz et al. (2015) categorize Hybrid BCI systems (or hBCI) as those which integrate other physiological or technical signals into the BCI. Several prototype user-interface modules for CEBL3 are currently in development; the first of which implements a P300-based BCI system, also known as the Rapid Serial Visual Presentation (RSVP) speller. The process of user/RSVP system interaction presents the user with an array of characters that flash quickly, as the user looks for the character they wish to use for spelling. While this is happening, the machine learning algorithm embedded within the RSVP system highlights changes in the EEG signals whenever the correct character flashes upon screen.

An alternate type of user-interface incorporates Motor Imagery (MI) to help the user select one of two items showcased in a circular "pie" menu (see Fig. 12.5). To do so, the machine learning algorithm identifies alterations of power in the mu band within the contralateral motor region of the brain as the user imagines moving their left or right arm. The paradigm within the MI interface may also incorporate an Event-Related Potential (Error-Related Negativity) in order to identify incorrect selections made by the BCI system. A third type of interface we wish to highlight is very similar to that of the MI paradigm; however, a key difference is that it allows for imagined mental tasks to produce changes in recognizable activity across different parts of the brain." With practice, this paradigm may yield multiple degrees of freedom and fluid, asynchronous control" (Forney and Anderson 2015).

This interface was designed to unlock new possibilities for user agency and increased degrees of freedom by way of feedback control, thus functioning primarily as a reward-based system. Using the CEBL system as the module menu (created by members of the Colorado State University Electroencephalography and Brain-Computer Interfaces Laboratory Colorado State University, see Forney et al. 2013a, b), in which imagined movement by the user provides the navigational control functions, such efforts would have the potential to help expand upon avenues for both training and learning within the system itself by way of the artistic-creative process (Fig. 12.6). By using the CAR to determine mean frequency for control, it integrates training procedures with a rewards-based gaming system to motivate users and boost confidence for their usability of the interface. Sense of personal agency and control over the system would stand to increase as users would be able to exercise artistic freedom without the impeding use of icons or a series of directive menus.

In considering the components of engagement and usability as integral prerequisites of new technology acceptance and adoption, the aforementioned interfaces stands to expand user possibilities for creative agency by reexamining not only the language typically inherent within BCI devices, but also the mediated user/interface interaction as well as the potential of art as a social-communicative medium. As noted, previous research has demonstrated the need for user-centric BCI design, and we believe that continued advancements within the field would stand to further development of technologies for specific users on a wider scale. An important next step could be working to clarify existing definitions for hBCI systems, as well as furthering design of systems that permit combined calibration activities as opposed to strict reliance upon the use of mental tasks. Also, continuing to develop BCI technology with rewards-based systems in mind may encourage users to unlock new modules for self-expression, which in turn could assist researchers with enhanced biofeedback training methodology and user control practices. Finally, wide-scale acceptance of these technologies may assist in promoting therapeutic intervention as a rehabilitative method that is especially beneficial for the individual user, and not just for the usability outcomes derived from the process.

Ideally, researchers from a variety of disciplines would utilize this approach in the future to promote art therapy as a viable means for therapeutic rehabilitation. This could best be achieved by including a dialogue box to enable patient/therapist discussion as well as a new form of visual language to assist users with expression of specific emotions. This type of interface may also be applied to other technological mediums, including virtual reality (VR) headsets and musical applications, in order to provide a complete omnisensory experience and allow for a wider range of self-expression for users. Ultimately, such advancements may enable the development of patient-centered art spaces to serve as virtual communities and support groups for sharing and discussing not just the artwork itself, but user stories that has previously gone untold.

12.13 Discussion

As the progression of technological development coincides with the increased presence of and dependency upon communicative devices and platforms, there is a visible and measurable cultural shift towards communication practices mediated by a specific interface. Orlikowski (2000) argues that altering user participation and the ecology of communication practices and social interactions could result in redefining the communicative status quo, while Shoemaker (1991) notes that there are consequences of these mediated processes which can be extremely difficult to predict. As BCI systems are technologically capable of affording various populations of users the ability to communicate via the combination of neurofeedback and biofeedback through physical, digital and biological components, it is likely that further development of these systems will lead to complex changes that extend far beyond physical alterations towards those which include reconfigurations of existing digital interfaces in order to mediate subsequent user interactions. As such, research that promotes continued development of BCI technologies in ways that place emphasis on improving communication means through these technologies may further the notion that the true potential of these systems exist not only for serving as a conduit for technical creation, but also as a vital component of one's ability for self-expression and self-fulfillment.

12.14   Conclusion

Though we believe this approach to BCI therapy has a tremendous amount of potential to improve the lives of its users, there are still concerns that must be addressed when implementing such a system on a wider scale. For one, the process of technology design brings with it major challenges related to mediated interactions including ethical concerns. In discussing ethical considerations of new technologies, Budinger and Budinger (2006) write that the designers of these technologies must consider all aspects of technology use, including the negative consequences that could occur from their use in the future. As mediated forms of social communication continue to shape the practices and behaviors of digital spaces and their respective populaces, it becomes crucial to evaluate not only the media-related choices that shape user self-expression and self-presentation within these environments, but the behaviors of and choices made by the corporations and organizations that control them as well (Yunn-Yu Sun 2012). Couldry (2013), similarly, emphasizes that ethical considerations regarding individual and corporate behavior need to be addressed when constructing an appropriate set of social norms within online environments. Concerning funding for future projects of this nature, the notion that advertisers instead of users may in fact be the greatest beneficiaries of this virtual structural adjustment suggests that these interfaces may play the role of key influencers towards the determination of ethical boundaries within these spaces. This should raise concern for both media scholars and medical professionals working to further the development and exploration of new modes of communication for disabled and healing populations, as the users may not always be the primary beneficiaries of these technological innovations.

As commercial interest within advancing technology continues to grow, resources may shift towards for-profit endeavors and away from technologies serving disabled populations. In order to continue accommodating BCI users, there is a need to take accountability and accommodate for the logic and application of new communication technologies. When specific technical constraints are placed upon the communication processes, it may lead to dependency and heteronomy amongst users (Schulz 2004). These problems are especially apparent in health science fields, as users who are in the most need for a particular technology rarely get the benefit of development and integration for maximum application to the unique physical and psychological needs that would enable communication and self-identification. Also, changing the choice architecture or creating interface features that "nudge" users towards certain behaviors contributes to a process that prompts unethical online behavioral practices that are responsible for reshaping the social norms of social mediated interactions (Dijck 2013). The notion that for-profit entities and not users may in fact be the greatest beneficiaries of this virtual structural adjustment suggests that these interfaces may play the role of key influencers towards the determination of ethical boundaries within these spaces. This should raise concern for both media scholars and medical professionals working to further the development and exploration of new modes of communication for disabled and healing populations.

Another issue that calls for greater attention within the healthcare technology field is that which arises from SNS research abilities to "provide the individual with efficient and convenient tools for maintaining contact with potential social resources based on personal needs and interests, rather than superordinate needs of a community" (Manago 2014, p. 3). This process creates a phenomenon that involves "commodifying" one's self, which may be destructive in the sense of an individual prioritizing her actions based on personal interests rather than evaluating those actions based on the greater needs of the community. In a drive to stay continuously connected with one another, maintaining a unified and socialized sense of self and defining ethical social norms within BCI hardware and interface design will prove fundamental in preserving our ability to make rational decisions and conduct individual behavior in an ethical manner (Gergen 2000).

Helping patients must remain the top priority of the field, and with new communication technologies at the forefront of a digital revolution, it is no longer enough "to demonstrate that a simple BCI might allow one specific function, most of the time, for some patients" (Allison 2009, p. 558). Instead, we must prioritize both accessibility and customization to enable individualized communication in the development of future BCI technologies. Through combining new technologies with visual educational strategies alongside the integration of knowledge from disciplines such as communication, sociology, and phenomenology, more innovative strategies towards communicating complex information about new health communication technologies can be developed and implemented in order to increase value for its highly-diverse user base. Nisbet and Scheufele (2009) suggest that science communication should be based on the interpersonal and social concepts including their preferred media sources and communication channels most relevant to individuals, and we believe that these types of advancements have the potential to achieve such a goal. By having health information occupy more visually-interactive and experiential spaces, we are more likely to see the communication enabled by these interactions work in a positive manner for the diverse user base; one that can generate awareness and positively influence attitudes towards new health and science innovations.

References

Allen PB (1995) Art is a way of knowing. Shambhala, Boston

Allison B (2009) The I of BCIs: next generation interfaces for brain–computer interface systems that adapt to individual users. In: Jacko JA (ed) Human-computer interaction. Novel interaction methods and techniques. Lecture notes in computer science, vol 5611. Springer, Heidelberg, pp 558–568

Anderson C, Besio W, Alzahrani S (2018) Comparison of conventional and tripolar EEG electrodes in BCI paradigms. In: Proceedings of the 7th international brain-computer interface meeting: BCIs, not getting lost in translation, Asilomar Conference Center, Pacific Grove, California, 21–25 May 2018. http://bcisociety.org/wpcontent/uploads/2018/05/BCI2018AbstractBook.pdf. Accessed 2 Jan 2019

Angheluta A, Lee BK (2011) Art therapy for chronic pain: applications and future directions. Can J Couns Psychother 45(2):112–131

Ball T, Kern M, Mutschler I, Aertsen A, Schulze-Bonhage A (2009) Signal quality of simultaneously recorded invasive and non-invasive EEG. NeuroImage 46(3):708–716. https://doi.org/10.1016/j.neuroimage.2009.02.028

Banakou D, Kishore S, Slater M (2018) Virtually being Einstein results in an improvement in cognitive task performance and a decrease in age bias. Front Psychol 9:1–14. https://doi.org/10.3389/fpsyg.2018.00917

Besio WG, Martinez-Juarez IE, Makeyev O, Gaitanis JN, Blum AS, Fisher RS, Medvedev AV (2014) High-frequency oscillations recorded on the scalp of patients with epilepsy using tripolar concentric ring electrodes. IEEE J Transl Eng Health Med 2:1–11. https://doi.org/10.1109/jtehm.2014.2332994

Blain-Moraes S, Schaff R, Gruis KL, Huggins JE, Wren P (2012) Barriers to and mediators of brain-computer interface user acceptance: focus group findings. Ergonomics 55(5):516–525. https://doi.org/10.1080/00140139.2012.661082

Brunner C, Birbaumer N, Blankertz B (2015) BNCI horizon 2020: towards a roadmap for the BCI community. Brain Comput Interfaces 2(1):1–10. https://doi.org/10.1080/2326263X.2015.1008956

Budinger T, Budinger MD (2006) Ethics of emerging technologies: scientific facts and moral challenges. Wiley, New Jersey

Cohen MX (2017) Comparison of linear spatial filters for identifying oscillatory activity in multi-channel data. J Neurosci Methods 278:1–12. https://doi.org/10.1016/j.jneumeth

Couldry N (2013) Why media ethics still matters. In: Ward SJA (ed) Global media ethics: problems and perspectives. Blackwell Publishing Ltd., pp 13–30

Cox R, Depoe S (2015) Emergence and growth of the "field" of environmental communications. In: Hansen A, Cox R (eds) The Routledge handbook of environment and communication. Routledge, London, pp 13–25

Czamanski-Cohen J, Weihs KL (2016) The body mind model: a platform for studying the mechanisms of change induced by art therapy. Arts Psychother 51:63–71. https://doi.org/10.1016/j.aip.2016.08.006

Daly JJ, Huggins JE (2015) Brain-computer interface: current and emerging rehabilitation applications. Arch Phys Med Rehabil 96(3):S1–S7. https://doi.org/10.1016/j.apmr.2015.01.007

Davis K (2011) Tensions of identity in a networked era: young people's perspectives on the risks and rewards of online self-expression. New Media Soc 14(4):634–651

Dijck JV (2013) You have one identity': performing the self on Facebook and Linkedin. Media Cult Soc 35:199–215

Dix A, Gill S (2018) Physical computing: when digital systems meet the real world. In: Filimowicz M, Tzankova V (eds) New directions in third-wave human computer interaction, vol 1. Springer, Gewerbestrasse, pp 123–144

Dourish P (2001) Where the action is: the foundations of embodied interaction. The MIT Press, Massachusetts

Du Gay P, Hall S, Janes L, Mackay H, Negus K (1997) Doing cultural studies: the story of the Sony Walkman. Sage Publications, London

Eco U (1986) Travels in hyperreality: essays. Harcourt Brace Jovanovich, San Diego

Ehresman C (2014) From rendering to remembering: art therapy for people with Alzheimer's disease. Int J Art Ther Former Inscape 19(1):43–51. https://doi.org/10.1080/17454832.2013.819023

Farr W, Price S, Jewitt C (2012) An introduction to embodiment and digital technology research: interdisciplinary themes and perspectives. National Centre for Research Methods Working Paper. MODE node, Institute of Education, London

Folgieri R, Dei Cas L, Dei Cas D, Vanutelli ME, Lucchiari C (2017) The creative mind-dracle. Paper presented at the Electronic visualisation and the arts, London, United Kingdom, 11–13 July 2017. http://dx.doi.org/10.14236/ewic/EVA2017.59. Accessed 20 Nov 2018

Forney E, Anderson C (2015) CEBL3: a modular platform for EEG signal analysis and real-time brain-computer interfaces. Poster presented at the 12th annual meeting of the front range neuroscience group, Fort Collins, Colorado, 10 Dec 2014. http://www.cs.colostate.edu/eeg/main/sites/default/files/forney_poster_cebl_FRNG-2015.pdf. Accessed 8 May 2016

Forney E, Anderson C, Gavin W and Davies P (2013a) A direct brain-computer interface for multimedia and environmental controls. Poster presented at the Colorado State University ventures innovation symposium, Fort Collins, Colorado, 12 April 2013. http://www.cs.colostate.edu/eeg/main/sites/default/files/poster_BCIPlayer.pdf. Accessed 16 May 2016

Forney E, Anderson C, Davies P, Gavin W, Taylor B, Roll M (2013b) A comparison of EEG systems for use in P300 spellers by users with motor impairments in real-world environments. In: Proceedings of the 7th international brain-computer interface meeting: BCIs, not getting lost in translation, Asilomar Conference Center, Pacific Grove, California, 3–7 June 2013. http://www.cs.colostate.edu/eeg/main/sites/default/files/forney_poster_p300_bcimeeting-2013.pdf. Accessed 27 Nov 2018

Forney E, Anderson C, Gavin W, Davies P (2015) Convolutional networks for EEG signal classification in non-invasive brain-computer interfaces. Poster presented at the 12th annual meeting of the front range neuroscience group, Fort Collins, Colorado, 9 Dec 2015. http://www.cs.colostate.edu/eeg/main/sites/default/files/frng_conv_poster2015.pdf. Accessed 8 May 2016

Forney E, Anderson C, Gavin W, Davies P (2018) Mental-task BCIs using convolutional networks with label aggregation and transfer learning. In: Proceedings of the 7th international brain-computer interface meeting: BCIs, not getting lost in translation, Asilomar Conference Center, Pacific Grove, California, 21–25 May 2018. http://bcisociety.org/wpcontent/uploads/2018/05/BCI2018AbstractBook.pdf. Accessed 20 Oct 2018

Gergen KJ (2000) The saturated self: dilemmas of identity in contemporary life. Basic Books, New York

Gharabaghi A, Naros G, Khademi F, Jesser J, Spüler M, Walter A, Bogdan M, Rosenstiel W, Birbaumer N (2014) Learned self-regulation of the lesioned brain with epidural electrocorticography. Front Behav Neurosci 8:429. https://doi.org/10.3389/fnbeh.2014.00429

Gruzelier J (2013) EEG-neurofeedback for optimizing performance. II: Creativity, the performing arts and ecological validity. Neurosci Biobehav Rev 44(2014):142–158

Gruzelier J, Egner T (2004) Physiological self-regulation: biofeedback and neurofeedback. In: Williamon A (ed) Musical excellence: strategies and techniques to enhance performance. Oxford University Press, London, pp 197–219

Guba ES, Lincoln YS (1994) Competing paradigms in qualitative research. In: Denzin NK, Lincoln YS (eds) Handbook of qualitative research. Sage Publications, Thousand Oaks, pp 105–117

Harrison B (2002) Seeing health and illness worlds—using visual methodologies in a sociology of health and illness: a methodological review. Sociol Health Illn 24(6):856–872

Houts PS, Doak Cecilia C, Doak Leonard G, Loscalzo MJ (2006) The role of pictures in improving health communication: a review of research on attention, comprehension, recall, and adherence. Patient Educ Couns 61(2):173–190

Huggins JE, Wren PA, Gruis KL (2011) What would brain-computer interface users want? Opinions and priorities of potential users with amyotrophic lateral sclerosis. Amyotroph Lateral Scler 12(5):318–324. https://doi.org/10.3109/17482968.2011.572978

Hwang Y, Southwell BG (2009) Science TV news exposure predicts science beliefs: real world effects among a national sample. Commun Res 36:724–742

Ito M (2008) Introduction. In: Varnelis K (ed) Networked publics. MIT Press, London, pp 1–14

Jones JP, Walker MS, Drass JM, Kaimal G (2018) Art therapy interventions for active duty military service members with post-traumatic stress disorder and traumatic brain injury. Int J Art Ther 23(2):70–85. https://doi.org/10.1080/17454832.2017.1388263

Kos A, Milutinovi CV, Umek A (2018) Challenges in wireless communication for connected sensors and wearable devices used in sport biofeedback applications. Future Gener Comput Syst. https://doi.org/10.1016/j.future.2018.03.032

Kübler A, Müller KR (2007) An introduction to brain computer interfacing. In: Dornhege G, Millán JR, Hinterberger T, McFarland D, Müller KR (eds) Toward brain–computer interfacing. MIT Press, Cambridge, pp 1–25

Kübler A, Holz E, Kaufmann T, Zickler C (2013) A user centered approach for bringing BCI controlled applications to end-users reviewed. In: Fazel-Rezai R (ed) Brain-computer interface systems-recent progress and future prospects. InTechOpen Limited, London. https://doi.org/10.5772/55802

Kübler A, Holz EM, Sellers EW, Vaughan TM (2015) Toward independent home use of brain-computer interfaces: a decision algorithm for selection of potential end-users. Arch Phys Med Rehabil 96(3):S27–S32. https://doi.org/10.1016/j.apmr.2014.03.036

Liebermann DG, Katz L, Hughes MD, Bartlett RM, McClements J, Franks IM (2002) Advances in the application of information technology to sport performance. J Sports Sci 20(10):755–769

Lizio R, Vecchio F, Frisoni GB, Rodriguez and Babiloni (2011) Electroencephalographic rhythms in Alzheimer's disease. Int J Alzheimer's Dis. https://doi.org/10.4061/2011/927573

Lowery SA, De Fleur ML (1983) Developing frameworks for studying mass communication. In: Milestones in mass communication research. Longman, New York, pp 1–29

Manago A (2014) Identity development in the digital age: the case of social networking sites. Oxford handbooks of identity online. https://doi.org/10.1093/oxfordhb/9780199936564.013.031

McLuhan M, Fiore Q (1967) The medium is the message. Random House, New York

Midalia S (1999) Textualizing gender. Interpretations 32(1):27–32

Monge-Pereira E, Molina-Rueda F, Rivas-Montero FM, Ibáñez J, Serrano JI, Alquacil-Diego IM, Miangolarra-Page JC et al (2017) Electroencephalography as a post-stroke assessment method: an updated review. Neurología (English Ed) 32(1):40–49. https://doi.org/10.1016/j.nrleng.2014.07.004

Morone G, Pisotta I, Pichiorri F, Kleih S, Paolucci S, Molinari M, Cincotti F, Kübler A, Mattia D (2015) Proof of principle of a brain-computer interface approach to support poststroke arm rehabilitation in hospitalized patients: design, acceptability, and usability. Arch Phys Med Rehabil 96(3):S71–S78. https://doi.org/10.1016/j.apmr.2014.05.026

Müller-Putz G, Leeb R, Tangermann M, Höhne J, Kübler A, Cincotti F, Mattia D, Rupp R, Müller KR, Millán JDR (2015) Towards noninvasive hybrid brain–computer interfaces: framework, practice, clinical application, and beyond. Proc IEEE 103(6):926–943. https://doi.org/10.1109/JPROC.2015.2411333

Nam C, Nijholt A, Lotte F (eds) (2018) Brain-computer interfaces handbook: technological and theoretical advances. CRC Press, Taylor & Francis Group, Boca Raton

National Organization for Arts in Health (2017) Arts, health, and well-being in America. https://thenoah.net/. Accessed 20 Nov 2017

Nijholt A, Jacob JKR, Andujar M, Yuksel BF, Leslie G (2018) Brain computer interfaces for artistic expression. In: Extended abstracts of the 2018 CHI conference on human factors in computing systems, Montreal, QC, Canada, 21–26 April 2018. https://doi.org/10.1145/3170427.3170618. Accessed 24 Sept 2018

Nisbet MC, Scheufele DA (2009) What's next for science communication? Promising directions and lingering distractions. Am J Bot 97:1767–1778

Ordikhani-Seyedlar M, Lebedev MA, Sorensen Pufthusserypady S (2016) Neurofeedback therapy for enhancing visual attention: state-of-the-art and challenges. Front Neurosci 10(352):1–15. https://doi.org/10.3389/fnins.2016.00352

Orlikowski W (2000) Using technology and constituting structures: a practice lens for studying technology in organizations. Organ Sci 11(4):404–428

Orndorff-Plunkett F, Singh F, Aragón OR, Pineda JA (2017) Assessing the effectiveness of neuro-feedback training in the context of clinical and social neuroscience. Brain Sci 7(8):95. https://doi.org/10.3390/brainsci7080095

Papacharissi Z (2002) The presentation of self in virtual life: characteristics of personal home pages. Journal Mass Commun Q 79(3):643–660

Papacharissi Z (2005) The real-virtual dichotomy in online interaction: new media uses and consequences revisited. Commun Yearb 29:215–237

Pfurtscheller G, Leeb R, Keinrath R, Friedman D, Neuper C, Guger C, Slater M (2006) Walking from thought. Brain Res 1071(1):145–152. https://doi.org/10.1016/j.brainres.2005.11.083

Postman N (1970) The reformed English curriculum. In: Eurich AC (ed) High school 1980: the shape of the future in american secondary education. Pitman, New York, pp 160–168

Postmes T, Haslam SA, Swaab R (2005) Social influence in small groups: an interactive model of social identity formation. Eur Rev Soc Psychol 16(1):1–42

Potash JS (2010) Guided relational viewing: art therapy for empathy and social change to increase understanding of people living with mental illness. Dissertation, University of Hong Kong

Ryan ML (2010) Narratology and cognitive science: a problematic relation. Style 44(4):469–495

Schulz W (2004) Reconstructing mediatization as an analytical concept. Eur J Commun 19(1):87–101. https://doi.org/10.1177/0267323104040696

Schweizer S, Thompson JL, Teel T, Bruyere B (2009) Strategies for communicating about climate change impacts on public lands. Sci Commun 31(2):266–274. https://doi.org/10.1177/1075547009352971

Scott SM, Gehrke L (2019) Neurofeedback during creative expression as a therapeutic tool. In: Contreras-Vidal JL, Robleto D, Cruz-Garza JG, Azorin JM, Nam CS (eds) Mobile brain–body imaging and the neuroscience of art, innovation and creativity. Bio- and neurosystems. Springer Nature (in press)

Shoemaker P (1991) Gatekeeping. Sage Publications, Newbury Park

Sigrist R, Rauter G, Riener R, Wolf P (2013) Augmented visual, auditory, haptic, and multimodal feedback in motor learning: a review. Psychon Bull Rev 20(1):21–25. https://doi.org/10.3758/s13423-012-0333-8

Star SL (1999) The ethnography of infrastructure. Am Behav Sci 43:377–391

Stewart EG (2004) Art therapy and neuroscience blend: working with patients who have dementia. Art Ther J Am Art Ther Assoc 21:148–155

Stromer-Galley J, Martey RM (2009) Visual spaces, norm governed places: the influence of spatial context online. New Media Soc 11(6):1041–1060

Umek A, Kos A, Tomažič S (2016) Sensor signal processing for biofeedback applications in sport. In: Proceedings of the 6th international conference on information society and technology (2), pp 327–330, Kopaonik, Serbia, 28 February – 2 March 2016. http://www.eventiotic.com/eventiotic/files/books/icist2016.pdf. Accessed 3 Jan 2019

Walther JB (1992) Interpersonal effects in computer-mediated interaction: a relational perspective. Commun Res 19:52–90

Walther JB, Van Der Heide B, Hamel LM, Shulman HC (2009) Self-generated versus other-generated statements and impressions in computer-mediated communication: a test of warranting theory using facebook. Commun Res 36(2):229–253

Webster A (2002) Innovative health technologies and the social: redefining health, medicine and the body. Curr Sociol 50(3):443–445

Wolpaw J, Wolpaw EW (2012) Brain-computer interfaces: principles and practice. Oxford University Press, New York

Wolpaw J, Birbaumer N, McFarland D, Pfurtscheller G, Vaughan TM (2002) Brain-computer interfaces for communication and control. Clin Neurophysiol 113(6):767–791. https://doi.org/10.1016/S1388-2457(02)00057-3

Wood A (2007) Digital encounters. Routledge, New York

Yunn-Yu Sun C (2012) Something old, something new, something borrowed, something blue: the construction of online identity and its consequences. Int J People Oriented Program 2(1):53–73

Zickler C, Halder S, Kleih SC, Herbert C, Kübler A (2013) Brain painting: usability testing according to the user-centered design in end users with severe motor paralysis. Artif Intell Med 59(2):99–110. https://doi.org/10.1016/j.artmed.2013.08.003

Chapter 13
BCI-Based Expressive Arts: Moving Toward Mind-Body Alignment

Rainbow Tin Hung Ho, Sunee H. Markosov, Nathan Sanders and Chang S. Nam

Abstract The aim of this chapter is to review the state of the art of BCI-based expressive arts, and review the possibilities as well as challenges involved in artistic expression and therapeutic applications of BCIs. We introduce the field of artistic BCI, its history, most common taxonomies and points of intersection with expressive arts-based therapies. We then discuss matching the artistic BCI technologies with different modalities of art-based interventions, and with different client categories, with the focus on mind-body alignment. We will conclude with a list of open problems and recommendations crucial for establishing a beneficial impact of BCI technology on artistic expression and therapeutic efforts.

Keywords BCI · Artistic BCI · Affective BCI · Expressive arts-based interventions · Mind-body alignment · Therapy · Creative arts · Digital media technologies

13.1 Introduction

In this section, we will introduce the field of artistic BCI, presenting a taxonomy focused on user control and modality of inputs and outputs, followed by and a brief history of artistic BCI approaches and applications. This section will also describe three categories of artistic BCI, those of medium innovation, facilitation of artistic expression for people with neuromuscular injury or illness, and as a form of art

R. T. H. Ho (✉)
Department of Social Work and Social Administration, University of Hong Kong, Pokfulam, Hong Kong
e-mail: tinho@hku.hk

R. T. H. Ho · S. H. Markosov
Centre on Behavioral Health, University of Hong Kong, Pokfulam, Hong Kong

N. Sanders · C. S. Nam
Department of Industrial and Systems Engineering, North Carolina State University, Raleigh, USA

© Springer Nature Switzerland AG 2019
A. Nijholt (ed.), *Brain Art*,
https://doi.org/10.1007/978-3-030-14323-7_13

therapy. The section will conclude by describing several examples of artistic BCI within the visual art and music modalities of artistic expression.

13.1.1 Taxonomy of Artistic BCIs

Artistic BCIs can be classified into four groups based on the type of user control: passive, selective, direct, and collaborative (Wadeson et al. 2015). Passive systems are heavily computer-based. The process is very algorithmic and does not require user interaction. Selective systems are similar to passive systems, but users can purposefully modulate their own biosignals that are being used as input. This gives users some control over the art they produce, but the results are still largely what Brian Eno might call 'generative'. In contrast, direct-control systems are toolboxes designed to facilitate deliberate artistic expression by giving the artist as much control as possible. There can still be severe constraints, but these are not a result of the control philosophy, but from the medium of BCI itself and its inherent technical limitations. Collaborative systems are usually forms of passive or selective systems in which multiple users can simultaneously influence the artwork.

13.1.2 Applications of Artistic BCIs

The applications of brain-computer interfaces in arts can be broadly grouped into three categories. The first is innovation, and the desire to explore the limits of new mediums. Following Joe Kamiya's discovery in 1962 that people could 'learn to discriminate the presence versus absence of EEG alpha activity' and his subsequent work on neurofeedback, there has been a rich vein of artistic exploration which drew upon and expanded his work (Nijholt 2015; Kamiya 2011). In 1976, David Rosenboom published the book "Biofeedback and the Arts—Results of Early Experiments" which documented eight years' worth of artistic experiments with 'biofeedback', as he called it (Blum 1989). Most of the works in Rosenbloom's back catalogue are to do with what we might now call 'generative music'. They describe systems in which participants were fitted with biophysical sensors so their brainwaves, electromyographic activity, respiration rate, or dermal conductivity could be used as inputs to audio synthesizers. Other experiments involved collaborations with artists and engineers such as John Cage and Robert Moog in efforts to blend Eastern spirituality and meditative practices with ambient music. Thom Blum, however, stated that these explorations were doomed to fail because these performances based on alpha rhythms turned out to be excruciatingly boring to everyone but the performers. This is an issue that is addressed by some of the more contemporary excursions into performance-based artistic BCI (Eaton et al. 2013, 2015).

The second application for artistic BCI is to provide an outlet for artistic expression for people with neuromuscular injury or illness, such as ALS or locked-in syndrome.

One of the first implementations of this idea was produced at the University of Tub-ingen in 2008 (Münßinger et al. 2010). At that point in time, P300 spellers had been around for about twenty years and had allowed locked-in patients to communicate with ever-increasing speed and accuracy. Applications were just beginning to be developed to allow creative expression and engagement in leisure activities, such as surfing the internet (Münßinger et al. 2010). Since that time, a few different teams have been pushing forward with BCI-based painting programs.

A third application of artistic BCI has been in the field of art therapy. This is perhaps a spiritual successor to the experimentation done in the '60s and '70s with alpha-wave modulation and neurofeedback. The idea is to engage users in an artistic experience, either directly or selectively, which can either teach them to enter an alpha wave state characterized by feelings of calmness and peace or provide feedback about their affective emotional state, thereby allowing them to control it, almost as a form of cognitive behavioral therapy. These types of works can take the form of a sort of choose-your-own adventure musical composition, where your brainwaves determine the tempo or character of a piece of music or virtual environment; or they may be more creative, where alpha-band power can be used to affect the color of virtual paint, or the shape of a virtual brush.

13.1.2.1 Examples in Visual Art

A raster graphics editor called Brain Painting was developed in Germany around 2010 (Münßinger et al. 2010). Unlike some multi-modal painting interfaces that emerged at the same time like BrainBrush (Van De Laar et al. 2013), Brain Painting was completely controlled via a P300-speller-style interface—cursor movement, object shapes, sizes, colors, brush styles, undo, redo, and so on. One of the unique aspects of this project is that the team adopted a user-centered design philosophy and tested the prototype on ALS patients. Even though users complained that the system was slow (imagine waiting 60 seconds to take any action in MS Paint) and uncomfortable (no one likes conductive gel in their hair), they still rated it a four out of five, and it did succeed in its mission to enable brain-based free painting (Zickler et al. 2013). The team went on to add more features to the toolbox, and found that one subject continued to paint regularly with the system at home. Despite the relatively low level of control, the user derived satisfaction from painting, and the program actually increased their quality of life (Holz et al. 2015a, b).

Neuro Brush is a collaborative system used for art therapy (Crawford et al. 2018). This is a form of competitive abstract painting, similar in theme to the "environment-demonstration-participation-performance event" described by Rosenboom (Blum 1989). In Neuro Brush, several people are connected to a BCI system as they control the brush of a simple paint program with a mouse. The size of their brush is modulated by their alpha-band power—the more alpha, the larger the brush. Color is controlled based on beta power—the more beta, the more subdued the tones. Each participant can see all the other participants' cursors. The idea is to paint Rothko-style color fields, and whoever is able to fill up the majority of the canvas with their own color wins. The feedback provided by paint color helps users enter the relaxed, meditative state needed to increase their brush size and thereby win the game.

An installation called "Mind the Chair" is an example of post-modern interactive performance art. This simple selective-control BCI comprised a chair affixed with lights and a sound system. Museum visitors were invited to sit in a chair and don a NeuroSky Mindwave, a single-channel wireless EEG. Lights and sound were modulated by brain activity (Folgieri and Lucciari 2016). NeuroSnap is a purely passive system developed in 2017. Modelled after Snapchat, it uses an Emotiv Insight which is a five-channel wireless EEG for registering alpha and beta waves which are used to select facial filters. The BCI is meant to infer the emotional state of the wearer and display an appropriate overlay (Lieblein et al. 2017).

13.1.2.2   Examples in Music

Some of the earliest pioneers of BCIs in music were Alvin Lucier, Richard Teitelbaum, and David Rosenboom (Miranda and Castet 2014). Alvin Lucier was the first person to compose a live EEG-based musical performance in 1965 with his piece 'Music for Solo Performer' (Pinegger et al. 2017; Miranda and Castet 2014). In this piece, he used amplified alpha rhythms from two electrodes attached theatrically to his forehead at the start of his performance to drive transducers attached to a plethora of percussive instruments on stage. Lucier was able to control the length of his alpha-wave bursts at will, and along with a mixer was able to control the volume of the percussive sounds, as well as the dominant instrument (Rosenboom 1997). David Rosenboom followed in 1970 with his "environment demonstration-participation-performance event" entitled 'Ecology of the Skin'. The performance made use of a modular synthesizer and ten musicians, each operating one of the modules. EEG signals were bandpass filtered and fed into Schmitt triggers to create alpha-wave detectors. Programmable logic controllers were used to track the amount of alpha activity generated by each performer, and the amount of time spent per minute producing alpha waves determined how much control they were given over their particular synthesizer module (Rosenboom 1997). In effect, the degree to which individual musicians were able to maintain states of wakeful relaxation determined what notes were played as well as their timbre.

More recently, Mick Grierson and collaborators at Goldsmith's, University of London created a toolkit for music brain-computer interfacing which uses a P300 speller and new ERP detection algorithms to allow musicians to make near real-time choices during a performance (Grierson 2008). This system was used in 'Braindrop', an algorithmically generated musical composition in which the performer, using the P300, could select which of three algorithms was at play (Grierson et al. 2011). Even more recently, a team of the University of Graz utilized user-centered design principles to develop a P300-based application that allows people to write sheet music and compose entire musical scores (Pinegger et al. 2017). Their goal was to create a system which could be utilized by the severely disabled, and early tests of the system resulted in very high selection accuracies and positive feedback from participants.

    The next section of this chapter will discuss the topic of expressive arts-based interventions and the broad range of possibilities that exist for applying BCI technologies in that field. One of the "precursor" approaches discussed here is the use of electroencephalogram (EEG), and other monitoring modalities to measure clients' affective state in the course of therapy. Section 13.2 will then describe the initial attempts at using BCI as a medium of creative arts based interventions, using the area of affective brain–computer music interface (aBCMI) as an example. The discussion will then proceed to the question of matching artistic BCI technologies with different modalities of art-based interventions, exploring the specificities of the various affective and artistic BCI technologies in the process. The section concludes with a taxonomy of different client categories, based on their level and type of impairment, suggesting different types of BCI-enabled expressive therapies for each category.

    The final section of this chapter discusses the challenges and difficulties facing the young and rapidly evolving field of BCI-enabled expressive therapies. The main, and perhaps obvious, challenge, which stems from the youth of the field, is that of the relative dearth of relevant research, and the lack of BCI-based studies involving practicing expressive therapists. Additional challenges are discussed in the framework where BCI field lies at the intersection of neuroscience and digital media, thus inheriting many of the challenges from those two fields. Drawing parallels with the relatively more explored field of the digital art-based therapy, Sect. 13.3 emphasizes the need for evidence-based research and encourages expressive therapy practitioners to be better acquainted with the intricacies of the BCI technologies, so as to avoid the most common pitfalls encountered in this field.

13.2   BCI and Expressive Arts Based Interventions

13.2.1   Expressive Arts-Based Interventions

While cognitive scientists have been investigating and trying to answer the question of whether creativity is innate or can be acquired (Folgieri and Lucciari 2016); the starting point of all therapeutic processes in expressive arts-based interventions is based on the humanistic belief that every individual has the innate ability to be creative and that the creative process is healing (Rogers 1993, p. 7). Rogers (1993), the founder of person-centered expressive arts, has stated that "part of the psychotherapeutic process is to awaken the creative life-force energy" (p. 1), and "what is creative is frequently therapeutic" (Roger 1993, p. 1). The expressive arts-based interventions utilize creative modalities such as visual art, music, drama, dance/movement, poetry, play, and other modalities individually or combined in the context of psychotherapy and counseling. It is an action oriented therapeutic approach in which both verbal and non-verbal communications will be used during the process. Acknowledging that every individual's preferred mode of expression is different (some may be more tactile or kinesthetic while others may be more visual, for example), the expressive arts-based interventions augment modes of expression in the psychotherapeutic process. In addition, while the term "psychotherapy" potentially "perpetuates a mind-body split with its exclusive reference to psyche" (McNiff 1981, p. viii), the inclusion of diverse modes of expression in the expressive arts-based interventions strives for mind-body alignment by introducing action in the healing process through inclusion of the body and the mind.

The advancement of technology, digital technology in particular, has expanded the scope of arts as well as the expressive arts-based interventions. Although some of the practitioners of expressive arts-based interventions continue to rely on traditional "hands-on" creative media, in the past two decades, many practitioners began to explore the therapeutic applications of technology media, its specificity, and its applicable populations and settings (Moon 2010). However, needless to say, the creative process and somatic experience of song writing differs greatly between these approaches and media. For example, one may argue that the act of creating a song in the context of a music-based intervention by using an upright piano or digital technology via electronic keyboard connected to a computer or an application in a smart-phone or even though the eye-tracking device may bring about the same outcome, a song. However, the creative process and somatic experience of song writing differs greatly among musical instruments and through the use of technology. The specificity of each medium, then, needs to be examined and applied carefully in order to fulfill the needs of target populations and clinical settings.

13.2.2 Applications of BCI-Enabled Technologies in Expressive Arts-Based Interventions

While the majority of BCI-enabled expressive arts based interventions remain hypothetical and not yet fully practiced or researched, the range of possibilities afforded by the recent technological developments is quite broad; and the conditions are ripe for expanding the use of BCI in creative therapies.

13.2.2.1 BCI for Monitoring Clients' Affective State During Expressive-Arts Based Therapies

A field related to BCI and Expressive Arts-based intervention that has been used relatively extensively is the use of electroencephalogram (EEG), qualitative electroencephalogram (qEEG), as well as other monitoring modalities to measure clients' affective state and physical response to therapy. Among the recent examples of research activity in this area is Belkofer's (2012) use of qualitative electroencephalogram (qEEG) to study the neurological effects of art-making. Another example of application of qEEG is drawing and clay sculpting studied by Kruk and the team (2014). Combined with progress made in the field of affective BCI (Daly 2018), this area of research produced some of the recent examples of BCI-enabled creative therapies.

13.2.2.2   Initial Introduction of BCI as a Medium of Creative Arts Based Interventions

As described by Daly (2018), a key feature of an affective BCI system is its ability to identify a user's affective (emotional) state, and change the expressive output in response to the user's changing emotions (Daly 2018). An affective brain–computer music interface (aBCMI) is designed to detect a user's affective state and use that information to control a music generator in order to adjust the user's affective state to the desired level. Using a group of 20 healthy subjects in the course of 5 sessions, the researchers were able to achieve the desired changes in two affective state dimensions (increase valence and reduce arousal) at statistically significant levels (see the Fig. 13.1 for illustration of the closed-loop aBCMI and Fig. 13.2 for the observed changes in the two affective state scores).

In the above example the clients' control over the music creation process is passive and relatively limited. The clients' current affective state, relative to the target affective state in the experiment, triggers continuous changes in the musical sub-features used by the music generator logic: tempo, mode, pitch range, timbre, and amplitude envelope, as described in the Williams et al. (2015). However, there are several examples of artistic aBCMIs where the clients have greater control over other areas of the creative process (composition, choice of instruments, etc.). Williams and Miranda (2018) offered a comprehensive review of the state of the art in BCMI field in "BCI

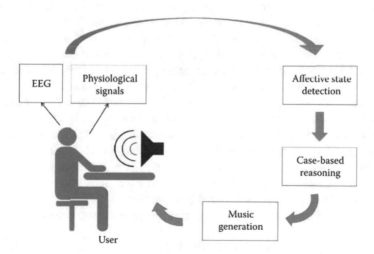

Fig. 13.1 An affective brain–computer music interface—aBCMI. Adopted from "Affective Brain–Computer Interfacing and Methods for Affective State Detection," by Daly (2018 p. 157) (reproduced with permission)

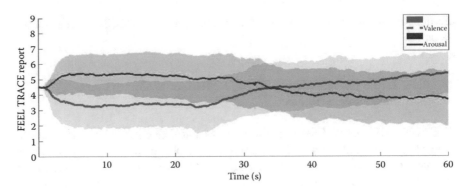

Fig. 13.2 Changes observed in two affective state scores of aBCMI users, under the "make happier" condition. Adopted from "Affective Brain–Computer Interfacing and Methods for Affective State Detection," by Daly (2018, p. 158) (reproduced with permission)

for Music Making: Then, Now, and Next." In addition to further exploring BCI-enabled real-time control and affective state feedback, the authors also described the potential for collaborative music-making, for clients with various degrees of physical impairment.

13.2.2.3 Matching Artistic BCI Technologies with Different Modalities of Arts Based Interventions: Exploration of Possibilities

Williams and Miranda (2018) described the taxonomy of BCMI's in terms of control modality. Active control refers to mapping the user's cognitive choices to musical features, while passive control refers to the use of BCI for detecting the subconscious mental states of the user and then informing the musical feature mapping. A hybrid system combines both control systems and allows simultaneous application of both (p. 199). Such a split between active and passive (and a hybrid combination of the two) is just one of the dimensions of the artistic BCI taxonomy. Wadeson's team (2015) and Prpa's team (2018) (see also Chap. 3 of this book) described an additional dimension of artistic BCI which deals with input modality (unimodal EEG vs. multimodal/hybrid of EEG/ECG/GSR, etc.), output modality (1D music/sonification, 2D painting/video, 3D audiovisual/virtual environment), and individual versus collaborative, open-loop versus closed-loop (i.e., with affective state feedback). Each combination of these different dimensions lends itself more naturally to different modalities of creative expression-based intervention, as well as to different client categories, including those with physical impairment or defective neurological functioning.

Music therapy would be the modality that is better compatible with the BCMI technology. For example, "Cortical Songs" is a music performance with the output as musical signals triggered by detecting user's neural activities (Matthias and Ryan 2007). "Affective brain-computer music interfacing" (Daly et al. 2016) allows for (passive) control over the music output through evaluating the user's affective state in responding to music. "The space between us" (Eaton et al. 2015) evaluates a multi-user BCMI design whereby the affective states of both a performer and listener are measured during a live musical performance. This last study explores the potential for interaction and for reaching a common emotional ground between the two participants, which offers interesting possibilities in therapeutic setting.

For Art therapy, visually-oriented artistic BCI such as Brain Painting is feasible and has been found suitable for applying to participants with different abilities. Botrel and co-authors' study is one of the examples. They showed that participants with amyotrophic lateral sclerosis used Brain Painting had an improved quality of life after the intervention (Botrel et al. 2015).

For the BCI-enabled drama therapy, some potentially relevant projects have been documented (Aparicio 2015; Aparicio and Cádiz 2017). For example, the study authors have been working with an actor with locked-in Syndrome and developed BCI-based communication system using the eye-tracking device for facilitating play writing and music and drama performance (Aparicio and Cádiz 2017). They also established the conceptual framework for how BCI technology can enable a person with physical challenges to be actor and/or director in a theatrical performance (Aparicio 2015). While Aparicio's research focused on theatre as art rather than as a tool in drama therapy, and thus falls into the domain of "artistic BCI", it provided a good foundation for expanding the research into BCI-enabled drama therapy. Similarly, the work "Enheduanna—A Manifesto of Falling: Live Brain-Computer Cinema Performance: Performer and Audience Participation, Cognition and Emotional Engagement Using Multi-Brain BCI Interaction" done by Zioga and co-authors is also an example for BCI based-performing arts which involved both the artist and the audience. In Zioga's work, an EEG-based BCI system was used for the simultaneous real-time multi-brain interaction (Zioga et al. 2018, p. 1). The authors collected qualitative (questionnaires) and quantitative (EEG) data from seven participants (one performer and two different audience members for each of 3 performances), and concluded that "the results reveal that the majority of the participants were able to successfully identify whether their brain-activity was interacting with the live video projections or not. A correlation has been found between their answers to the questionnaires, the elements of the performance that they identified as most special, and the audience's indicators of attention and emotional engagement" (Zioga et al. 2018, p. 1).

Although real BCI-based dance therapy has not yet fully established, a "precursor" for BCI-enabled dance/movement therapy research has been developed using portable EEG technology with the sensors sufficiently robust to movement so as not to distort the EEG signals. Bearss and the team used EEG sensors to detect changes in the alpha band in conjunction with lower depression scores after just one dance class, as well as over the course of the 12-week program, in people with Parkinson's disease (Bearss et al. 2017). In a another project named "Movement in Mind: Dance, Self-Awareness and Sociality—An investigation of dance as treatment/therapy", the same EEG protocol was used for participants who were diagnosed with depression and similar results were found (Barnstaple 2017). While these studies are not situated in the area of BCI-enabled dance therapy, they do offer a potential starting point, where the "input" side of the BCI technology (EEG signals) is linked to the dance and movement therapy practice, and can serve as a basis for neurofeedback as well as for assessment of therapeutic efficacy.

Another example of potentially relevant research is provided in "Measuring brain mechanisms underlying dance-therapy: past & future directions" (Shafir 2018), where the author discusses using Laban Movement Analysis to identify sets of movement components (characteristics) whose execution enhances different emotions. The author further discussed using Kinect and machine learning techniques as a biofeedback system able to identify these movement components from people's movements based on their 3D data, and offers future research directions that would correlate the subject's affective state data from EEG to the 3D data of their movements. There are numerous other studies in this domain, some focusing on measuring the expressive components of the dance and movement, e.g. "Neural decoding of expressive human movement from scalp electroencephalography (EEG)" (Cruz-Garza 2014) where the authors record brain activity and movement of certified Laban Movement Analysis (LMA) dancers and are able to isolate the expressive component of the movement. This theme of isolating the expressive movement from EEG data in the BCI context was further explored in "Towards a whole body brain-machine interface system for decoding expressive movement intent: Challenges and Opportunities" (Contreras-Vidal et al. 2017), where the authors explore "applications to artistic brain-computer interfaces (BCIs), movement aesthetics, and gait neuroprostheses endowed with expressive qualities" (para 1). These articles, taken together, offer a "mosaic" of information which can be synthesized and built upon in developing BCI-enabled dance and movement Therapy practices and research studies.

Finally, multi-modal Expressive Arts Therapy, which relies on the interrelatedness and polyesthetic nature of arts (Estrella 2007; Malchiodi 2007; Knill 2010), lends itself more naturally to BCI-based approaches as the Expressive Arts Therapy encourages shifting from one art form to another as well as multiple outputs (1D, 2D, and 3D, or combination of all). Yet, no literature has been found in this area. Nevertheless, the current BCI output modalities (1D, 2D, 3D, or combination of all) may be matched with those currently practiced in creative arts interventions by augmenting their therapeutic tools or making therapeutic tools available to those clients who could not otherwise be reached without BCI-enabled technologies. It is apparent that selecting the method of BCI input modality (EEG, ECG, other sensors, multimodal) and the level of BCI control modality (passive, active, hybrid between the two, collaborative vs. individual) should depend on client's conditions, comfort level with technology, psychosomatic needs, therapeutic objectives and settings as well as creative therapists' competency with the process of BCI technologies.

		I.	II.
Physical Impairment	High	Stroke Survivors Parkinson's Disease "Locked-in" Syndrome	Intellectual disability with physical impairment
	Low	IV. Neurotypical individuals without physical impairment	III. Dementia Autism Spectrum Disorder Intellectual Disability
		Low	High
		Neurological Impairment	

Fig. 13.3 Quadrants for exploration of BCI-enabled expressive arts based interventions

13.2.3 Possibility of BCI-Based Expressive Arts for Therapeutic Purposes

The possible therapeutic applications of BCI-based expressive arts can be categorized into four types based on the needs of clients with different physical and intellectual challenges (Fig. 13.3): (I) Clients with high physical impairment and low neurological impairment, (II) Clients with high physical impairment and high neurological impairment, (III) Clients with low physical impairment and high neurological impairment, and (IV) Clients with low physical impairment and low neurological impairment.

13.2.3.1 Clients with High Physical Impairment and Low Neurological Impairment

Clients in quadrant I would likely be benefited from the pure EEG (non-hybrid) input modality, and may have to start their therapeutic sessions with a more passive control (and then gradually move up to more active control), due to their physical limitations and the learning curve involved in mastering the active-control modalities. Through currently existing BCI output modalities such as "NeuroBrush," and "Brain painting" stated in the studies by Botrel et al. (2015), BCI may make it possible for the clients in this quadrant to express their creativity and help the mind of an immobile client reconnect with his/her body.

The use of body sensations is critical to many forms of therapy, e.g. Expressive Therapy Continuum in its Kinesthetic/Sensory (K/S) Level (Hinz 2009) or Sensorimotor Psychotherapy (Ogden and Minton 2000). Furthermore, the idea that the body plays an important role in storing and processing emotions has found empirical support in neuroscience (Franks 2006). While this represents a challenge for working with clients who became immobile or lost a part of their body, the bodily sensations are often not lost, and even if the motor control is impaired, the connection between the mind and the body continues. In the case of clients who lost a body part, up to 80% of amputees experience a phantom limb, usually in a form of a painful sensation (Sherman et al. 1984). A significant number of patients with Locked-in syndrome also maintain body sensations as well (Pistoia 2017). Such clients could still benefit from the BCI-enabled expressive arts therapies that would tap into those bodily sensations.

13.2.3.2 Clients with High Physical Impairment and High Neurological Impairment

For clients in quadrant II, the options for BCI-enabled expressive arts based interventions may be more limited, with heavier reliance on passive control and hybrid input modality, to maximize the information available to the therapist (the possibilities for collaborative BCI modalities would likely be limited as well). However, there are still some avenues available for this category, e.g. the use of affective state monitoring and feedback (Daly 2018; Williams and Miranda 2018) can be useful for clients with limited verbal communication abilities. Furthermore, there is existing body of research in helping children with autism spectrum disorder (ASD) in developing neuroplastic changes of their mirror neuron system with the use of a BCI-based game with neurofeedback features (Pineda et al. 2012; Friedrich et al. 2014). Such methods could in theory be extended to clients suffering from both neurological and physical impairments.

13.2.3.3 Clients with Low Physical Impairment and High Neurological Impairment

Clients in quadrant III may benefit more from the hybrid and mobile input modalities (mobile—to accommodate dance and movement therapy, since the client's mobility is not restricted), and may have to rely more heavily on passive control and affective state monitoring and feedback, due to their potential inability to verbalize their affective state. The neurofeedback-based methods described in the previous section, e.g. BCI-based games for clients with ASD (Pineda et al. 2012; Friedrich et al. 2014) can be combined with the artistic BCI approaches described earlier, such as Brain Painting (Botrel et al. 2015) for art therapy or as "Cortical Songs" (Matthias and Ryan 2007) and Brain-Computer Music Interface (BCMI) approaches (Eaton et al. 2015) for music therapy, as a complimentary form to the existing expressive therapy modalities based on traditional media. The emphasis for this group would likely need to be on minimizing the intrusiveness of the BCI devices, and on making proper use of the clients' EEG feedback by the therapists, which would require the therapists to receive additional training in both the theoretical and the practical aspects of BCI technologies and related neuroscience concepts.

13.2.3.4 Clients with Low Physical Impairment and Low Neurological Impairment

For clients in quadrant IV, the choice of BCI modalities would likely be dictated by expanding functionality beyond the traditional arts modalities (e.g., 3D output like virtual environment), as well as by affective state feedback (as described in the Daly 2018, and Williams and Miranda 2018, above), and moderated by the convenience factors (i.e., the BCI devices would need to be sufficiently unobtrusive in order to justify the continued use). While some aspects of this client category (e.g., affective BCI, neurofeedback) are partly shared with the quadrant III, the use of BCI would need to be further justified, since the clients in this category are not impaired either physically or neurologically.

One possible avenue where the BCI-based approaches may have appeal is the younger clients who tend to be "digital natives" and thus comfortable with the new media and computer technologies. In that respect, the growing interest in digital art as a therapeutic modality (Garner 2017) is likely to stimulate interest in BCI-enabled expressive arts interventions as well. In terms of the BCI taxonomy based on input and output modalities (Prpa and Pasquier 2018), the emphasis with the "digital native" clients may well end up being on the 3D output BCI, i.e. Virtual Reality, Augmented Reality, and perhaps also on collaborative virtual environments, where both clients and therapists are present and interact in the form of virtual avatars rather than physically. Such modalities, while interesting to consider, and potentially powerful, would need to be thoroughly tested in practice, to make sure that the important relationship and sensory aspects of expressive therapies are not compromised in the process.

Another interesting application of the collaborative BCI-based expressive therapy is the possibility of both the client's and the therapist's affective state being available for observation by both sides of the therapeutic relationship, represented either visually or sonically (Eaton et al. 2015) or via a combination of modalities. Such an approach could serve in building emotional attunement and trust between the client and the therapist (as opposed to the unidirectional application where only the therapist can observe the client's affective state). However, such applications would necessarily raise issues of both client and therapist's confidentiality. Also, some circumstances may require the therapist to contain their emotional state in client's presence, in order to better accomplish the therapeutic goals, so a careful balance would need to be struck with respect to the level of "emotional transparency" to be made available in such a collaborative affective BCI-based therapeutic practice. It is important to note that the applications described in this section are just examples of what's possible based on the technologies that exist at this moment. The BCI space is evolving quite rapidly and many new possibilities are likely to become available to researchers and practitioners in the near future.

13.3 Difficulties/Challenges/Ethical Concerns

The challenges that the BCI-enabled expressive therapies are facing in fact are inter-twined with and driven by the general challenges faced by the BCI technologies. The most obvious challenge for BCI-enabled expressive therapies stems from the fact that it is a young and rapidly evolving field, with very little empirical research implemented to date. While there are thousands of research articles on the field of BCI, very few of them directly address the topic of artistic and/or affective BCI-enabled expressive therapies. Two examples of such directly relevant research were mentioned previously, the affective brain–computer music interface (aBCMI) used to regulate subjects' affective state (Daly 2018), and the BCI-enabled real-time collaborative music-making with affective state feedback, for clients with various degrees of body impairment (Williams and Miranda 2018). Even these experiments can be considered as only tangentially relevant to BCI-enabled expressive therapies, and perhaps more properly situated in the fields of "affective BCI" (Daly 2018) and "artistic BCI" (Williams and Miranda 2018), because, despite their stated therapeutic effect, there doesn't appear to be a practicing expressive therapist involved in either study. Until the researchers and practitioners in the expressive therapies field make the interdisciplinary leap and start conducting directly relevant studies, the concept of "BCI-enabled expressive therapies" will largely remain in the realm of potentiality.

Despite such dearth of directly relevant research, the situation remains very promising, as significant parallels can be drawn between the challenges faced by the emerging field of BCI-enabled expressive therapies and the challenges (and opportunities) which lie at the points of contact between the fields of expressive therapies, neuroscience, and digital media technologies. This is due to the fact that the BCI field itself can be said to lie at the intersection of neuroscience on one hand, and digital media on the other (Fig. 13.4). Put differently, whereas neuroscience intersects BCI at its "input" and "control" phases, to use the taxonomy discussed in Sect. 13.2.2.3 above, borrowed from Prpa and Pasquier (2018), the digital media field intersects with BCI at the latter's "output" phase.

Therefore, the challenges faced by the expressive arts therapies practitioners and researchers in their encounters with neuroscience and with digital media, and the lessons learned from those encounters, are directly relevant to the exploration of challenges and opportunities in the emerging field of BCI-based expressive therapies. Significant research and practical experiences have already been accumulated at expressive therapies' intersections with neuroscience and digital media, and should be brought to bear in the study of BCI-enabled expressive therapies, in addition to the previously discussed affective BCI and artistic BCI research fields.

As stated in Juliet King's "Art Therapy: A Brain-based Profession" (2016), "Current technologies that measure brain activity during art-making hold promise for art therapists by gathering the empirical evidence needed to substantiate the efficacy of interventions…. The use of technology such as the qEEG might help in identifying what media will work best with certain conditions, and might also help identify brain

Fig. 13.4 Intersections of expressive therapies, neuroscience, and digital media

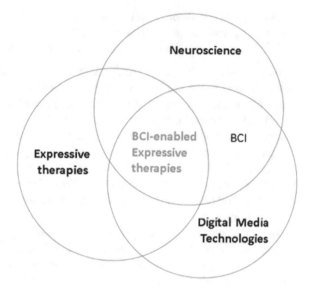

functions in relation to basic media used in art therapy. From this, art therapy protocols for certain populations might be developed and tested, providing the field with much needed evidence-based and outcomes research" (p. 85). The author reinforces the last point by quoting a study conducted by Kaiser and Deaver (2013), where the top priority for art therapy researchers is to provide "outcomes, efficacy, and evidence-based research" (King 2016, p. 87). The same holds true for all expressive arts based interventions.

Thus, the challenge for researchers and practitioners of expressive therapies alike is to sufficiently acquaint themselves with the intricacies of neuroscience, and by extension—of the BCI technologies, so as to be able to deliver evidence-based research and avoid the common pitfalls inherent in those disciplines. In addition to these general BCI challenges, additional recommendations specific to affective BCI (aBCI) were discussed in Daly's (2018) "Affective Brain–Computer Interfacing and Methods for Affective State Detection," and are summarized below:

- Identify the application area and the purpose of the aBCI.
- Identify the category the aBCI being used falls into (active, passive, or a hybrid approach that combines both).
- Identify how affect will be used within the aBCI (e.g., aiding with communication/control accuracy, providing a form of therapeutic feedback to the user, or used in combination with other input modalities).
- Define how affect will be categorized and measured within the aBCI.
- Identify the affective state detection method to use in the aBCI and verify that it works correctly.
- Take into account the inter-trial and inter-user variability, as well as intra-session non-stationarity of affective state responses.

- Remove artifacts from the neurological data used to control the aBCI.
- Plan an appropriate testing strategy, taking into account the inherent challenges in identifying a user's "ground truth," a measure of their actual affective state.

While some of the above recommendations echo the general BCI challenges, they also point to the need for special care when dealing with affective BCI, which is one of the most promising BCI areas for expressive therapies, due to its broad appeal and the potential to deliver benefits for clients in all categories, including the "normal" (non-body-impaired neurotypical) clients. These pitfalls have already been extensively explored by neuroscience researchers, e.g., in "Using neurophysiological signals that reflect cognitive or affective state: six recommendations to avoid common pitfalls" (Brouwer et al. 2015) and the resulting recommendations should be used by researchers and practitioners in their future experiments with affective BCI-enabled expressive therapy.

Similarly, insights gained from the pre-existing experience with digital art therapy can be used in the BCI-based expressive therapies, with respect to the BCI's "output" phase, as discussed earlier in this section. The field of digital art therapy, while still relatively young, has accumulated a significant body of research, much of which is directly relevant to BCI. The need for new/modified therapy frameworks to accommodate the specificities of the digital media also extends itself to dealing with the BCI-enabled expressive therapies: each modality and choice of materials brings specific characteristics, and therefore, may influence the therapeutic dynamics and outcomes. As described in the introduction to "Digital Art Therapy," "Art therapists have cited the sensuous quality of traditional media as something that is lost with digital media…But for all its cold, non-fluid qualities, there is a seductive quality to digital media that comes from experience with the medium…over time as familiarity is gained, the experience of interacting with it changes to something that can be similar to entering the creative process with traditional media" (Garner 2017, p. 11). The same principle applies to the output of the artistic BCI technologies—there is a need for both clients and practitioners to overcome the initial learning curve, so as to be able to appreciate the specificities of this new medium. An additional factor to take into account is the fact that a large proportion of the expressive therapists are "analogue natives" (pre-Internet generation), whereas a growing number of clients are "digital natives," and that divide needs to be addressed by additional research into digital art therapy, including its applications in the BCI-enabled expressive therapies.

There are many other challenges and concerns that have haven not been discussed yet, which usually accompany new media and technologies. For example, new ethical concerns arising from BCI usage in therapy are covered by the topic of "neurosecurity" and additional privacy and confidentiality issues in the context where both the therapist and client are connected via BCI. The new level of access to the client's inner images and emotions afforded by the BCI technology comes with new challenges and ethical responsibilities which will need to be addressed by practitioners and researchers in the field. Nevertheless, these numerous new challenges will only become fully apparent when direct research and practice in the BCI-enabled expressive therapies commences in earnest.

References

Aparicio A (2015) Immobilis in mobili: performing arts, BCI, and locked-in syndrome. Brain Comput Interfaces 2(2–3):150–159. https://doi.org/10.1080/2326263x.2015.1100366

Aparicio A, Cádiz RF (2017) Wheels within wheels: brain-computer interfaces as tools for artistic practice as research. Augmented cognition. In: Schmorrow D., Fidopiastis C (eds) Augmented cognition. Enhancing cognition and behavior in complex human environments. AC 2017. Lecture notes in computer science, vol 10285. Springer, Cham, pp 266–281

Barnstaple RE (2017) Movement in mind: dance, self-awareness and sociality—an investigation of dance as treatment/therapy. Dissertation, York University

Bearss KA et al (2017) Improvements in balance and gait speed after a 12 week dance intervention for Parkinsons disease. Adv Integr Med 4(1):10–13. https://doi.org/10.1016/j.aimed.2017.02.002

Belkofer CM (2012) The impact of visual art making on the brain. Dissertation, Lesley University

Blum T (1989) Review: biofeedback and the arts: results of early experiments by David Rosenboom. Comput Music J 13:86–88

Botrel L et al (2015) Brain painting V2: evaluation of P300-based brain-computer interface for creative expression by an end-user following the user-centered design. Brain Comput Interfaces 2(2–3):135–149. https://doi.org/10.1080/2326263x.2015.1100038

Brouwer A et al (2015) Using neurophysiological signals that reflect cognitive or affective state: six recommendations to avoid common pitfalls. Front Neurosci 9. https://doi.org/10.3389/fnins.2015.00136

Contreras-Vidal JL et al (2017) Towards a whole body brain-machine interface system for decoding expressive movement intent challenges and opportunities. In: 5th international winter conference on brain-computer interface, BCI 2017. Institute of Electrical and Electronics Engineers Inc., pp 1–4. https://doi.org/10.1109/iww-bci.2017.7858142

Crawford C, Cioli N, Holloman A (2018) NeuroBrush: a competitive, artistic multi-modal BCI application. In: Conference: CHI: artistic BCI workshop

Cruz-Garza JG (2014) Neural decoding of expressive human movement from scalp electroencephalography (EEG). Front Hum Neurosci 8. https://doi.org/10.3389/fnhum.2014.00188

Daly I (2018) Affective brain–computer interfacing and methods for affective state detection. In: Nam C et al (eds) Brain-computer interfaces handbook: technological and theoretical advances. CRC Press, Taylor & Francis Group, Boca Raton, FL, pp 147–163

Daly I et al (2016) Affective brain-computer music interfacing. J Neural Eng 13(4):046022. https://doi.org/10.1088/1741-2560/13/4/046022

Eaton J, Miranda E (2013) BCMI systems for musical performance. In: 10th international symposium on computer music multidisciplinary research, sound, music motion

Eaton J et al (2015) The space between us: evaluating a multi-user affective brain-computer music interface. Brain Comput Interfaces 2(2–3):103–116. https://doi.org/10.1080/2326263x.2015.1101922

Estrella K (2007) Expressive therapy: an integrated arts approach. In: Malchiodi C (ed) Expressive therapies. The Guildford Press, New York, pp 183–209

Folgieri R, Lucciari C (2016) Creative thinking: a brain computer interface of art. In: International conference on live interfaces. https://doi.org/10.13140/rg.2.2.31683.40489

Franks DD (2006) The neuroscience of emotions. In: Stets JE, Turner JH (eds) Handbook of the sociology of emotions. Springer, New York, pp 38–62

Friedrich EV et al (2014) Brain-computer interface game applications for combined neurofeedback and biofeedback treatment for children on the autism spectrum. Front Neuroeng 7. https://doi.org/10.3389/fneng.2014.00021

Garner RL (2017) Introduction. Digital art therapy: material, methods and applications. Jessica Kingsley, London, pp 9–20

Grierson M (2008) Composing with brainwaves: minimal trial P300 recognition as an indication of subjective preference for the control of a musical. In: International computer music conference

Grierson M, Kiefer C, Yee-King M (2011) Progress report on the EAVI BCI toolkit for music: musical applications of algorithms for use with consumer brain computer interfaces. In: Proceedings of the international computer music conference 2011, pp 110–113

Hinz LD (2009) Overview of the expressive therapies continuum. In: Expressive therapies continuum: a framework for using art in therapy. Routledge, New York, pp 3–19

Holz EM et al (2015a) Independent home use of Brain Painting improves quality of life of two artists in the locked-in state diagnosed with amyotrophic lateral sclerosis. Brain Comput Interfaces 2(2–3):117–134

Holz EM, Botrel L, Kaufmann T, Kübler A (2015b) Long-term independent brain-computer interface home use improves quality of life of a patient in the locked-in state: a case study. Arch Phys Med Rehabil. https://doi.org/10.1016/j.apmr.2014.03.035

Kaiser D, Deaver S (2013) Establishing a research agenda for art therapy: a Delphi study. Art Ther 30(3):114–121. https://doi.org/10.1080/07421656.2013.819281

Kamiya J (2011) The first communications about operant conditioning of the EEG. J Neurother 15:65–73. https://doi.org/10.1080/10874208.2011.545764

King JL (2016) Art therapy: a brain-based profession. In: The Wiley Blackwell handbook of art therapy. Wiley, Chichester, pp 77–89

Knill PJ (2010) Foundations of a theory of practice. In: Knill PJ et al (eds) Principles and practice of expressive arts therapy toward a therapeutic aesthetics. Jessica Kingsley, London, pp 79–170

Kruk KA et al (2014) Comparison of brain activity during drawing and clay sculpting: a preliminary qEEG study. Art Ther. 31(2):52–60. https://doi.org/10.1080/07421656.2014.903826

Lieblein R, Hunter C, Garcia S et al (2017) Augmented cognition. Enhancing Cogn Behav Complex Hum Environ 10285:345–353. https://doi.org/10.1007/978-3-319-58625-0

Malchiodi CA (2007) Expressive therapies: history, theory, and practice. In: Expressive therapies. The Guildford Press, New York, pp 1–15

Matthias J, Ryan N (2007) Cortical songs: musical performance events triggered by artificial spiking neurons. Body Space Technol 7(1). http://doi.org/10.16995/bst.157

McNiff S (1981) Overview. In: The arts and psychotherapy. Thomas, Springfield, IL, pp v–xxiii

Miranda R, Castet J (eds) (2014) Guide to brain-computer music interfacing. Springer, London

Moon CH (2010) A history of materials and media in art therapy. In: Materials & media in art therapy: critical understandings of diverse artistic vocabularies. Routledge Taylor & Francis Group, New York, pp 3–47

Münßinger JI, Halder S, Kleih SC et al (2010) Brain painting: first evaluation of a new brain-computer interface application with ALS-patients and healthy volunteers. Front Neurosci 4:182. https://doi.org/10.3389/fnins.2010.00182

Nijholt A, Nam CS (2015) Arts and brain-computer interfaces (BCIs). Brain Comput Interfaces 2:57–59. https://doi.org/10.1080/2326263X.2015.1100514

Ogden P, Minton K (2000) Sensorimotor psychotherapy: one method for processing traumatic memory. Traumatology 6(3):149–173. https://doi.org/10.1177/153476560000600302

Pineda J et al (2012) Self-regulation of brain oscillations as a treatment for aberrant brain connections in children with autism. Med Hypotheses 79(6):790–798. https://doi.org/10.1016/j.mehy.2012.08.031

Pinegger A, Hiebel H, Wriessnegger SC, Müller-Putz GR (2017) Composing only by thought: novel application of the P300 brain-computer interface. PLoS One 12:1–20. https://doi.org/10.1371/journal.pone.0181584

Pistoia F et al (2017) Commentary: embodied medicine: mens sana in corpore virtuale sano. Front Hum Neurosci 11. https://doi.org/10.3389/fnhum.2017.00381

Prpa M, Pasquier P (2018) BCI art: brain computer interfaces in contemporary art. Available via ResearchGate, https://www.researchgate.net/publication/324705672_BCI_art_brain-computer_interfaces_in_contemporary_art. Accessed 30 Oct 2018

Rogers N (1993) A path to wholeness: person-centered expressive arts therapy. In: The creative connection: expressive arts as healing. Science and Behavior Books, Palo Alto, CA, pp 1–9

Rosenboom D (1997) Extended musical interface with the human nervous system. Leonardo mono-graph series, supplemental issue

Shafir T (2018) Measuring brain mechanisms underlying dance-therapy: past & future directions. In: Proceedings of the 3rd international mobile brain/body imaging. https://depositonce.tu-berlin.de/bitstream/11303/8075/3/proceedings_MoBi2018.pdf. Accessed 10 Nov 2018

Sherman RA et al (1984) Chronic phantom and stump pain among American veterans: results of a survey. Pain 18(1):83–95. https://doi.org/10.1016/0304-3959(84)90128-3

Van De Laar B, Brugman I, Nijboer F et al (2013) BrainBrush, a multimodal application for creative expressivity. In: ACHI 2013 sixth international conference on advances in computer-human interactions, pp 62–67

Wadeson A, Nijholt A, Nam CS (2015) Artistic brain-computer interfaces: state-of-the-art control mechanisms. Brain Comput Interfaces 2(2–3):70–75. https://doi.org/10.1080/2326263x.2015.1103155

Williams D et al (2015) Affective calibration of a computer-aided composition system by listener evaluation. In: Proceedings of the ninth triennial conference of the European Society for the cognitive sciences of music (ESCOM2015). http://cmr.soc.plymouth.ac.uk/publications/ESCOM15-DW.pdf. Accessed 04 Jan 2019

Williams DA, Miranda ER (2018) BCI for music making: then, now, and next. In: Nam C et al (eds) Brain-computer interfaces handbook: technological and theoretical advances. CRC Press, Taylor & Francis Group, Boca Raton, FL, pp 193–205

Zickler C, Halder S, Kleih SC et al (2013) Brain painting: usability testing according to the user-centered design in end users with severe motor paralysis. Artif Intell Med 59:99–110. https://doi.org/10.1016/j.artmed.2013.08.003

Zioga P et al (2018) Enheduanna—a manifesto of falling. Live brain-computer cinema performance: performer and audience participation cognition and emotional engagement using multi-brain BCI interaction. Front Neurosci 12. https://doi.org/10.3389/fnins.2018.00191

Part V
Brain Art: Control, Tools, Technology, and Hacking

Chapter 14
Brain-Controlled Cinema

Richard Ramchurn, Sarah Martindale, Max L. Wilson, Steve Benford
and Alan Chamberlain

Abstract This chapter explores the space of Brain-Computer Interaction as a tool to enhance storytelling within cinema, as a means to overcome some of the main critiques of interactive film in terms of interaction and immersion in the media. Using the Performance-led Research in the Wild methodology, we create complete professionally-made experiences to explore possible brain-computer interactions with film, and exhibit them in-the-wild to the public. As well as reviewing the findings of these investigations, this chapter primarily contributes an exposition on artistic practices, and draws conclusions for future developments in brain-controlled film. We present two case studies of BCI films that have been made and publicly screened, *The Disadvantages of Time Travel* (2015) and *The MOMENT* (2018). For each we (a) describe the experience we produced, (b) present detail about how it was made regarding different elements of the filmmaking process, (c) give an overview of how they were received by audiences, and (d) summarise key lessons learned about filmmaking practice. In discussing how they were made and received, therefore, we set out implications for scripting, storyboarding, sound design, cinematography, directing and editing as well as interactive platform development. The chapter concludes by discussing possible techniques, processes, mappings, and BCI implementations that could be put together to make future films, as well as presenting both opportunities for industry and open challenges that remain for brain-controlled film.

R. Ramchurn (✉)
Horizon Centre for Doctoral Training, University of Nottingham, Nottingham, UK
e-mail: richard.ramchurn@nottingham.ac.uk

S. Martindale
Horizon Digital Economy Research, University of Nottingham, Nottingham, UK
e-mail: sarah.martindale@nottingham.ac.uk

M. L. Wilson · S. Benford · A. Chamberlain
Mixed Reality Lab, University of Nottingham, Nottingham, UK
e-mail: max.wilson@nottingham.ac.uk

S. Benford
e-mail: steve.benford@nottingham.ac.uk

A. Chamberlain
e-mail: alan.chamberlain@nottingham.ac.uk

© Springer Nature Switzerland AG 2019
A. Nijholt (ed.), *Brain Art*,
https://doi.org/10.1007/978-3-030-14323-7_14

Keywords Neurocinema · Interactive film · Affective narrative · BCI · Art ·
Cinema

14.1 Introduction

Interactive film, as an ambition, dates back to at least the 1920s, emerging historically
from the desire to transfer the engaging experiences of interactive theatre into the
medium of film. As we review further in Sect. 14.2, sporadic attempts over time have
had limited sustained effect on the industry, and it is widely concluded that this is
because the nature of interacting with a film conflicts with the immersive experiences
that films are designed to create (Shaul 2008; Polaine 2005). Our research has
focused on the design, development, and deployment of brain-controlled films, in
which conscious physical interaction is taken away, such that the film can be both
interactive and immersive.

To study brain-controlled films, we adopt a Performance-led Research in the
Wild methodology (Benford et al. 2013), in which artistic practice, studies, and
the exploration of theory are tightly interleaved. This methodology relies on the
deployment of real-world applications of artistic practice with the public. Reflection
on practice allows researchers to interrogate the process to make such works, and the
impact and creation of new theories. Similarly, studies of these artefacts are therefore
of real-world experiences, and reveal findings for both practice and theory.

So far, our research has allowed us to explore the benefits that ambiguous and
unconscious control bring to creating dynamic artistic experiences (Pike et al. 2016),
and to present data about how audiences, including those that were not in control
of the movie, have experienced these films, and how they want to experience them
again (Ramchurn et al. 2019). In this chapter, however, we bring together these
findings to contribute an exposition of artistic practice involved in making brain-
controlled films. We present the process of professionally creating the films, information about how they were exhibited, and key extracts of our previously published
findings (Pike et al. 2016; Ramchurn et al. 2019) that specifically contribute to a
developed understanding of artistic practice. Perhaps most importantly for artistic
practice, is that designing the intended final experience is necessary to shape nearly
all stages of creating the film, including script design, storyboarding, scheduling,
recording, editing, and production. In reviewing these cycles of practice, performance, and reflection (which build upon each other), we conclude by discussing
implications for future practice.

14.2   Related Work

14.2.1   Interactive Cinema

Interactive film has had sporadic development through the years, so much so that every generation claims to be the 'first' to have done it. The earliest example to be found goes back to 1926, *As You Like It, Not Shakespeare*. Directors Leventhal and Norling, had been experimenting with anaglyphic 3D films for some time, one of their last films used this technique to project the climax of their film as two overlaid scenes, one in red one in blue (Zone 2014). The audience could choose what ending they wanted, or flip between them by closing the corresponding eye. Unfortunately, this film has been lost over time; we have no reports of how audiences responded and there is little to no evidence that it made an impact on filmmaking practice in the following decades. It wasn't until 1968 with Raduz Cincera's *Kinoautomat* that interactive film was once again in the spotlight (Willoughby 2007). At the Montreal world fair Cincera presented a film that offered choices for the audience vote on at key moments, which the projectionist would then choose and project the appropriate reel. It was made as a response to the political process in Czechoslovakia at the time and was banned on its return. What *Kinoautomat* did do is introduce a mode of interaction which has been repeated up to the present day. It used remote controls for the audience to make choices. Theatrical realises of *I'm Your Man*, *Mr Payback* (Hales 2015), *Choose Your Own Documentary* (Penlington et al. 2014), the app based *Late Shift* (CtrlMovie 2016) and *Bandersnatch* (Reynolds 2018) all use this approach of: 'a decision has to be made', the users make a choice and a corresponding scene is played.

It has been convincingly argued that interactive narratives have a central problem, that of methods of control disrupting narrative immersion (Shaul 2008; Polaine 2005). This conflict can be reasoned as the more you actively make efforts to interact, be it via a mouse, a remote control, app or even closing one eye, the more you are pulled out of a narrative. At one extreme the narrative may actually stop and wait for you to make a choice before continuing and at best you will still have to divert your attention to the choice options as the movie plays, make a choice under a time constraint and find the button to press to continue with the experience.

To say that interaction and narrative immersion are incompatible would be a simplification, however. The video game is becoming a place players do experience well told stories. But in video games the aforementioned loop is ongoing; the fun is in the ludic experience of doing. And while great stories can be told inside video games, they tend to happen over many hours of play, where controls have become second nature to the player, and often involve non or limited interactive cut scenes.

When players have learned how to control their character this control shares some attributes with a passive BCI interaction, in that there is ongoing feedback, and one does not need to think about control within the interaction loop. Likewise within a passive BCI system one does not need to know how or what one's input is doing for interaction to occur.

A key divergence from the above forms of interactive movies are the database movies made by Kinder's *Labyrinth Project* (Kinder 2003) and Manovic's *Soft Cinema* (Anderson 2006). Made in the late 1990s and early 2000s these have roots in hypertext fiction and the aesthetic of the early internet. Manovic's interactive films algorithmically recombine on each play-through, and thematically respond to the multimedia non-linear medium of the pre-YouTube internet. These films have no ongoing control from users, but rather evolve a narrative from a real-time assembly of images, voices and music. We see the role of artist shifting here, no longer the auteur creator, but giving away editorial control to a pre-defined algorithm:

> *"Manovich and Kratky have modelled a compositional mode in which the work of the 'artist' is shifted from encoding desired meanings into a montage structure to establishing the rules and metadata by which the SoftCinema engine will create its own combinations of media elements (video, text, sound, composition), the meaning of which will ultimately be produced through reception and interpretation by a viewer."* (Manovich et al. 2001)

Some have taken this artistic method further and included the viewer's physiological responses as data in the algorithmic system. The following three pieces of interactive narrative were created by artists and researchers utilising physiological sensors including BCI and subconscious control to affect narrative films.

The first is by Pia Tikka. She created a filmic installation called *Obsession*, which consists of four screens centred around viewers who are continuously monitored for heart rate and skin conductance by biosensors. *"A real time feed-back loop is established when the authored cinematic system influences the perception and cognition of the viewer, whose bodily feedback, in turn, affects the montage"* (Tikka 2010). Tikka's work is based around the concept of Enative Media, which draws on theories of embodiment and Eisenstein's concept of affecting the viewer with montage by *"deliberately experiment[ing] with the unconscious emotional responses of [the] audience"*. Tikka argues for cinema as an extended cognitive externalisation of consciousness. *"In the enactive setting, the notion of interface becomes implicit, perhaps even to the degree of being non-conscious."*

As a second example, Alexis Kirke's *Many Worlds* is a branching narrative story. Sensors measuring Electrocardiograph (EKG), Electromyograph (EMG), Electroencephalograph (EEG) and Galvanic Skin Responce (GSR) are applied to four audience members. The system uses this physiological input to assess levels of arousal over time, an evaluation which is then used to make binary choices at two decision points of the film (Kirke et al. 2018).

Finally, *The Angry River*[1] is an interactive film by Armen Perian, which uses eye tracking to follow one of four characters *"transforming you into the unreliable narrator"*. *The Angry River* was inspired by the book *'The House of Leaves'*, which is a layered and networked narrative with multiple narrative perspectives (Pressman 2006). One of Perian's questions was: how do you film the unfilmable novel? By answering this question through interactive film, Perian provides hints at the artform's potential to tell unfamiliar, complex and non-linear narratives.

---

[1] https://www.theangryriver.com.

These three interactive movie experiences deviate from the trend of offering conscious decision making as an interactive method while watching cinematic content. Rather, they rely on unconscious interactions, which deviate from the tried and tested standards of human-computer interaction. Within the field of HCI, usually the emphasis is on direct manipulation (Shneiderman et al. 2016). However, others have put forward that ambiguity can be a resource that designers can exploit (Sengers and Gaver 2006). This concept of co-construction between users and systems is somewhat analogous to the very nature of film and could therefore be a powerful resource when applying interactivity. As we are the creators of our own meaning, this empowers the user to take ownership of their own interpretation.

14.2.2 Neurocinematics

The disciplines of film and neuropsychology have become entangled. In 2008, Uri Hasson coined the term Neurocinematics whereby he set out a way of judging the effectiveness of how much a film controls a viewer's mind by correlation of subjects' fMRI and eye tracking data (Hasson et al. 2008). In the paper, Inter Subject Correlation (ISC) is argued to be a measure for the effectiveness of film. Hasson showed participants sections of *The Good, The Bad And The Ugly* by Sergio Leone, *Bang! You're Dead* by Alfred Hitchcock, an episode of *Curb your Enthusiasm* by Larry David, and an unedited shot of Washington square park. He found that ISC was highest with Hitchcock, then Leone, then David and then finally the park shot. Based on this ISC Hasson suggests a single continuum going from reality, to documentary, art house, Hollywood finally to propaganda, where the further to the right the higher the ISC. Hasson does offer a disclaimer as to which is the better film:

> "while the ISC cannot provide an aesthetic judgment as to the right cinematic style to be taken, it may serve as an objective scientific measurement for assessing the effect of distinctive styles of film-making upon the brain, and therefore substantiate theoretical claims made in relation to them" (Hasson et al. 2008)

In Hasson's line of aesthetic, see Fig. 14.1, he moves from reality to propaganda with rising correlation between subjects. Finally, taken to an extreme, the possibility

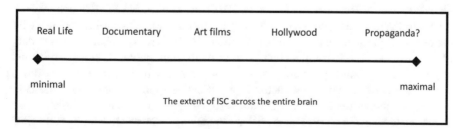

Fig. 14.1 Hasson's line of filmic aesthetic in relation to inter-subject correlation of brain data (Hasson et al. 2008)

to achieve a tight grip on the viewers' minds can be used for creating an unethical form of propaganda or brainwashing (Hasson et al. 2008). It is not surprising that marketing and advertising companies have adopted these techniques. Furthermore, Neurocinematics has been taken up by neuro-marketers and movie studios to judge the effectiveness of how much a piece of content "controls viewer's neural responses". In fact, some have taking it further, using these techniques to assess scripts, characters and casting (Randall 2011).

There are some who critique Hasson's definition of effectiveness based on high ISC as it by definition excludes 'noisy' cases where more complex readings of film occur (Poulaki 2014). Poulakia even suggests exploring the opposite direction to ISC, exploring the possibilities when minds diverge; this is echoed by the ex-CEO of a neural marketing company who was quoted in 2011 saying:

> "*Multiple if not infinite versions of one film with myriad story twists and endings will be produced and consumed. Netflix and Facebook will play a big part in film 'personalization.' [...] Real-time instant consumer brain response-based personalization will create true dynamic modifications of the same movie and afford endless delight to consumers.*" (Randall 2011)

What is described in the above quote is an interactive film which is altered by neurological responses. As Hasson finds, as we watch movies, a brain-computer interface can be used to observe changes in brain data caused by that movie. To then go full circle, that data can be used in turn to affect the stream of visual and auditory information. Non-diegetic (editing, sound, cinematography), diegetic (the film world), characters and actions of the film can be used as feedback components. This will produce an ongoing, real time feedback loop, which has been explored in other mediums (Höök 2008). This is the central tenet of what Zander has termed Passive BCI (Zander and Kothe 2011) and this is the concept behind brain-controlled cinema.

14.3   Our Case Studies of Brain-Controlled Cinema

Below are two case studies of our brain-controlled movies, along with the resultant research produced using a Performance-led Research in the Wild (Benford et al. 2013) methodology. The case studies, compared in Table 14.1, are of two films: *The Disadvantages of Time Travel* (Pike et al. 2016) and *The MOMENT* (Ramchurn et al. 2019). Both are written, directed, and designed by co-author Richard Ramchurn with the help of developers and researchers from the Mixed Reality Lab at the University of Nottingham. The two films contrast in scale, design, genre, directing style, and intention. They explore a passive interaction design, which frames the narrative around the central human interactor in an affective loop. In describing the case studies, we set out the director's intentions for design and lay out their practice; we have tried to keep the observations relevant to making an interactive brain-controlled film rather than a general filmmaking process.

Table 14.1 Comparison of two our interactive brain-controlled film productions

	The Disadvantages of Time Travel	The MOMENT
Duration	16 min	23 min
Genre	Arthouse	Sci-fi Thriller
Narrative	4 layers: perspectives on events	3 layers: following characters
Headset	NeuroSky mindwave	NeuroSky mindwave 2
System	Quartz composter, custom python	Max MSP, brain wave OSC
Control	Blinking, Attention, Meditation	Attention
Actors	11 Non-professional child actors	9 Professional actors
Total crew	6	28
Pre-production	3	6
Production	6	25
Post-production	2	4
Budget	£19k	£54k
Funding	Kickstarter, Arts Council England	Arts Council England, EPSRC

14.4 Case Study 1—*The Disadvantages of Time Travel*

The director's intention when starting to make *The Disadvantages of Time Travel* (see Fig. 14.2) was to reproduce the psychic space of the central character. This was achieved by making a filmic world and interactive system that responded to the viewer and fell into sync with the viewer's physiological signs. The initial design for the film came from an artistic hunch based on years of editing and a concept written in Walter Murch's book *In the Bink of the Eye* (Murch 2001). In his book, Murch observes that his actors would blink just at the moment that he would want to cut and hypothesized that blinking happens as a punctuation of thought.

> *"If it is true that our rates and rhythms of blinking refer directly to the rhythm and sequence of our inner emotions and thoughts, then those rates and rhythms are insights to our inner selves and therefore as characteristic of each of us as our signatures."* (Murch 2001)

We explored this concept by mapping each blink data point to a hard cut creating the temporal montage of the film as shown in Fig. 14.3. The film was continually in a state of spatial montage as two layers of dream footage and two layers of reality footage blended together based on NeuroSky Attention and Meditation data respectively (see Fig. 14.4). This blending of layers produced what Manovic terms Spatial Montage in his book *The Language of New Media* when describing interactive film (Manovich et al. 2001). While he defines temporal montage as the changing of images in time—the predominant cultural embodiment of film—he characterises spatial montage as the compositional arrangement of multiple images on a single screen. While there is historical precedent for this, such as Vertov's *Man With A Movie Camera* (Petric 1987), his direct analogy is from our computer mediated experience of windows open on a desktop.

Fig. 14.2 Poster for The Disadvantages of Time Travel

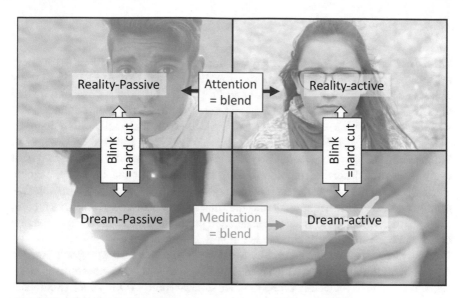

Fig. 14.3 mapping of BCI and blink data and to temporal and spatial montage

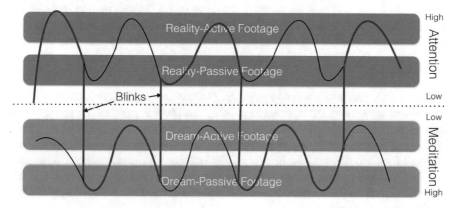

Fig. 14.4 Interactive Design of *The Disadvantages of Time Travel* , from Pike et al. (2016)

We used NeuroSky's MindWave headset as the BCI for the film. The single, dry sensor Bluetooth device outputs raw EEG signals and pre-processed Attention, Meditation and blink data. While NeuroSky uses proprietary algorithms in their chip to calculate these measures, some work has been done to validate these as well as the raw EEG (Kim et al. 2012; Lim et al. 2012). For our primary concern of usability, however, the MindWave has advantages of being inexpensive, very quick to fit, and does not require a training period, all of which is advantageous when exhibiting to the public.

14.4.1 How It Was Made

The Disadvantages of Time Travel was shot over 9 days in Scotland by a small professional crew with young non-professional actors aged between 12 and 17 years old. It was made between August 2014 and July 2015 with a budget of £19k, funded via Kickstarter and Arts Council England Grants For The Arts public funding, and was supported by B3 Media, AND festival, FACT and the University of Nottingham.

The design of *The Disadvantages of Time Travel* was inspired by dream and imagination; the director wanted a system to respond to the viewer in a way that every frame was different from one viewing to the next, to recreate the hypnagogic space we encounter between wakefulness and sleep, that of the lucid dream. This was reinforced in the cinematography and the music composition and implemented in the system design. This looseness translated to production as the crew experimented on set and used storyboards as a guide rather than a strict plan.

The system comprised a NeuroSky MindWave headset and a MacBook Pro. We developed a Python program which reads the Bluetooth messages sent from the headset, which are then routed internally as Open Sound Control (OSC) data (Schmeder et al. 2010) to a patch written in Quartz Composer, which is responsible for playback.

Script The script was written as short vignettes of events and memories from the directors childhood, which were reduced and sorted into a narrative structure, then roughly storyboarded as key moments. The cinematographer and director drew out more detailed storyboards and worked out shots during the evenings before each day's shoot.

Sound Design The music was composed by Hallvardur Ásgeirsson. He worked with mood-films created by the director to develop the main themes of the score before the scenes were completed. The composer worked remotely from before production started and throughout post-production. The composed tracks were designed in a way that they would never be heard in their entirety. The compositions were split into 4 stems; two of these stems play together and then fade to the other two stems with blinking data. This produces a soundtrack that has a continuous tempo and key but flips between instruments.

Cinematography The filmic design was comprised of four levels of narrative. These were delineated by the qualities of the shots for each scene, which were active, passive, dream and reality, and related to lenses, frames per second and grip equipment of each scene. When shooting the film, the cinematographer and director developed a shorthand for each scene designation, which defined the tool-set to be used. A few examples of this are: active shots would be handheld, passive on a tripod; dream shots would use a tilt shift lens and reality shots would use clean optics; reality shots were filmed at 25 frames per second, dream at 50fps. The crew also shot on two cameras, which allowed the same action to be recorded at different speeds and on different lenses. This allowed for a lot of spontaneity and inventiveness on set while continuing to get the coverage needed.

Directing A key concern in making an interactive film is the added effort it will take, which has implications for production time and budget. Even with traditional film a large amount of usable footage ends up unused on the 'cutting room floor'. Directing this film was very close to a traditional process; it involved working with actors and the cinematographer as a director would usually do. However, there are a few small differences to take into account. In essence we were shooting four times the duration as we would in a linear narrative film, yet that does not mean that we needed quadruple the time to do it. In *The Disadvantages of Time Travel* , shots were designed to be longer than what would normally be expected, and there were not four times as many camera or actor set-ups. It is these set-ups that can take the time in filmmaking. However, the director needed to keep a close eye on the coverage of the four layers.

Editing Perhaps the most divergent practice from the traditional filmmaking model came in the editing of *The Disadvantages of Time Travel* . In a traditional editing process, one creates the rhythm of the film, and encodes meaning by juxtaposing images in temporal sequence. In this process, however, it became apparent that the work was in making the four layers sync and fit with each other; it was in making opportunities for meaning to be encoded. The decision to film the dream layer footage in 50 frames per second made that process much easier as these scenes could be changed in speed.

However, it became apparent in the editing process that the film could not be easily reviewed in the way it will be experienced by viewers, and so the editor created intermediate screeners where all four layers were placed in four quadrants of the screen. This worked extremely well as one could move from shot to shot by moving one's eyes. This technique is reminiscent of the layout used by filmmaker Mike Figgis in the movie *Timecode* (Fabe 2014). The grading was an element where the editor could be fully creative, as they used colour to express the emotion in the film. In the few moments where there was not complete coverage (where there were not four unique shots), the editor was able to use colour grades to define the different layers.

14.4.2 How It Was Received

The film was screened outside FACT (in Liverpool, UK) over 6 days, in a cinema converted from a vintage caravan (see Fig. 14.5). The cinema accommodated up to 6 audience members and one interactor. During these screenings, we conducted a research study; we recorded interviews with the interactors, logged the brain data from the NeuroSky device and recorded the resultant films that were created. The full results of this research have been published by Pike et al. (2016). The text below discusses the key issues that relate to the reception of, and interaction with, our brain-controlled cinema.

Fig. 14.5 Caravan Cinema used to exhibit *The Disadvantages of Time Travel*

As a cultural event people found the experience to be different to that of watching a traditional film. While there was no one typical reported experience, some interactors felt they could affect the film: *"I felt like I could slow it down, speed it up and I could move on"*. The content of the film was also challenging in nature, in that the narrative had an emotional charge. The added interactivity, for some, amplified emotional intensity: *"I found it, I don't know if scary is the word, but perhaps a little unnerving in places because usually when you see a film you see outside the character but because you could move to see from inside the characters perspective"*.

The blink control method became the prominent aspect of interaction. On average across all the interactors blink frequency was about 5 per minute, the same as in television watching; however some used their interaction carefully, *"I definitely had control over the way it was edited as I'd seen that version before, I kind of knew I could change the footage that I hadn't seen and see the other version."* This person blinked only 2 times per minute.

Where Hasson speculates that the power of movies *stems from their ability to take control of viewers minds* (Hasson et al. 2008) and was able to measure this by correlation between subjects, viewers experiencing *The Disadvantages of Time Travel* would find each film was unique. Each audience saw a different film because their physiological activity was divergent from each other.

From the perspective of control, we learned a lot from comparing the blinking method of influencing the film with the influence that brain data had on the film. Blinking is both an explicitly conscious action and an automatic unconscious response, while the NeuroSky brain data is entirely less explicit. For some the limited aspect of control was welcomed, *"I was happy with the amount of control, because I didn't know the parameters of how to affect it and trying to manually affect how your brain is reacting is really difficult."*

We found the control that viewers felt was sometimes at odds with the enjoyment of the film and immersion in the narrative: "*a lot of the time I found it difficult to remove myself from the thought of... the fact that I was changing it and I was controlling it, and I kept thinking like why is my mind doing that pace like what's going on in the film?*" We found this to be especially common in regard to the blinking method of interaction.

Some comments brought into relief the dissonance of narrative flow and interaction. This is backed up by the previous arguments regarding narrative flow: "*Sometimes you notice that you have the agency, and that flipped you out of flow. But sometimes you've really added to the dramatic effect.*" So, it seemed that we reached a boundary of the level of interaction. "*I think that, maybe more control would have taken away from the immersive elements of the film.*"

Some positive reports occurred when interactors stopped trying to actively control the film. For some it's a conscious decision: "*Sometimes I just let go. It's good*". For others the narrative immersion of the story overpowered their agency: "*I completely forgot, I was concentrating on the film.*"

14.4.3 Observations

Below are three of the main observations that we made during the making and screenings of *The Disadvantages of Time Travel*:

- **Some forms of physiological feedback are stronger than others; blinking as a mode of control overpowers that of attention and meditation levels**. We found that people interacting with this film derived more enjoyment when they let go of trying to actively control the film and immersed themselves in the narrative. And so, in future designs, departing from conscious control of the interaction was identified as grounds for further exploration of multi-media BCI experiences.
- **Using cinematic techniques to delineate narrative constructs was an effective strategy and has further potential for BCI cinema**. In this film temporal and spatial montage, thematic juxtaposition and deconstructed music was used as feedback. To use interactive feedback to build possibilities for narrative flow could go towards solving disruption caused by conscious interaction.
- **The traditional process of filmmaking need not be completely redesigned**. A lot of footage that is usually discarded can be used to good effect in an interactive film context. By subtly redesigning the process the production can be economical with the shot footage.

14.5   Case Study 2—*The MOMENT*

To continue the artistic and research journey into interactive brain-controlled cinema, the director set out to make a new film, with three main objectives when embarking on this project. Firstly, it had be coherent for non-interacting viewers, as well as the people wearing the brain scanner. To do this, the director made efforts to maximise narrative potential by cross referencing and parallel storytelling, so that on each retelling of the story it maximises the potential for meaning making. Secondly, it should encourage repeat viewings. The strategy for this was to maximise the potential versions of the film, so that people noticeably experience a new version of the film on each viewing. Finally, the director wanted to further explore subconscious control; and therefore decided to remove blinking and meditation as a control method and work only with different interpretations of attention data.

The MOMENT (see Fig. 14.6) is a sci-fi thriller, with an interactive narrative and a 24-min runtime. It was made between January 2017 and July 2018 with funding from the EPSRC, Arts Council England, and the University of Nottingham. Its final budget was £54,000 and it was shot in Manchester, England. The interactive system was re-developed in the six months leading up to pre-production. We modified BCI mappings and feedback, scene combination mechanisms, and narrative approach, based on what we found when presenting the previous film to audiences.

The director set out to create an interactive film which had a narrative that would be understandable, but have variations, and would be minimally disrupted by the act of interaction. As with the last film, it was not set out as a branching narrative, but a film that responds to the user by adapting the telling of the narrative in diverse manners.

Also central to the design was that it would be open to, and reward, repeat viewing. Thus, *The MOMENT* was designed to be rich in detail, to allow new narrative arcs and nuances, and to encourage new understandings with each version viewed. *The MOMENT* has 17 scenes, which can each be re-configured in 6 different ways (paired combinations of 3 narratives). At any second the film can edit between the two active narratives of the current configuration for that scene. In all, not taking into account differences in real-time editing, there are 16,926,659,444,736 possible combinations across the 17 scenes; it would take over seven hundred million years to watch them all back-to-back. Thus, there is little possibility of the same configuration being seen twice. A visualization of 29 recorded routes through the film can be seen in Fig. 14.7.

Given that the structure, environment, and interaction would be new and unfamiliar to the audience, the director decided to make a genre film, so that viewers would have a familiar element of the viewing experience to relate to. Having an anchor amongst the novelty also meant we were able to research how well the recombined narrative was understood by audiences.

Design In line with the iterative nature of the practice, the director/researcher was able to address and improve on some of the issues we observed in our previous film. Specifically, in terms of voluntary control, the first film was unbalanced. Blinking was linked to temporal montage and, while effective when interactors were able to

Fig. 14.6   The poster for *The MOMENT*

Fig. 14.7 Journeys through *The MOMENT* taken from the logs of 29 example experiences; from (Ramchurn et al. 2019)

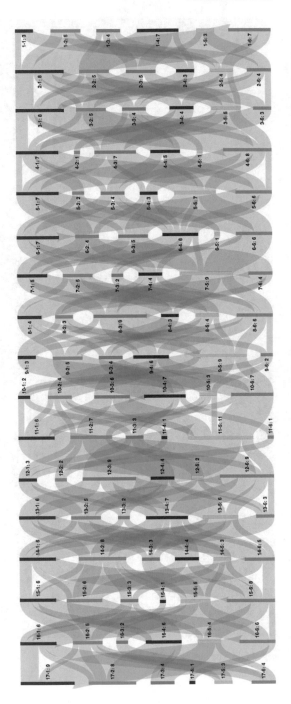

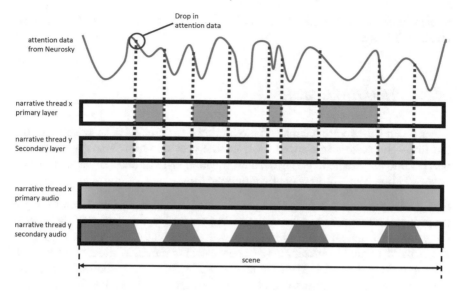

Fig. 14.8 How *The MOMENT* recombines

forget about it, it often overpowered the other Attention and Mediation mappings. Blinking encouraged play, which was at odds with immersion into the narrative, and it was apparent the subtle nature of feedback to Attention and Meditation was occluded by this stronger form of interaction.

Within the neuroscience and cinema literature, there is the concept of event segmentation (Cutting 2014). The director took this concept of event segmentation and considered whether a period of Attention could be mapped to a shot, and if that would feel natural to an interactor. The director took some of the Attention data from *The Disadvantages of Time Travel* to see what the period duration of Attention was; it averaged out at about 6 seconds,[2] which is within the range of average shot lengths (Bordwell 2002).

The film was designed to have 3 narrative layers, or threads. Each narrative thread has itself 2 alternative sound designs: a primary and a secondary. When a scene is played it visually cuts between 2 narrative threads, with the primary sound design from one thread and secondary from the other. A diagram showing this can be seen in Fig. 14.8. This makes a total of 6 possible combinations of each of the 17 scenes. These scene combinations are chosen based on interactions during the previous scene; how fast it cut and how long was spent on each narrative thread.

It was important to have the above design in place before starting to write the script. The design created a ready-made framework that came with its own rules, such as: narrative threads run in parallel time; each thread must relate to both other

---

[2]This was revisited after the premiere, by analysing the log files created while people watched *The MOMENT*; we found the average length of attention at 4.2 s.

threads at all times; threads can be co located; each narrative thread follows one character's story.

14.5.1 How It Was Made

When it came to pre-production and production—the process of practically making the film—the scale was stepped up from the previous film. Therefore, planning was paramount; detailed lighting designs, location layouts, and storyboards were all created in line with a professionally produced film.

Script As previously mentioned, the interactivity was designed prior to pre-production and so, when it came to writing the script, the nature of the interactivity informed the structure of the narrative and layout of the script. Figure 14.9 shows the diagram the director used to construct the narrative structure of the film, prior to writing the script. The original concept was to have a shorter looping film that would be watched once, twice or three times. In the process of writing the script the film became longer as narrative coherence was prioritised.

Usually, a script is written in the way it is to be seen, which becomes a plan for the filmmakers to follow, as it is a textual version of the film. With *The MOMENT* this is not the case; the scenes are malleable and cannot appear on the page as they do on screen. Several attempts at writing the script were aborted before an appropriate script structure was decided on. This involved putting each of the three narrative threads together within each scene so that writing across narrative threads was possible.

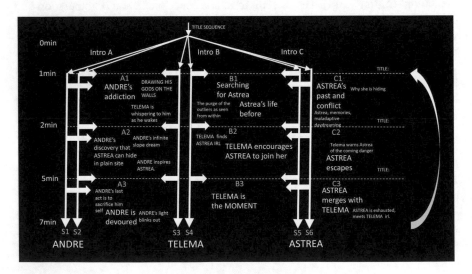

Fig. 14.9 Early 3-part narrative structure for *The MOMENT*

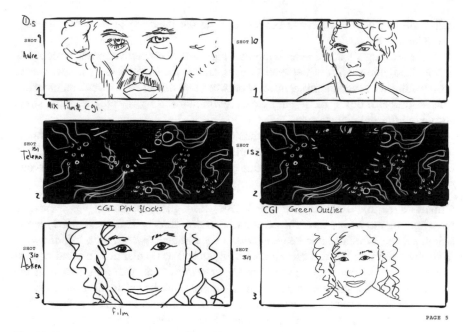

Fig. 14.10 Parallel storyboards from *The MOMENT*

Storyboards The storyboards, as seen in Fig. 14.10, were also drawn in parallel, with three boards to a column all depicting the same point in time for each narrative thread. It was here that visual continuity, possible match and reverse cuts, and places where the story is told between the narrative layers was visually planned. A total of 432 panels were drawn.

Sound Design The sound design and composition were handled by two long-term collaborators: Hallvardur Ásgeirsson and Scrubber Fox. The music and sound design are key to the combinatory aspect of the film. Each narrative thread consists of 2 separate compositions. The system is designed in such a way that primary and secondary audio tracks from each scene come together to form the soundtrack. This effectively means that the soundtrack is unique each time it is interacted with. Interestingly, each composer took a different approach to designing the soundtrack; while one composer took a more classical approach, the other used synth-based systems to develop more of a soundscape. One of the composers also noted that the viewer of the film, in his opinion, took on the role of the conductor, in the sense that they were 'actively' controlling the music. Due to the interactive nature of the media and the different ways that the tracks could be merged to develop a combined soundtrack, the composers had to create twice the amount of musical material.

The foley (the diegetic sound effects), the primary and secondary compositions, and the vocal track are discrete files, which are re-combined in real-time as the film plays, creating moments of synchronisation and audio-visual dissonance, both of which are found in cinema and have effects on the meaning of the story's reception (Chion 1994).

The audio segues become an important part of the interactive movie's recombination. The video will cut between threads and at the beginning of a new scene, however the audio can blend seamlessly between two threads or at scene boundaries. This is similar to the 'J' or 'L' cut used in traditional editing, where the sound starts before or after the picture cuts.

As the musicians were composing music for multiple visual threads, it was important that they could see what was happening at the same time in the film. Similar to the last film, the director produced a single video that shows all of the layers on one screen (see Fig. 14.11).

Cinematography When designing the lighting throughout the film, it was decided to have lighting motifs for each character and narrative thread. This created a dynamic colour palette, which was in keeping with a style known as tech noir. Lighting had to be designed into the set; certain colour lights had to be planned for, and needed to make sense in the diegetic world of the narrative. This was all planned in the months prior to the shoot with the cinematographer and the production designer. An example of this is a scene in a flat where characters from multiple threads appear. The character Astrea has a green lighting motif. She has a scene near a balcony where we designed the set so that a green light in the kitchen was positioned to light her near the end of her scene.

Directing Again, the process of directing on set was not far removed from that of a traditional film. A notable difference was that the scenes and shots had to be carefully timed and annotated in order to match with their counterpart narratives. As the film was designed to cut between narrative threads at any point, parallel shots had to be carefully composed so as to make meaning between them. For example, we would place similar compositions of shots between people to show a connection between the characters. If any changes to storyboards occurred these would be recorded to be referred to later. Scenes were shot out of order, which is typical, but it meant a lot of moving between storyboards or screenshots, to cross reference the current and the parallel shot. As in our previous production, shots were allowed to run on for longer than normal, to allow for these shots to be cut together in the final interactive configuration.

Editing When it came to prepare the footage for the interactive system, where one would usually edit the film, an entirely new process had to be devised. Here we will discuss where this project diverged from a typical editing process. We edited with Adobe Premier. Firstly an animatic (where the storyboard frames are put in place of the footage) was produced. All three narrative threads were placed on the same timeline. Scenes were then prepared separately (best takes were chosen and on set recorded sound was synced to video). These scenes were then placed on top of the animatic. In order to keep track of these layers and multiple takes it was necessary to be scrupulous about how the metadata was planned and organised. The video and audio tracks were renamed to reflect the narrative layers. Titles were used to visually

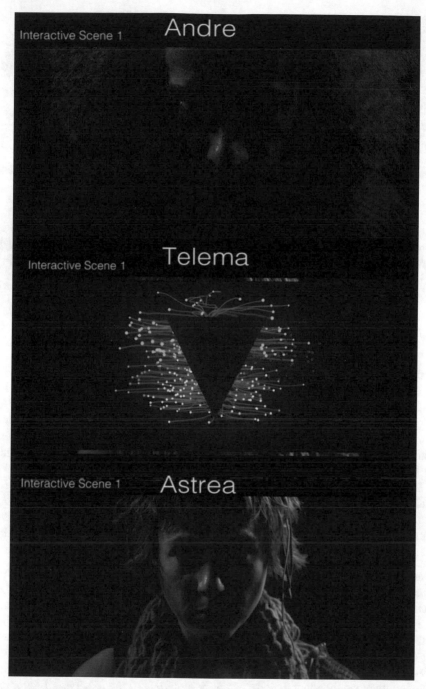

Fig. 14.11 A still from the assembly edit showing one point in time across three scenes

Fig. 14.12 Left—a view from inside the caravan. *Photo credit* Studio Softbox. Right—The mobile cinema on on site at Sheffield DOC/FEST

show which scene was currently playing both in the timeline and on the composition screen. Assembly edits were exported showing all three levels of narrative, which provided context to the sound and animation departments who were still creating content during the editing process (see Fig. 14.11).

14.5.2 How It Was Received

Before *The MOMENT* was made, we obtained a grant from the EPSRC to convert a caravan into a mobile cinema, as shown in Fig. 14.12. The cinema seats up to 7 people comfortably; one person controls with up to 6 non-controlling audience. *The MOMENT* was previewed at 2 venues for a total of 7 days. During this time interviews with audiences were conducted by researchers, director's notes were made, and press interviews took place. From the interviews and notes additional changes were made to the system and the content was refined, editing 4 mins from the runtime. *The MOMENT* then toured to Sheffield DOC/FEST, Kendal Calling, Arts by the Sea, and Blue Dot Festival in the UK and Ars Electronica in Linz, Austria. We conducted interviews, ran a questionnaire, recorded system logs and each film that was watched. More detail of our research into experiences of *The MOMENT* is presented by Ramchurn et al. (2019). Below are some interesting highlights from that research.

Controlling the Film From questionnaire feedback, some controllers reported trying to exert intentional control but experiencing a form of partial control. "*It felt less like control and more like influence*". Even although they did not experience direct control, they still found enjoyment in knowing that they are responsible for the control being exerted. "*It's a really exciting experience to think that on a very, almost unconscious or becoming conscious way, your brain is interacting with a piece of art someone else has made*". Some controllers found that when they couldn't consciously control the film, they would adopt a more lean back attitude. "*At the beginning I felt it cut a lot and I was like, oh, I'm cutting it a lot. At the beginning, I wanted to control it more because I wanted to figure out what's going on with the film. And then I was* *just like, right, I'm just going to just watch it and see what happens. And then that's when the third character came in and I think things made sense for me a bit more".* When drawing a distinction between video games and film, this perspective provided positive feedback on keeping it as an immersive film experience; we discuss issues of control and interaction further in Ramchurn et al. (2019). Some found that the process of trying to gauge their control resulted in an introspection and mindfulness where they honestly question what is happening in their own heads. *"Interactive film just is amazing to me, just because it's not only some medium that is entertaining but also can pose a lot of questions and make you think a lot".*

It was reported that the interactivity gave the film more value than that of a normal movie. *"It's like an art form where if you ask three different people to paint a vase of flowers, you get three really different outcomes. With this, there's a real sense of ownership I think. There is a sense that you do have a stake in what comes out at the end of this."* The uniqueness of each film caused a feeling of ownership. The experience of no narrative decision making was likened to that of watching a normal film too. One of the most common reference points that people have is that of *Choose Your own Adventure* books (Livingstone 1984). *"I used to read a lot of the adventure books where you got the choice as to what you did. And even though I read quite a lot of them, I didn't find them that satisfying because you could muddle it yourself, you know, once you knew what you were doing, and I much prefer to be out of control in that situation, rather than decide."*

Viewing without Controlling The majority of people who watch *The MOMENT* do not get to interact with it, as it has been designed to be screened to an audience with only one interactor. While it may have been harder to reconstruct the narrative for some, our questionnaire and interview data showed that it was enjoyed by non-interactors and that the work of comprehension was actually an enjoyable process. *"I think I just like the idea that you can watch it with many different people being the controller each time and see a different narrative of a different way of watching the film. For me, I think that's probably more important as just a regular cinema goer and somebody who doesn't want to try and pull too much out of that experience other than just going along for the enjoyment of it."* This is a good sign for future presentations in auditorium settings.

Repeat Experiences In relation to one of the primary aims for this second film, repeat viewings, people did express the desire to see the film again. This was true of controllers and non-controllers. Some who had seen the film first as a non-controller were able to use their prior knowledge as a framework for understanding the film whist controlling. People were keen to compare versions, either in discussions after a screening, where viewers would talk about what they saw, or by booking another ticket and coming back to see it again. *"I loved seeing it again and I want to see it again. Because I know for a fact that if I went in knowing nothing about it, and I saw my edit first, I'd be like, okay, I'm a bit confused. If I went in and saw the second one, I'd be able to put those two together which is amazing."*

14.5.3 Observations

Below are three of the main observations that we made during the making and screenings of *The MOMENT*:

- **Removing explicit voluntary interaction also reduced interruptions to immersion**. We removed blinking in order to minimise the controller effects we found in *The Disadvantages of Time Travel* and to encourage a letting go of control. The interaction operates by internalised responses rather than physical actions. We found that this film fits in to our previous taxonomy, that it moves people away from voluntary control and awareness of their control. We found that people were less likely to be removed from the narrative flow of the film by interactive side effects.
- **Creating smooth narratives from unknown dynamic possibilities is an open challenge**. There are a massively high number of potentially different films that can be created, but when does any one of these films achieve a distinct narrative? Where the parallel stories interact become touch points in mapping these narratives out. Exploring narrative structures made by interactors may give clues for future system designs that encourage cohesive narrative building. Where before we talked about temporal montage and spatial montage, in this case the system could be described as a form of narrative montage.
- **BCI in film needs to connect to cinematic vocabulary to relate to practice**. In *The Disadvantages of Time Travel* there was the original aim that every frame would be unique. This was done by using image blending to cross dissolve between visual layers. *The MOMENT* relied solely on cuts—temporal montage—as the visual interactive mode of feedback. It will be important for the evolution of the medium of brain-controlled cinema to have all cinema's visual vocabulary available to it.

14.6 Discussion

Below we discuss two main topics: (1) where BCI film, as we have explored it, sits in the broader picture of interactive film, and their notions of control, and (2) the possible opportunities for the film industry that brain-controlled film represents, alongside challenges that remain.

14.6.1 Brain-Controlled Cinema as Interactive Film

From the research of our work, we identified a taxonomy that characterized the variability in experience of *The Disadvantages of Time Travel*, in relation to control as typically considered in HCI (see Fig. 14.13). The taxonomy maps extent of voluntary

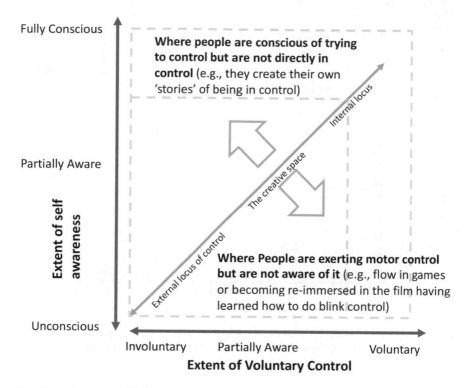

Fig. 14.13 Taken from #Scanners: exploring the control of adaptive films using brain-computer interaction (Pike et al. 2016)

control and extent of self-awareness to examine the trajectories through the film that people would take as they became familiar with the control mechanism. For full detail of this taxonomy, in the context of control in HCI, readers should refer to Pike et al. (2016).

14.6.1.1 Different Types of Control

It is interesting to consider how our case studies relate to the broader literature of interactive film experiences. Figure 14.14 plots our brain-controlled films with previously discussed interactive films on Pike et al.'s taxonomy. The data for the other interactive films is extrapolated based on their interactive characteristics. For example, we note that *As You Like It*, *Kino Automat* and *Mr Payback* all rely on a conscious interaction—closing one eye, clicking a button—that is both voluntary and has to be done consciously. We see, for example, that there is an overlap between *The Disadvantages of Time Travel* and *As You Like It*, which is specifically to do with manual eye control, although the autonomic nature of blinking is distinct from the manual control of one eye.

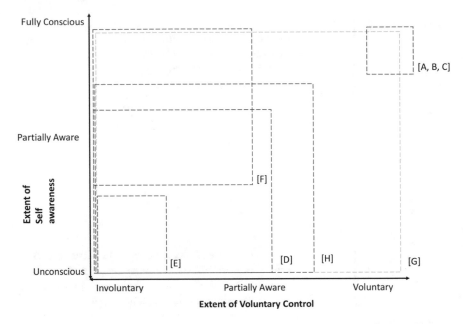

Fig. 14.14 (adapted from Pike et al.)—**a** *As You Like It*, **b** *Kinoautomat*, **c** *Mr Payback*, **d** *Obsession*, **e** *Many Worlds*, **f** *The Angry River*, **g** *The Disadvantages of Time Travel*, **h** *The MOMENT*

The Angry River uses eye tracking which, like the blinking attribute of *The Disadvantages of Time Travel* , can be both unconscious and directed by the user; what differentiates them is that in *The Disadvantages of Time Travel* blinking causes a noticeable cut, which brings an awareness of the user's agency. The audience in *The Angry River* cannot know at what point their influence has changed the playback of the film. The BCI aspect of *The Disadvantages of Time Travel* and *The MOMENT* is characterised by user's being both unconscious of their interactions and unaware of the extent of control they have. *Obsession* and *The MOMENT* share a lot of characteristics, as they both embody unconscious control. *Many Worlds* has even less feedback; only 2 points in the film, that 'give away' the interactive aspect of the film leading to a conscious awareness of the interaction.

14.6.1.2   BCI Control as Interaction

The central critique of interactive film, as discussed earlier in the chapter, is that interaction undermines narrative immersion. Narrative works within video games when interaction is part of the diegetic world of the game; but this does not naturally extend to watching movies.

In attempts to remove the need for interruptive explicit interaction in brain-controlled film, our research (Pike et al. 2016; Ramchurn et al. 2018, 2019) has examined the nature of control that we have produced. We found that although participants could not easily exert direct control in our brain-controlled interactions, it was important for viewers to feel that they had controlled and influenced the film, and that this was sufficient for a good brain-controlled experience. Of all of our chosen interactions, blinking in *The Disadvantages of Time Travel* was the most explicit and conscious, and we found that having such direct control could quickly become disruptive. Consequently, in *The MOMENT*, we created a more implicit and passive interaction, both within and between scenes; essentially removing direct control from the interactive film. This perhaps does not fall into the typical definition of interaction and direct manipulation when speaking from a human-computer interaction standpoint. But as we have seen there is some form of interaction happening even if it is ambiguous. Gaver et al. (2003) has argued that ambiguity can be a useful and enlightening resource for design and interpretation, in this case the limits of control. Likewise in the context of the juxtaposition of images that is the very essence of montage, by designing opportunities for possible meaning making, the narrative flow can run unhindered.

Open research questions remain as to what can be considered as 'interaction' and 'control'; to ask whether our films are brain-responsive, rather than brain-controlled? Have we created an interactive film, or a responsive film? And have we gone too far away from an interactive film? Can interaction and control ever be achieved whilst making a good interactive film? Indeed, the nature of control and interaction is rather an open question for passive BCI systems more broadly.

14.6.2 The Future of BCI Film: Challenges and Opportunities

Ethical implications of media personalisation have already surfaced that make it more and more difficult to verify the authenticity of media (Chesney and Citron 2018). As current business models of surveillance capitalism (Zuboff 2015) remain unchallenged, interaction at scale using brain-computer interfaces may result in unforeseen consequences. Already our media consumption is curated by algorithms in our social media timelines, our watch-next queue, and in the commissioning process of new media giants such as Netflix and Amazon. It is not unlikely that brain data will be used to personalise aspects of consumable media on larger scales in the near future, which will tailor personal media to the individual, for good or for ill. In this regard, communities are already interested in the data produced about viewers and their preferences, by the options they select while watching *Bandersnatch*.

The release of *Bandersnatch*, and the trial run children's episodes released in the preceding 2 years, indicates that interactive films are back 'on trend' thanks to the possible interactions facilitated by viewing streamed media on tablets and phones. *Bandersnatch* has been broadcast in 27 languages to 137 million Netflix subscribers (Reynolds 2018). At the same time, simple BCI headsets are available now at a cost comparable to previous large-scale deployments of interactive movie apparatus such as the remote controls installed into theatres in the 1990s. BCI cinema has a unique potential within the interactive narrative medium but content needs to be produced and distributed. Artistic and scientific cross-pollination has the potential to explore this space before higher bandwidth ubiquitous technologies become the norm.

14.6.2.1 Scaling up

A key challenge for interactive film has been creating an experience for large audiences, rather than an individual. Some previous approaches rely on collective control and others simply don't scale up because they rely on the control of a single viewer at a computer. So far our brain-controlled cinematic experiences have been situated in small cinema spaces, with one controller and an audience of 5 or 6. For one hundred people to see the film this way will take a full day of screenings, with dedicated staff running the film and a booking system, and it can only be on in one place at a time. So, when considering scaling up the experience, although it seems natural to follow the tradition of cinematic presentation and screen it in an auditorium, the question is how to do that and still offer equivalent value to the audience. Limited people will get a chance to interact, and the film may seem less special when more people are experiencing it. On the other hand each audience would experience a unique edit, which may be perceived as an added value.

It was important for *The MOMENT* to work as a spectator experience, and for this to happen *The MOMENT* had to feel like a film. To achieve this we incorporated familiar story elements and a cohesive narrative structure. The edits produced by the brain-controlled interaction are made to feel natural and are incorporated into the non-diegetic story by the audience.

14.6.2.2 Shared Control

In our as yet unpublished work, we have explored how multiple audience members interact with a BCI film at the same time. There is still more to explore as we found mixed results. Some positive interactions observed were with groups that knew each other and were satisfied to cooperate in a non-hierarchical way. Some other groups suffered from individuals hijacking the experience, effectively taking over the control. The shared responsibility inherent in group control could both mitigate against shyness and address divergent behaviour. By designing for interactional behaviours in shared control we could create potentially interesting and unique filmic experiences.

14.6.2.3 Industry Opportunities for BCI Film

We found that controllers reported feeling a responsibility and privilege when making their version, and that other audience members felt the experience provided an insight into the controller's state of mind, which raises the possibility that these perceptions could be exploited by inviting famous or influential controllers. Although these versions could possibly be seen as authoritative, it may also give more value to the non-interacting viewers.

The high levels of variation between versions may actually encourage repeat viewing, as the rarefied privilege of controlling is made available to others. Encouraging repeat viewing is of particular interest to industry business models. As we have said, each viewing of *The MOMENT* is unique, never to be repeated, therefore personalised ownership could also provide additional revenue. We are currently exploring design requirements of an archive of versions of *The MOMENT*. User responses to this archive will give clues as to the viability of, and value to, viewers of such personalised versions of media.

The platform created for *The MOMENT* is versatile and could be used by other studios, filmmakers and artists to write and direct new narratives across genres.

14.6.2.4 Live Scoring for Dynamic Films

One option that addresses some challenges and opportunities facing BCI films involves looking back to the early days of cinema. Before the 'talkies', in the days of silent film, a piano player would compose their own music to the movie as it played. This tradition has re-emerged with well-known films being shown with live orchestras,[3] and classic and cult films being re-scored by modern bands (Ninja Tune 2003; Murray 2011). This tradition of the 'Live Score' has been updated for the new century and has become an understood form of experiential cinema. The film plays, and the orchestra or band will play along to it, perhaps with a written score, taking visual cues from the film. The brain-controlled films we have been discussing, however, are different every time they are played, and the score to the film is also altered in real time. In this case therefore it is not as simple as performing a live score for a film, or even re-scoring a known linear film. In order to produce a live scored brain-controlled film, therefore, one must work with musicians to discover how best to facilitate their performance.

We recently premiered the world's first live scored brain-controlled film at Reykjavik International Film Festival in October 2018. In order to make and perform a malleable score we had to develop a way of creating communication between the musicians, the interacting audience members' brain data and the playback system. To do this we amended the existing system for *The MOMENT* to provide automation to the musicians' software, and a user interface for the musicians that displayed pertinent and timely information about film content to them.

---

[3]http://moviesinconcert.nl.

The live score becomes an additional cultural experience that can be offered to film festivals and music venues, increasing access to the work and its appeal to audiences.

14.7 Conclusions

In this chapter, we have contributed two case studies of creating brain-controlled films, based upon a Performance-led Research in the Wild methodology, focusing particularly on the artistic practices involved and how closely they relate to core issues of interactive control in Human-Computer Interaction.

To do this, we reviewed the history of interactive film and neurocinematics in order to highlight the difference between brain-controlled and previous forms of interactive films that use either conscious or alternative involuntary modes of interaction. As part of this review, we highlight three examples of existing interactive films that use non-obvious modes of interaction. We then presented our two case studies, which build upon each other, and explore both different artistic practices and different modes of brain-controlled interaction. We reflect on these different modes of interaction further in our discussion, as well as presenting open issues and challenges for future work, for both filmmaking and brain-controlled interaction with dynamic media, specifically with the film industry in mind.

At the core of this work is the opportunity to move away from homogenised media culture. New forms of media can tell new types of stories and it is evident from media coverage that our brain-controlled cinema has captured the public's imagination, which perhaps points to a desire to break the mould of traditional storytelling. We hope the experiences and technology platforms that we have reported here can be used by other filmmakers and other artists more broadly. The findings of this research, however, are relevant to Human-Computer Interaction, as well as the design of interactive cinematic experiences. The issues of involuntary control that we have explored could be useful for future designs of a wide range of technological interactions—to add depth, layer context, and enrich our increasingly digital lives. Amongst the core of our own lessons learned, is that understanding these possibilities involves as much design and creativity as it does technological innovation, in order to make observations about the human behaviours involved, voluntary or otherwise.

Acknowledgements This work was supported by the Engineering and Physical Sciences Research Council [grant numbers EP/L015463/1, EP/M02315X/1]. We would like to acknowledge the financial support we received from the BFI Film Hub Propeller Scheme, Arts Council England Grants for the Arts, the EPSRC Telling Tales of Engagement competition, and those who contributed to the original Kickstarter campaign. Further, we would like to acknowledge the support provided by our partners: B3 Media and Live Cinema. Finally, we would like to acknowledge our long-time collaborators Mat Johns, Scrubber Fox, Hallvardur Asgierson, and Rachel Ramchurn, as well as the cast and crew for both our productions.
Data access: As per the previously published works from this project, datasets cannot be made available due to concerns over protecting the anonymity of our participants.

References

Anderson S (2006) Soft cinema: navigating the database. Mov Image 6(1):136–139. https://doi. org/10.1353/mov.2006.0002

Benford S, Greenhalgh C, Crabtree A, Flintham M, Walker B, Marshall J, Koleva B, Rennick Eggle- stone S, Giannachi G, Adams M, Tandavanitj N, Row Farr J (2013) Performance-led research in the wild. ACM Trans Comput-Hum Interact 20(3):14:1–14:22. http://doi.acm.org/10.1145/ 2491500.2491502

Bordwell D (2002) Intensified continuity visual style in contemporary american film. Film Quart 55(3):16–28

Chesney R, Citron DK (2018) Deep fakes: a looming challenge for privacy, democracy, and national security

Chion M (1994) Audio-vision: sound on screen. Columbia University Press

CtrlMovie (2016) Ctrlmovie. http://www.ctrlmovie.com/. Accessed 19 Sept 2018

Cutting JE (2014) Event segmentation and seven types of narrative discontinuity in popular movies. Acta Psychologica 149:69–77. https://doi.org/10.1016/j.actpsy.2014.03.003

Fabe M (2014) Closely watched films: an introduction to the art of narrative film technique. Univ of California Press

Gaver WW, Beaver J, Benford S (2003) Ambiguity as a resource for design. In: Proceedings of the SIGCHI conference on human factors in computing systems. ACM, New York, NY, USA, CHI '03, pp 233–240. http://doi.acm.org/10.1145/642611.642653

Hales C (2015) Interactive cinema in the digital age. In: Koenitz H, Ferri G, Haahr M, Sezen D, Sezen TI (eds) Interactive digital narrative: history. Theory and Practice, Routledge, London, UK, pp 36–50

Hasson U, Landesman O, Knappmeyer B, Vallines I, Rubin N, Heeger DJ (2008) Neurocinematics: the neuroscience of film. Projections 2(1):1–26. https://doi.org/10.3167/proj.2008.020102

Höök K (2008) Affective loop experiences—what are they? In: Oinas-Kukkonen H, Hasle P, Har- jumaa M, Segerståhl K, Øhrstrøm P (eds) Persuasive technology. Springer, Berlin, Heidelberg, pp 1–12

Kim Y, Moon J, Lee H, Bae C, Sohn S (2012) Integration of electroencephalography based services into consumer electronics. In: 2012 IEEE 16th international symposium on consumer electronics, pp 1–2. https://doi.org/10.1109/ISCE.2012.6241692

Kinder M (2003) Designing a database cinema. The cinematic imaginary after film, Future cinema, pp 346–353

Kirke A, Williams D, Miranda E, Bluglass A, Whyte C, Pruthi R, Eccleston A (2018) Unconsciously interactive films in a cinema environment—a demonstrative case study. Digital Creativity 29(2– 3):165–181. https://doi.org/10.1080/14626268.2017.1407344

Lim CG, Lee TS, Guan C, Fung DSS, Zhao Y, Teng SSW, Zhang H, Krishnan KRR (2012) A brain- computer interface based attention training program for treating attention deficit hyperactivity disorder. PLOS One 7(10):1–8. https://doi.org/10.1371/journal.pone.0046692

Livingstone I (1984) Deathtrap dungeon. Wizard Books

Manovich L, Malina RF, Cubitt S (2001) The language of new media. MIT press

Murch W (2001) In the blink of an eye: a perspective on film editing. Silman-James Press

Murray R (2011) 65daysofstatic score 'silent running'. https://www.clashmusic.com/news/ 65daysofstatic-score-silent-running. Accessed 29 Jan 2019

Ninja Tune (2003) Man with a movie camera by the cinematic orchestra. https://www.ninjatune. net/release/the-cinematic-orchestra/man-with-a-movie-camera-2. Accessed 29 Jan 2019

Penlington N, Guiterrez De Jesus F, Smail S, Waston N (2014) Choose your own documentary, theatre performance, 16th oct–18th oct 2014. Giant Dwarf Theatre, Sydney, viewed 18th Oct

Petric V (1987) Constructivism in film: the Man with the Movie Camera: a cinematic analysis. Cambridge University Press Cambridge

Pike M, Ramchurn R, Benford S, Wilson ML (2016) #scanners: Exploring the control of adap- tive films using brain-computer interaction. In: Proceedings of the 2016 CHI Conference on

Human Factors in Computing Systems, ACM, New York, NY, USA, CHI '16, pp 5385–5396, http://doi.acm.org/10.1145/2858036.2858276

Polaine A (2005) The flow principle in interactivity. In: Proceedings of the second Australasian conference on interactive entertainment. Creativity & Cognition Studios Press, Sydney, Australia, Australia, IE '05, pp 151–158. http://dl.acm.org/citation.cfm?id=1109180.1109204

Poulaki M (2014) Neurocinematics and the discourse of control: towards a critical neurofilmology. Cinéma & Cie: Int Film stud J 22–23

Pressman J (2006) House of leaves: reading the networked novel. Stud Am Fict 34(1):107–128

Ramchurn R, Chamberlain A, Benford S (2018) Designing musical soundtracks for brain controlled interface BCI systems. In: Proceedings of AM'18. ACM, New York, NY, USA, pp 28:1–28:8. http://doi.acm.org/10.1145/3243274.3243288

Ramchurn R, Martindale S, Wilson ML, Benford S (2019) From director's cut to user's cut: to watch a brain-controlled film is to edit it. In: Proceedings of the 2019 CHI conference on human factors in computing systems. ACM, New York, NY, USA. https://doi.org/10.1145/3290605.3300378, (In Press)

Randall K (2011) Rise of neurocinema: how hollywood studios harness your brainwaves to win oscars. https://www.fastcompany.com/1731055/rise-neurocinema-how-hollywood-studios-harness-your-brainwaves-win-oscars

Reynolds M (2018) The inside story of bandersnatch, the weirdest black mirror tale yet. https://www.wired.co.uk/article/bandersnatch-black-mirror-episode-explained

Schmeder A, Freed A, Wessel D (2010) Best practices for open sound control. In: Linux audio conference, vol 10

Sengers P, Gaver B (2006) Staying open to interpretation: engaging multiple meanings in design and evaluation. In: Proceedings of the 6th conference on designing interactive systems. ACM, New York, NY, USA, DIS '06, pp 99–108. http://doi.acm.org/10.1145/1142405.1142422

Shaul NB (2008) Hyper-narrative interactive cinema: problems and solutions. BRILL

Shneiderman B, Plaisant C, Cohen M, Jacobs S, Elmqvist N, Diakopoulos N (2016) Designing the user interface: strategies for effective human-computer interaction. Pearson

Tikka P (2010) Enactive media—generalising from enactive cinema. Digital Creativity 21(4):205–214. https://doi.org/10.1080/14626268.2011.550028

Willoughby I (2007) Groundbreaking czechoslovak interactive film system revived 40 years later. https://www.radio.cz/en/section/panorama/groundbreaking-czechoslovak-interactive-film-system-revived-40-years-later

Zander TO, Kothe C (2011) Towards passive brain-computer interfaces: applying brain-computer interface technology to human-machine systems in general. J Neural Eng 8(2):025005

Zone R (2014) Stereoscopic cinema and the origins of 3-D film, 1838–1952. University Press of Kentucky

Zuboff S (2015) Big other: surveillance capitalism and the prospects of an information civilization. J Inf Technol 30(1):75–89

Chapter 15
The Making of Brain Painting—From the Idea to Daily Life Use by People in the Locked-in State

Andrea Kübler and Loic Botrel

Abstract We describe how the classic P300 based BCI-controlled spelling paradigm was transformed into a painting tool. Initiated by an artist, we ("we" refers to all people who were involved in the making of Brain Painting at any one time. "I", "me" or "AK" refers to the first author. Individual names/initials are used to identify a specific person being involved in a specific step of the history including end-users with disease) developed Brain Painting in close interaction with two end-users of the technology in the locked-in state due to amyotrophic lateral sclerosis. We enabled them to use Brain Painting at their home independent of the researchers being present. Throughout the use of Brain Painting we applied measures of effectiveness, efficiency, and end-user satisfaction leading to a unique long-term evaluation covering hundreds of sessions. We demonstrated a positive effect on quality of life, self-esteem, and other measures of well-being. The artists created a plethora of Brain Paintings, which have been shown at several exhibitions. In this chapter, we describe firstly, the history of Brain Painting. Secondly, we explain in detail how the P300-BCI controlled application works. Thirdly, we dwell on how we enabled use of Brain Painting in daily life at the end-users' home. We, fourthly, dedicate a separate section to the impact of Brain Painting on quality of life, and show finally some of the Brain Paintings and summarize the artists' Brain Painting exhibitions. We end with a summary, emphasising that if BCI developers follow the user-centred design, i.e., listen to and interact closely with the targeted end-users of technology, the transfer of BCI-controlled applications out of the laboratory to the end-users' home becomes more likely, feasible and hopefully more real.

Keywords P300-BCI · Painting · Creativity · Quality of life · ALS · Exhibition

A. Kübler (✉) · L. Botrel
Department of Psychology I, Institute of Psychology, University of Würzburg, Marcusstr. 9-11, 97070 Würzburg, Germany
e-mail: andrea.kuebler@uni-wuerzburg.de

L. Botrel
e-mail: loic.botrel@gmail.com

© Springer Nature Switzerland AG 2019
A. Nijholt (ed.), *Brain Art*,
https://doi.org/10.1007/978-3-030-14323-7_15

15.1   The History

Two independent, but linked happenings occurred in 2005. Firstly, the at the time most famous German artist of modern art Jörg Immendorff (JI) was in the advanced state of amyotrophic lateral sclerosis (ALS), a neurodegenerative, terminal disease, which caused tetraplegia and necessitated constant artificial ventilation by means of tracheotomy. The artist was thus, no longer able to paint and could only advice his scholars to paint what was on his mind. Secondly, the development of the BCI controlled P300 letter matrix was advanced such that communication was possible in exemplary patients in the locked-in state due to ALS (Hoffmann et al. 2008; Nijboer et al. 2008). However, no BCI controlled applications for communication and interaction was used by any patient at home, worldwide. To dwell on the reasons for their non-use is beyond the scope of this chapter (the reader is referred to for example Kübler et al. 2014; Nijboer 2015), but one reason is important for the making of Brain Painting: the wish of the patients to use the P300-BCI for other purposes than communication, more precisely for entertainment, and the lack of respective offers. It then happened accidentally that the previous supervisor of the present author Andrea Kübler (AK), Niels Birbaumer (NB), and the artist Adalbert–Adi–Hösle (AH), who created an oeuvre with the brain as central topic (www.retrogradist.de), met at an exhibition at the "Kunsthalle Tübingen", Germany.[1] Of course, the leading topic of their conversation was to link the art of the brain with BCI. Further, it turned out that AH knew JI. The artist hoped to get some funding for future brain-art projects with the aid of NB, and NB speculated to get some funding by JI via AH for the further development of BCI. Neither the one nor the other worked out, but in 2008 AK received funding for Brain Painting within a large scale European project (see acknowledgements).

As often, when NB met people and found them interesting, these people knocked at the door of my (AK) office and told me that I should take care of them. This was also true for AH, and of course I had to tell him that, at the time, we had no funding to invest in Brain Painting. Nevertheless, I liked the idea and the vision that we could offer some entertainment to our patients. So I brought together my PhD student and postdoctoral fellow (at the time) Sebastian Halder and Femke Nijboer, who had the knowledge of programming the P300 matrix, and testing BCI controlled applications with end-users at their home, and AH who had the vision of how to turn the P300 matrix into a painting tool.

In the remainder of this chapter the authors will tell how Brain Painting developed from the idea to its daily use at home by ALS patients in the locked-in state. The making of Brain Painting is an example of following consequently the user-centred design approach (UCD, ISO 9241-210 2008), which defines "usability" and stipulates the early involvement of the end-user of the respective technology. With respect to adopting a new assistive technology (AT)—the category to which belongs BCI-controlled Brain Painting—functionality, possibility of independent use, and ease of use were rated most important by potential end-users of this technology (Zickler et al.

---

[1] https://kunsthalle-tuebingen.de/.

2009). The UCD has been adapted to BCI controlled applications (Kübler et al. 2014) and the proof of its usefulness and practicability also when working with patients in the locked-state is revealed below within the realm of the Brain Painting application. For the definition of usability in the BCI context, see Sect. 15.5 "Daily life use by people in the locked-in state".

In 2007, the first exhibition that included Brain Painting was organised by AH at the Künstlerbund Tübingen (artist association,[2] Fig. 15.1). Besides his brain related work, the visitors could watch the artist while he was Brain Painting and try the application themselves. The first proof-of-principle study used a P300 matrix that included all functions deemed important by the artist AH (see Fig. 15.5, Sect. 15.2 "The BCI controlled application"). Besides himself (Fig. 15.2), ALS patient LEK[3] in the locked-in state was also presented with the Brain Painting matrix (Fig. 15.3). Both were able to use the matrix and, thus, create the first ever Brain Paintings (see Painting 15.1 of Sect. 15.7 "The Gallery") during training in the laboratory, at home and during the exhibition (Kübler et al. 2008).

Fig. 15.1 Vernissage at the Künstlerbund Tübingen (local art exhibition hall), November 2007. In the foreground ALS patient LEK who was the very first who created Brain Paintings in 2007 (see Sect. 15.7 "The Gallery" Painting 15.1 for one of her paintings)

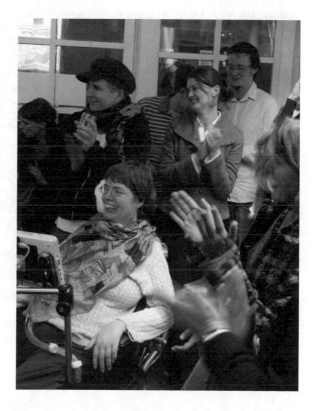

---

[3]Initials of Brain Painting end-users with permission.

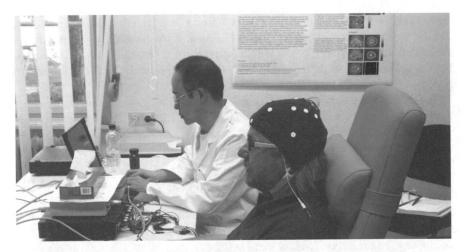

Fig. 15.2 Brain Painting initiator Adi Hösle is practicing Brain Painting in the lab

Fig. 15.3 Proof-of-principle with the Brain Painting prototype by LEK with ALS in the locked-in state at her home; for the painting visible on the "virtual canvas" see Sect. 15.7 "The Gallery" Painting 15.1

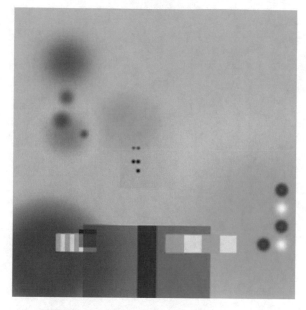

Painting 15.1 The first ever Brain Painting by an end-user with ALS (see also Fig. 15.3; ©Liane Krauss and Andrea Kübler)

Painting 15.2 "Wandering of soul" by Sonja Balmer. "I, as an impressionist painter", she said in 2012, "tried to depict a sunset at the horizon of a sea with a flying bird in the twilight. However, during the painting a memory of my near-death experience came up of the being between life and death and I tried to depict what I felt." She re-made the impressionist painting with Brain Painting (©Sonja Balmer and Andrea Kübler)

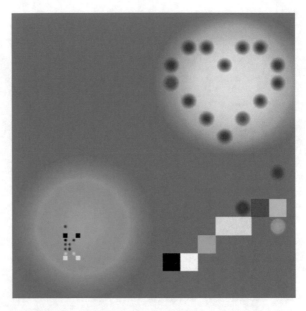

Painting 15.3 Made by an end-user with ALS who participated in the Zickler et al evaluation study (Zickler et al. 2013). He dedicated the painting to his wife who was also his primary caregiver. The heart on a yellow circle expresses his love and the sunshine his wife brings into his life. The "K" on the lower left hand side corner is the initial of his last name (©Andrea Kübler)

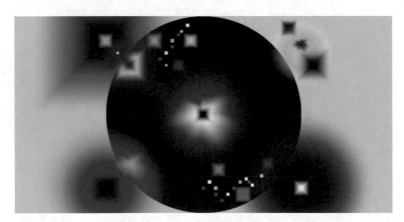

Painting 15.4 Melly exploding by HHEM. Dedicated to her caregiver who was furious that day resulting in a tantrum. This example shows impressively the emotional power of Brain Painting (©Heide Pfützner, with permission)

Münßinger and colleagues then performed the first scientific evaluation of the Brain Painting application prototype (Münßinger et al. 2010). Ten healthy subjects and 3 with ALS participated and compared the classic P300 spelling matrix (Farwell and Donchin 1988) to the Brain Painting matrix. The Brain Painting matrix was rated significantly more complex and more difficult to use which was reflected in the drop of average accuracy and target P300 amplitude. Taking into account the feedback of the participants, the matrix was changed to black and white and the commands were sorted according to their functions (Fig. 15.6, Sect. 15.2 "The BCI controlled application"). Ten newly recruited subjects participated and—as compared to the classic P300 letter matrix—no drop in performance was found, neither for accuracy nor for the target P300 amplitude.

These were encouraging results and we, thus, considered BCI-controlled Brain Painting ready for evaluation by end-users with disease. Four such end-users were included, two with ALS, one with stroke (pontine, arteria basilaris), and one with Duchenne muscular dystrophy. All were severely paralysed and all but one with restricted or no speech. They participated in seven BCI sessions, including copy-spelling, copy-painting, and free-painting (Zickler et al. 2013). It was the first time that a BCI application was evaluated with measures suggested by the UCD (Kübler et al. 2014). Average performance of all patients was above 80% accuracy. Satisfaction with the device as measured with the BCI-related items of the Extended Quest questionnaire (Zickler et al. 2013), a measure for general satisfaction with assistive technology (Demers et al. 2002), was rated between 4.1 and 4.9 (scale from 1 to 5, with 5 corresponding to highest satisfaction). Importantly, learnability was rated 5 by all participants. They were also highly satisfied with reliability, and, surprisingly, all but one also with speed.

Following the UCD, four different BCI-controlled applications—two for communication, two for entertainment—were evaluated by altogether 19 patients with severe paralysis due to different diseases (Kübler et al. 2014). Brain Painting received the highest general and BCI specific satisfaction ratings and it was the only application that end-users could imagine to use in daily life (see Table 5 of Kübler et al. 2014). Thus, the Brain Painting application was ready to be transferred out of the lab to the end-users' home.

In the meantime AH organised several Brain Painting sessions with German artists, one of those being Neo Rauch.[4] At the same time, AH came into contact, via the Charité Hospital in Berlin, with Heide Pfützner (HHEM[5]) from Leipzig, Germany, an artist devoted to impressionist painting and diagnosed with ALS in 2007. In 2012 when we were introducing Brain Painting to her, she was in the locked-in state. Providing her with the Brain Painting application required some adaptations to facilitate home use without researchers being present, instruction of caregivers, and a remote access for saving data and long-distance support (see Fig. 15.8 and Sect. 15.5 "Daily life use by people in the locked-in state"). One year later in sum-

---

[4]https://www.faz.net/aktuell/feuilleton/maschinelle-hirnmalerei-einmal-neo-rauch-sein-1984215. html.

[5]HHEM is Heide Pfützner's artist name, so we use it instead of her initials.

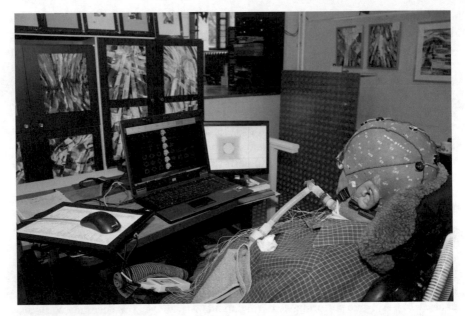

Fig. 15.4 JT during Brain Painting in his atelier; the overlay of a row with the Einstein face is visible (Photograph is shown with permission by Jürgen Thiele, ©Andrea Kübler)

mer 2013, we were contacted by another artist diagnosed with ALS in 2006, Jürgen Thiele (JT). Retired architect and renowned aquarelle painter, he spent most of his time in his workroom. JT was so passionate about painting that he kept painting by putting a pencil between his teeth while his art assistant would mix the colours. His condition allowing him to speak, he enjoyed to tell with a smile his Leitmotiv: "If I can no longer paint with my hands, I'll paint with my head". Two-days in Königs-Wusterhausen, near Berlin, were enough to provide JT with Brain Painting at home. Learning from experience, we installed a dedicated laptop and monitor in the middle of his atelier, surrounded by all his creations (Fig. 15.4).

For him it was the first time to use a computer! Both artists had devoted caregivers and family who empowered them to use Brain Painting. Both artists were visited by our team three times only, and their entire oeuvre was created with remote supervision and intervention when necessary (for a documentation of this process see Holz et al. 2015a, b). Until the end of 2016 HHEM had been painting for $N = 484$ sessions with an average duration of 93.7 min ($SD^6 = 50.6$). Satisfaction on a 10 point Likert-type scale (1 worst, 10 best) was on average $M = 6.94$, ($SD = 2.60$). JT had been painting until summer 2016 for $N = 225$ sessions with an average duration of 59.6 min, ($SD = 3.1$). Satisfaction was on average $M = 7.48$ ($SD = 2.59$). After many painting sessions both artist wished more options; specifically JT asked for the possibility of drawing lines. Thus, again interacting closely with the artist, Brain Painting Version

---

[6]Standard deviation.

2 was developed (Botrel et al. 2015) and allowed for quite another type of paintings (see Paintings 15.8 and 15.9 Sect. 15.7 "The Gallery"). However, this second version came at a price: as it had two matrices and considerably more options, it was more complicated to operate, and only JT used it extensively ever since then. With this long-term use of BCI, the proof-of-principle exists that BCI can be used at home by locked-in patients without on-site support by researchers, provided we offer the end-users something valuable for their quality of life.

15.2 The BCI Controlled Application

In the past years, the development of BCIs toward independent home use by locked-in patients made quite some progress with respect to speed and choice of applications. The speed of letter selection with a BCI is now equal to that achieved by other assistive devices for communication that operate in the so-called scanning mode, i.e. about 10–20 letters a minute (Brady et al. 2016; Kaufmann et al. 2011). In the scanning mode, options for selection are presented sequentially to the end-user of the technology and a command is transferred either by pressing a button with residual finger movement, or other muscle twitch or eye movement via an eye tracker (for a combination with such technology see e.g., Müller-Putz et al. 2011; Riccio et al. 2015). For Brain Painting, we used the so-called P300-BCI because of its ease of use and reliability. In fact, the P300-BCI is, to date, the most reliable BCI in the short (Cheng et al. 2017; Kaufmann et al. 2013; Ryan et al. 2018) and long run (Holz et al. 2015b; Sellers et al. 2010; Wolpaw et al. 2018) and evaluation studies exist (Guy et al. 2018; Kübler et al. 2014; Mc Cane et al. 2014). Almost all end-users including those in the locked-in state achieve high an effectiveness and efficiency already in the first or few follow-up sessions (Halder et al. 2016; Severens et al. 2014), and can maintain this performance in the long-term (Holz et al. 2014; Sellers et al. 2010). Importantly for patients with neurodegenerative disease, the P300-BCI has been used by late stage ALS patients for several years without a drop in performance (Holz et al. 2014; Sellers et al. 2010). One of the striking advantages of the P300 as input signal for BCI is that a decline in effectiveness can be easily compensated by increasing the number of repetitions, i.e. stimulus presentations. Albeit at the cost of efficiency, because more time is needed for selections, the P300-BCI still enables the end-user to select items at high accuracy, which may be favoured over speed.

The Brain Painting application uses the well-known P300 component of the event-related potential as input signal. Consequently, all options for painting were inserted in the cells of the P300-BCI matrix, firstly by the artist AH (Fig. 15.5). The subsequent black and white matrix was the application offered to all end-users (Fig. 15.6). At the very beginning of Brain Painting the rows and columns of the matrix were highlighted (flashed) in random order as in the classic P300-BCI spelling application. In 2011 Kaufmann and colleagues introduced a face overlay instead of simply flashing rows and columns (Jin et al. 2012; Kaufmann et al. 2011). In the P300-BCI the signal-to-noise ratio depends on the ability of the classifier to identify target-

L	Q	⬜	⚫	75	W	C	31
B	GR		∩		3	7	15
25	50	↻	⚪	➲	M	63	127
S	100		↺		255	511	R
1	2	**A**	M	Z +	Z -		S
4	8	G	T	H	UD	RD	STOP

Fig. 15.5 The original Brain Painting matrix which comprised commands for cursor position, colour, zooming in and out of the P300 matrix grid. The numbers on the right side indicated the grid size (odd numbers between 3 and 511) on which the painting cursor could be moved (see Sect. 15.3 "Brain Painting software and matrices" for details). The size of the shape to be printed was dependent on the grid size, and determined by a brush size factor (1, 2, 4 or 8). The percentages (25, 50 and 100) allowed for modulating the color opacity. End-users can undo or redo actions by using UD and RD. The A character activates the "amorph" command which applies distortion on the picture. Letters L, Q, S, T and H are fillers only with no meaning

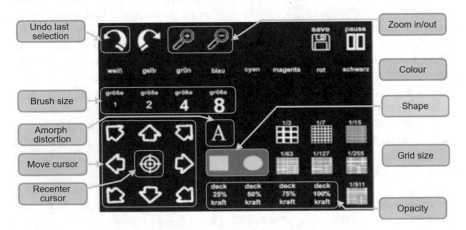

Fig. 15.6 Adapted black and white matrix of the P300-Brain Painting application. Items for selection were sorted according to their function. 1st line: redo, undo, zoom in, zoom out; 2nd line: colours; 3rd line: size of the brushstroke; 4th line: cursor position up and diagonal up, using the amorph filter (strong image distortion), grid sizes for determining available cursor position and brush reference size; 5th line: putting the cursor back to the center, left and right, square, circle, grid sizes; 6th line: cursor down, diagonal down, colour intensity, grid size

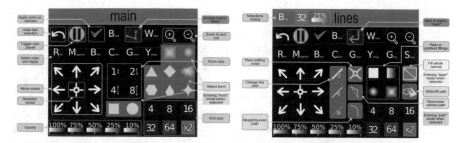

Fig. 15.7 P300 matrices labeled with their function or commands. Left: the 'main' matrix is the first matrix presented to the user and is based on Brain Painting V1 matrix (see Fig. 15.5). Right: the 'line' matrix can be selected from the main matrix and allows for drawing lines, selecting color gradients, or filling the canvas with a colored layer. Detailed descriptions of the commands are in the Sect. 15.3 "Brain Painting software and matrices")

specific event-related potentials, and thus, every measure to increase the number or the amplitude of such potentials leads to a better classification result. It has long been shown that processing of familiar faces evoke specific potentials, namely the N170, P300, and N400f.[7] In 2013 Kaufmann and colleagues could show that the overlaying rows and columns with the famous Einstein face lead to a tremendous increase in effectiveness and efficiency of the P300-BCI in patients with neurodegenerative disease (Kaufmann et al. 2013). In the meantime many other authors confirmed the increase of performance when using faces as stimuli, though mostly with healthy subjects (Cheng et al. 2017; Jin et al. 2014; Yeom et al. 2014). The Einstein face was, thus, integrated as flashing stimulus in the Brain Painting application. For HHEM the number of sequences could be reduced from 10 to 5 for 100% accuracy (Holz et al. 2015a).

In the next section we will describe in detail the functions of the Brain Painting matrices versions V1 and V2 (Figs. 15.6 and 15.7).

15.3   Brain Painting Software and Matrices

Brain Painting requires the use of two monitors. The first displays the matrices for selection of functions. This monitor is placed in front of the end-users since this is where their attention needs to be focused on most of the time. A second monitor serves as a digital canvas on which the painting is composed. Painting often involves brush strokes and smooth hand movements in a controlled motion. Instead of a continuous motion, the P300 paradigm provides one command every few seconds. The Brain Painting interface had to be constructed toward bridging this gap.

---

[7]The "f" stands for "faces" as in such paradigms an N400 is elicited in response to face perception (Eimer 2000).

An initial grid size of 3 by 3 provides nine positions toward which the "brush" or cursor for painting can be moved. The end-user moves the cursor to the desired location, selects either a square or a circle, the opacity, and then a colour. The colour selection "prints" the shape on the initially empty canvas. To reduce the number of steps for the next selection, the cursor position and any parameters change are kept, hence facilitating the next step: for example, if a red square was chosen at first, and the objective is to paint a blue square, it is not necessary to "click" again on the square. By choosing cursor positions, forms, and colours, the painting is composed step by step. The grid size can be increased up to 511 lines and columns, decreasing proportionately the size of the shape in preparation. However, a brush size from 1 to 8 allows for increasing this grid-dependent shape size. It is also possible to zoom in and out of the painting using cursor location as a reference, without changing any other grid and brush sizes parameters, thus allowing for painting very small details (see e.g., Painting 15.5). As all primary and secondary colours are available, it is possible to additively mix colours by picking the respective colours and opacities. The undo and redo features can revert or repeat any selection that was voluntarily or involuntarily made, and were often used. The frequent use of the redo and undo buttons demonstrates that the artists did not accept false selections, which is often assumed by outsiders. A pause function allows the artist to rest. "Save" conserves the current version of the painting ready to be re-loaded whenever wished. More shapes, a fuzzy square, fuzzy circle and a cloud-type shape were added for selection to extend the artistic possibilities. Above the matrix, there is a horizontal status bar that shows all the current parameters (e.g., grid size, brush size, opacity). As soon as a colour is chosen, the shape—defined by all the current parameters—is printed to the canvas at the position of the cursor.

With the request of more shapes and line drawing, specifically by JT, a new version of Brain Painting had to be developed, including two matrices. The first "main" matrix of Brain Painting V2 (Botrel et al. 2015) comprises the same painting functions, with the exception of shape preview. Instead of selecting all parameters and only printing the shape after colour selection, Brain Painting V2 displays the current shape below the cursor as soon as a selection is made. The shape preview remains attached to the cursor until its colour is defined or it is applied using the new "apply" selection, which prints it on the canvas. This matrix also contains more shapes such as diamond, star, waterdrop (see Painting 15.8 of "The Gallery") and triangle. The second "lines" matrix is available by selecting its specific icon. This matrix replaces all shape-related options (green in Fig. 15.7) by line related ones (blue in Fig. 15.7). Simply said, drawing custom shapes or lines can be done by placing crosses on the canvas, and the end-user can then choose what to do with those crosses, either connect them with a line, a curve or a polygon. To do so, the end-user has to choose "editing crosses". When two crosses are connected, a line is created. Three and more crosses result into a path. Selection of "open/close" determines whether the first and the last crosses are linked, selection of "straight/curve" transforms straight lines into curves and "toggle fill" transforms the path into a filled polygon. Selecting "apply" prints the previewed path on the canvas. The editing crosses disappear after this step. This second matrix further comprises linear, circular and X-shaped gradient modes

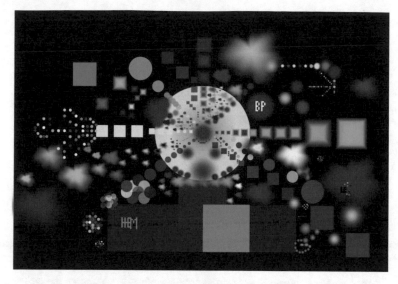

Painting 15.5 "Spökenkiekers Brägen"—"The Brain of the Clairvoyant" by HHEM painted for the Annual Meeting "Psychologie und Gehirn" of the German Society for Psychology Section Biological Psychology in Würzburg 2013. This is probably the most complex and detailed Brain Painting ever prodcued (©Heide Pfützner, with permission)

Painting 15.6 a By end-user D of Zickler et al. (2013) was actually the first, which aimed at realistic as opposed to abstract content (an apple tree and skyscraper, ©Andrea Kübler). Such efforts were also undertaken by artist HHEM, "Der Schürzenjäger"—colloquial expression for a man who approaches women extensively. "Schürze" translates to "pinafore" usually worn by women and "Jäger" is a "hunter". The snowman in the middle wears such a pinafore identifying him as the "Schürzenjäger" surrounded by snow-women (©Heide Pfützner, with permission). The painting demonstrates the humor often present in HHEM's paintings

Painting 15.7 "Magazin" (stack) by JT. The painting nicely captures JT's preference of structures, clear contours and forms (©Jürgen Thiele and Andrea Kübler)

Painting 15.8 "Blut und Gold" (Blood and Gold). The painting demonstrates the new options and potential of Brain Painting V2. The new shape "waterdrop" was used extensively by JT (©Jürgen Thiele and Andrea Kübler)

for the filled polygons with the additional gold and silver colours requested by end-users. The "fill canvas" selection allows for filling the entire canvas with a colour, enriched by opacity and gradient modifiers. Paintings 15.8 and 15.9 of Sect. 15.7 "The Gallery" demonstrate vividly the potential of the Brain Painting matrix V2 compared to its predecessor.

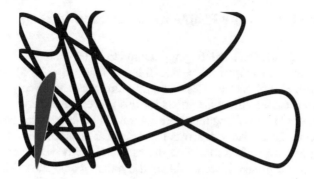

Painting 15.9 "Gestrüpp" (brushwood). Also made with Brain Painting V2. It demonstrates the potential of the new option to draw lines and curves and to connect them. The lines/curves were drawn by placing "crosses" on the canvas and then choosing the respective command for their connection. The fun and joy JT had with lines can be seen vividly (©Jürgen Thiele and Andrea Kübler)

15.4   Other Approaches to Painting with the Brain

P300 based BCI paradigms can only send one command about every 6–20 s depending on the individual signal-to-noise ratio. BCI also provides paradigms that enable continuous motor imagery based control over binary actions. Such paradigm was translated into "Braindrawing" in a pilot study (Holz et al. 2012). A cursor would trace a line while moving at a constant pace on an empty canvas. By imagining right or left hand movement, the end-user was able to move the cursor left or right. Despite attempting with a better than average healthy participant, fine and non-stop continuous control was inefficient, and, thus, not further pursued.

Many artists have used brain signals as a source of creativity. An eponym project exploited brain signals into a paint emitting robot arm (Funk and Raschke 2013). The control signal was based on four "thought" patterns directly trained via the Emotiv Epoc Toolbox, and produced "artistic" outputs. The proof of concept device is, however, not provided with any estimate of statistical significance with respect to the degree of control. This inevitably leads to the question of how much random activity or false detections could be integrated into brain generated arts. Our experience with Brain Painting artists demonstrates, that they were very intolerant toward random selections may they arise from system or "brain" failure. Thus, a toolbox such as Emotiv Epoc would serve the aspect of "ease of use" but not that of meaningful "communication", here via Brain Painting art. Another way of enhancing a no-hands painting experience with brain signals was demonstrated by van de Laar and colleagues with BrainBrush (van de Laar et al. 2013). A P300 interface would allow for tool selection while the gyrosensor of an Emotive Epoc steered the brush on the canvas. To our knowledge, none of these approaches were implemented and developed such that end-users with disease could use it independently in their home environment.

15.5   Daily Life Use by People in the Locked-in State

Before we brought the Brain Painting application to HHEM and JT, we ensured that the environment could support frequent BCI use (Kübler et al. 2015). Signal acquisition was realized with 8 electrodes and calibration was only necessary few times across the years and we visited the artists for this procedure (Holz et al. 2015b). As the P300 was stable across long time periods, we did not introduce auto-calibration (Kaufmann et al. 2012). As both artists lived at great distance from the University of Würzburg and we wished to enable the artists to paint whenever they wished independent of our presence, we established a remote connection (Fig. 15.8). We set up the Brain Painting application such that every start of a session was recorded including time, number of selections, and the painting result. The artists could end a session by themselves and were then asked several questions for evaluation. Following the UCD, we assessed all aspects of usability, i.e., effectiveness, efficiency, and satisfaction. Effectiveness is a measure of how accurate and complete the end-user of technology can operate the application, i.e. how often the intended selection can be achieved (=accuracy). The often-used measure of accuracy defined as "percentage of correct responses", requires a template against which the achieved results can be compared. Such a template is not available during free-painting. Thus, we asked the artists, at the end of each session, to indicate their perceived level of control, and whether the ability to control the application changed across the session. Efficiency relates the invested costs to effectiveness. A common such measure is the information transfer rate (ITR), which, however, also requires accuracy. One can assume an accuracy between 70 and 100%, because below 70%, no meaningful communication or painting is possible (Kübler et al. 2001) and the artists would terminate the session. As a subjective measure of efficiency, we assessed the level of exhaustion after each session. Satisfaction while using the device was also assessed after each session with visual analogue scales ranging from zero to ten for satisfaction, joy, and frustration. For the entire evaluation procedure, the respective measures, and the results thereof, the reader is referred to Holz and colleagues (2015b).

During Brain Painting hard- and software related problems occurred. HHEM stated, for example, that the program did not react although everything looked fine; the PC crashed; EEG signals looked strange although enough gel was being used. Likewise, JT experienced crashing of the Brain Painting application and bad EEG signals. Generally, the artists tried to fix the problems themselves (re-start or cancellation of the session) before they contacted us. They also experienced sessions of "no control" which was commented with having "no fun". Also subject-specific or environmental factors interfered with performance, such as being too tired, coughing, experiencing respiratory problems or too many interruptions by others (see Holz et al. 2015b). Despite these problems, the end-users enjoyed painting and it positively influenced their quality of life, which we will detail in the next section.

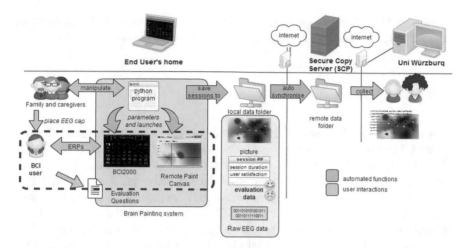

Fig. 15.8 Scetch represents the different devices and actors for Brain Painting at home. Researchers do not interact directly with the end-user, who is assisted by caregivers in using the Brain Painting application. Evaluation data of each session is retrieved with automated processes via secure cloud storage. Red dotted lines indicate the unity of end-user, the BCI, and the Brain Painting application

15.6   The Impact on Quality of Life

To assess the impact of the BCI-controlled Brain Painting application on the life of HHEM and JT, Holz and colleagues (2015a, b) applied the Psychosocial Impact of Assistive Device Scale (PIADS) (Jutai and Day 2002). It consists of 26 items assessing the three dimensions *competence*, *adaptability*, and *self-esteem*. Competence (twelve items) measures feelings of competence and efficacy, and is sensitive to the impact of AT on performance and productivity. Adaptability (six items) refers to the willingness to try out new things and to take risks, and is sensitive to the enabling and liberating aspects of AT. Self-esteem (eight items) captures feelings of emotional health and happiness and is sensitive to the impact of AT on self-confidence and emotional well-being. HHEM and JT indicated whether the AT had a positive or negative impact and the degree of this impact on the specified dimension (range, -3 to 3; -3 indicates maximum negative impact, 3 maximum positive impact, and zero no perceived impact [the description of the PIADS is from Holz et al. 2015b]). For HHEM Brain Painting had the highest impact on "happiness, self-esteem, productivity, usefulness, self-confidence, expertise, well-being, and performance"; for JT on "happiness, efficiency, self-esteem, productivity, security, usefulness, self-confidence, expertise, skilfulness, wellbeing, and quality of life". For both artists Brain Painting had the highest impact on the self-esteem dimension. The only negative impact was stated by JT (-1) on "independence", probably because he depended on his caregivers for using Brain Painting. Taken together, we can state that BCI controlled Brain Painting affects positively the quality of life of locked-in patients diagnosed with ALS in many aspects and it demonstrates the potential of BCI provided we offer an application that is important and valuable for the end-users.

15.7   The Gallery

In this section, we show several Brain Paintings created during the past 12 years by several end-users, lay painters and artists alike (see Paintings 15.1, 15.2, 15.3, 15.4, 15.5, 15.6, 15.7, 15.8, 15.9). Explanations about the making and stories associated with the Brain Paintings are given in the respective captions. Painting 15.5 "Spökenkiekers Brägen" by HHEM made it to the front page of the scientific Journal Brain in June 2013.[8] Artist SB (Painting 15.2), for who we could not establish Brain Painting at home due to a lack of support by caregivers, told us her vision for BCI technology: "In a year's time", she said in 2012, "I would like to move from the nursing home to an apartment with 24 h care. I am imagining an art-room in which I conduct my brain painting, write my books, communicate, twitter, send e-mails, listen to music, relax etc. Everything is steered through BCI. A video conferencing connects me to the outside world. My visitors will also be involved. Perhaps we can even play a game. Being confined to bed or wheelchair, this is the only possibility to connect to the outside world. It allows me to act independently and I feel empowered". Now, six years later, the BCI community made important steps toward SB's vision, but is not there yet.

15.8   Exhibitions

Paintings by the end-users were exhibited for the first time in 2013 at the Town Hall of Easdale, a Scottish island where HHEM's daughter works.[9] HHEM herself was there for the opening. The journey and all the associated necessities were costly and, thus, HHEM and her family started a crowd funding action via startnext.[10] The envisaged sum of 5,000 € was indeed achieved and HHEM went to Easdale with her team. The first exhibition in her home country Germany was at the Annual Meeting of the German Biological Psychology Society in Würzburg, Germany, 2013 for which she created Painting 15.5 of Sect. 15.7 "The Gallery". Thereafter her Brain Paintings were shown at the International BCI Meeting 2013 in Asilomar, Pacific Grove, California.[11] In 2014 we were approached by the Deputy of Cultural Affairs of the "Forum trifft Kunst", an Art Gallery in the main building of the Volksbank

---

[8]https://academic.oup.com/brain/issue/136/6.
[9]http://www.easdaleislandhall.org/gigs/index.php?id=396.
[10]https://www.startnext.com/de/mein-hirn-brennt.
[11]http://meetings.bcisociety.org/2013/.

Fig. 15.9 HHEM visiting her exhibition at the "Forum trifft Kunst" of the Volksbank Würzburg. Left Loic Botrel and right Elisa Holz from the Brain Painting team at the University of Würzburg (photograph shown with permission by Heide Pfützner, ©Andrea Kübler)

Würzburg (a bank of Germany), and her paintings were shown there for half a year.[12] Figure 15.9 shows HHEM at the entrance of the Volksbank where the exhibition was advertised. In 2016 her so far largest exhibition "Brain on Fire" was held in Hamburg at the Warth & Klein Grant Thornton AG.[13]

JT started a little later with Brain Painting. He used to paint with watercolours and as his disease progressed he could only paint with the help of his art assistant, who mixed the water colours according to JT's order and then put the brush in JT's mouth. When speaking became more difficult, this was no longer possible, but Brain Painting came right in time, such that he could continue painting albeit in a quite different manner. His 2014 exhibition "AQUA-RELL" in the "Bürgerhaus Hanns Eisler" in his hometown Königs Wusterhausen had already included some of his Brain Paintings.[14] In 2016 JT had his first "Brain Paintings only" exhibition "malerei digital" (painting digital) in the "Vertikale Galerie" (vertical gallery) in Lübben, Spreewald.[15]

---

[12]http://www.i1.psychologie.uni-wuerzburg.de/int/aktuelles/einzelansicht-startseite/news/brainpaint/.

[13]https://www.wkgt.com/themen/hamburg-ausstellung/.

[14]https://www.kulturbund-dahme-spreewald.de/events/aqua-rell-juergen-thiele/.

[15]https://www.dahme-spreewald.info/de/seite/34812.html.

Fig. 15.10 Entrance to the exhibition "Brain Painting" in the "Naturkundemuseum Münster". Here finally Brain Paintings of both artists HHEM and JT are jointly shown. Left Lisa Klepfer who organised the exhibition together with Andrea Kübler, head of the Brain Painting team at the University of Würzbur (©LWL/Puschmann, Münster, Germany, with permission)

Finally, and more than one year after JT had deceased, Brain Paintings of HHEM and JT were shown together from November 9, 2018, to January 20, 2019[16] in the "Naturkunde Museum" (Natural History Museum) during the biannual exhibition "The Brain"[17] (Fig. 15.10). Both artists have sold some of their paintings.

15.9   Summary and Conclusion

Taken together, the making of Brain Painting has now covered more than 15 years and took place in close interaction with the end-users of the P300-BCI technology following the user-centred design. We demonstrated with two end-users in the locked-in state due to ALS that the regular use at home of a BCI controlled application is feasible and leads to increased quality of life provided we can offer something to the end-users to which they do not have access otherwise. This having said, it is important to mention that we asked HHEM whether we should integrate Brain Painting into her software controlled by the eye-tracker, but she refused. She stated

---

[16]https://www.lwl-naturkundemuseum-muenster.de/de/naturkundemuseum/ausstellungen/gehirn/brain-paintings/.

[17]https://www.lwl-naturkundemuseum-muenster.de/de/naturkundemuseum/ausstellungen/gehirn/.

that painting with the BCI is less exhausting because the eye movement does not have to be as precise as for the eye tracker. End-users HHEM and JT evaluated Brain Painting for around 4 years of use and thus, contributed to the so far longest ever evaluation of a BCI controlled application. The main results of this evaluation are the now quantified positive effect on quality of life, the insight that a BCI-controlled application for entertainment may be used in daily life even if the perceived control is not always high, and that a cap with gel electrodes does not prevent daily use. This now published knowledge is encouraging for patient end-users with disease and BCI researchers alike as it demonstrates a path for BCI-controlled applications from the lab to the end-users' home. Brain Painting also demonstrates that BCI can provide more than communication and interaction, but also the possibility to express oneself creatively and emotionally far beyond verbalization of basic needs. Such an achievement, however, is only possible if we listen closely to those people for who we are developing such technology. The UCD tells us how to do this.

Acknowledgements We thank all the patient and healthy end-users who participated in short- and long-term studies throughout the years. Thanks to Dr. Elisa Holz, Dr. Sonja Kleih, Dr. Femke Nijboer, Dipl.-Psych. Claudia Zickler, Dr. Adrian Furdea, Dr. Sebastian Halder who contributed to the making of Brain Painting. Dr. Elisa Holz together with Dr. Loïc Botrel conducted the long-term studies with HHEM and JT. We thank Adi Hösle for the idea, his endeavour and contribution to the making of Brain Painting. This work was supported by the European ICT Program ProjectFP7-224631 (TOBI). This manuscript only reflects the authors' views and funding agencies are not liable for any use that may be made of the information contained herein. HHEM is funded by gtec, Linz, Austria (EEG cap and electrodes).

References

Botrel L, Holz EM, Kübler A (2015) Brain Painting V2: evaluation of P300-based brain-computer interface for creative expression by an end-user following the user-centered design. Brain Comput Interfaces 2(2–3):135–149. https://doi.org/10.1080/2326263x.2015.1100038

Brady NC, Bruce S, Goldman A, Erickson K, Mineo B, Ogletree BT, Paul D, Romski MA, Sevcik R, Siegel E, Schoonover J, Snell M, Sylvester L, Wilkinson K (2016) Communication services and supports for individuals with severe disabilities: guidance for assessment and intervention. Am J Intellect Dev Disabil 121(2):121–138. https://doi.org/10.1352/1944-7558-121.2.121

Cheng J, Jin J, Wang X (2017) Comparison of the BCI performance between the semitransparent face pattern and the traditional face pattern. Comput Intell Neurosci 2017:1323985. https://doi.org/10.1155/2017/1323985

Demers L, Weiss-Lambrou R, Ska B (2002) The Quebec user evaluation of satisfaction with assistive technology (QUEST 2.0): an overview and recent progress. Technol Disabil 14(3):101–105

Eimer M (2000) Event-related brain potentials distinguish processing stages involved in face perception and recognition. Clin Neurophysiol 111(4):694–705

Farwell LA, Donchin E (1988) Talking off the top of your head: toward a mental prosthesis utilizing event-related brain potentials. Electroencephalogr Clin Neurophysiol 70(6):510–523

Funk M, Raschke M (2013) Brain Painting: action paintings based on BCI-input. University of Bremen, Paper presented at the Mensch & Computer

Guy V, Soriani MH, Bruno M, Papadopoulo T, Desnuelle C, Clerc M (2018) Brain computer interface with the P300 speller: usability for disabled people with amyotrophic lateral sclerosis. Ann Phys Rehabil Med 61(1):5–11. https://doi.org/10.1016/j.rehab.2017.09.004

Halder S, Kathner I, Kubler A (2016) Training leads to increased auditory brain-computer interface performance of end-users with motor impairments. Clin Neurophysiol 127(2):1288–1296. https://doi.org/10.1016/j.clinph.2015.08.007

Hoffmann U, Vesin JM, Ebrahimi T, Diserens K (2008) An efficient P300-based brain-computer interface for disabled subjects. J Neurosci Methods 167(1):115–125. https://doi.org/10.1016/j.jneumeth.2007.03.005

Holz EM, Kaufmann T, Franz D, Hösle A, Kübler A (2012) Brain drawing: first evaluation results. Paper presented at the TOBI workshop III, Würzburg, Germany

Holz EM, Botrel L, Kübler A (2014) Independent BCI use in two patients diagnosed with amyotrophic lateral sclerosis. Paper presented at the 6th international BCI conference, Graz, Austria, 16–19 September 2014

Holz EM, Botrel L, Kaufmann T, Kubler A (2015a) Long-term independent brain-computer interface home use improves quality of life of a patient in the locked-in state: a case study. Arch Phys Med Rehab 96(3 Suppl):S16–26. https://doi.org/10.1016/j.apmr.2014.03.035

Holz EM, Botrel L, Kübler A (2015b) Independent home use of Brain Painting improves quality of life of two artists in the locked-in state diagnosed with amyotrophic lateral sclerosis. Brain Comput Interfaces 2(2–3):117–134. https://doi.org/10.1080/2326263x.2015.1100048

ISO 9241-210 (2008) Ergonomics of human system interaction—Part 210: Human-centred design for interactive systems (formerly known as 13407). International Organization for Standardization (ISO), Switzerland

Jin J, Allison BZ, Kaufmann T, Kübler A, Zhang Y, Wang X, Cichocki A (2012) The changing face of P300 BCIs: a comparison of stimulus changes in a P300 BCI involving faces, emotion, and movement. PLoS ONE 7(11):e49688. https://doi.org/10.1371/journal.pone.0049688

Jin J, Allison BZ, Zhang Y, Wang X, Cichocki A (2014) An ERP-based BCI using an oddball paradigm with different faces and reduced errors in critical functions. Int J Neural Syst 24(8):1450027. https://doi.org/10.1142/s0129065714500270

Jutai J, Day H (2002) Psychosocial impact of assistive devices scale (PIADS). Technol Disabil 14:107–111

Kaufmann T, Schulz SM, Grunzinger C, Kübler A (2011) Flashing characters with famous faces improves ERP-based brain-computer interface performance. J Neural Eng 8(5):056016. https://doi.org/10.1088/1741-2560/8/5/056016

Kaufmann T, Völker S, Gunesch L, Kübler A (2012) Spelling is just a click away—a user-centered brain-computer interface including auto-calibration and predictive text entry. Front Neurosci 6:72. https://doi.org/10.3389/fnins.2012.00072

Kaufmann T, Schulz SM, Koblitz A, Renner G, Wessig C, Kübler A (2013) Face stimuli effectively prevent brain-computer interface inefficiency in patients with neurodegenerative disease. Clin Neurophysiol 124(5):893–900. https://doi.org/10.1016/j.clinph.2012.11.006

Kübler A, Neumann N, Kaiser J, Kotchoubey B, Hinterberger T, Birbaumer N (2001) Brain-computer communication: self-regulation of slow cortical potentials for verbal communication. Arch Phys Med Rehabil 82(11):1533–1539

Kübler A, Halder S, Furdea A, Hösle A (2008) Brain Painting: BCI meets art. Paper presented at the 4th international brain-computer interface workshop and training course, Graz, Austria

Kübler A, Holz EM, Riccio A, Zickler C, Kaufmann T, Kleih SC, Staiger-Sälzer P, Desideri L, Hoogerwerf EJ, Mattia D (2014) The user-centered design as novel perspective for evaluating the usability of BCI-controlled applications. PLoS ONE 9(12):e112392. https://doi.org/10.1371/journal.pone.0112392

Kübler A, Holz EM, Sellers EW, Vaughan TM (2015) Toward independent home use of brain-computer interfaces: a decision algorithm for selection of potential end-users. Arch Phys Med Rehabil 96(3 Suppl):S27–32. https://doi.org/10.1016/j.apmr.2014.03.036

Mc Cane LM, Sellers EW, McFarland DJ, Mak JN, Carmack CS, Zeitlin D, Wolpaw JR, Vaughan TM (2014) Brain-computer interface (BCI) evaluation in people with amyotrophic lateral sclerosis. Amyotroph Lateral Scler Frontotemporal Degener 15(3–4):207–215. https://doi.org/10.3109/21678421.2013.865750

Müller-Putz GR, Breitwieser C, Cincotti F, Leeb R, Schreuder M, Leotta F, Tavella M, Bianchi L, Kreilinger A, Ramsay A, Rohm M, Sagebaum M, Tonin L, Neuper C, del Millan RJ (2011) Tools for brain-computer interaction: a general concept for a hybrid BCI. Front Neuroinform 5:30. https://doi.org/10.3389/fninf.2011.00030

Münßinger JI, Halder S, Kleih SC, Furdea A, Raco V, Hösle A, Kübler A (2010) Brain Painting: first evaluation of a new brain-bomputer interface application with ALS-patients and healthy volunteers. Front Neurosci 4:182. https://doi.org/10.3389/fnins.2010.00182

Nijboer F (2015) Technology transfer of brain-computer interfaces as assistive technology: barriers and opportunities. Ann Phys Rehabil Med 58(1):35–38. https://doi.org/10.1016/j.rehab.2014.11.001

Nijboer F, Sellers EW, Mellinger J, Jordan MA, Matuz T, Furdea A, Halder S, Mochty U, Krusienski DJ, Vaughan TM, Wolpaw JR, Birbaumer N, Kubler A (2008) A P300-based brain-computer interface for people with amyotrophic lateral sclerosis. Clin Neurophysiol 119(8):1909–1916. https://doi.org/10.1016/j.clinph.2008.03.034

Riccio A, Holz EM, Aricò P, Leotta F, Aloise F, Desideri L, Rimondini M, Kübler A, Mattia D, Cincotti F (2015) Hybrid P300-based brain-computer interface to improve usability for people with severe motor disability: electromyographic signals for error correction during a spelling task. Arch Phys Med Rehabil 96(3 Suppl):S54–S61. https://doi.org/10.1016/j.apmr.2014.05.029

Ryan DB, Colwell KA, Throckmorton CS, Collins LM, Caves K, Sellers EW (2018) Evaluating brain-computer interface performance in an ALS population: checkerboard and color paradigms. Clin EEG Neurosci 49(2):114–121. https://doi.org/10.1177/1550059417737443

Sellers EW, Vaughan TM, Wolpaw JR (2010) A brain-computer interface for long-term independent home use. Amyotroph Lateral Scler 11(5):449–455

Severens M, Van der Waal M, Farquhar J, Desain P (2014) Comparing tactile and visual gaze-independent brain-computer interfaces in patients with amyotrophic lateral sclerosis and healthy users. Clin Neurophysiol. https://doi.org/10.1016/j.clinph.2014.03.005

van de Laar B, Brugman I, Nijboer F, Poel M, Nijholt A (2013) BrainBrush, a multimodal application for creative expressivity. Paper presented at the the sixth international conference on advances in computer-human interactions, Nice, France, 24 February–1 March

Wolpaw JR, Bedlack RS, Reda DJ, Ringer RJ, Banks PG, Vaughan TM, Heckman SM, McCane LM, Carmack CS, Winden S, McFarland DJ, Sellers EW, Shi H, Paine T, Higgins DS, Lo AC, Patwa HS, Hill KJ, Huang GD, Ruff RL (2018) Independent home use of a brain-computer interface by people with amyotrophic lateral sclerosis. Neurology 91(3):e258–e267. https://doi.org/10.1212/WNL.0000000000005812

Yeom SK, Fazli S, Muller KR, Lee SW (2014) An efficient ERP-based brain-computer interface using random set presentation and face familiarity. PLoS ONE 9(11):e111157. https://doi.org/10.1371/journal.pone.0111157

Zickler C, Di Donna V, Kaiser V, Al-Khodairy A, Kleih S, Kübler A, Malavasi M, Mattia D, Mongardi S, Neuper C, Rohm M, Rupp R, Staiger-Sälzer P, Hoogerwerf E-J (2009) BCI applications for people with disabilities: defining user needs and user requirements. Paper presented at the 10th association of the advancement of assistive technology in Europe conference, Florence, Italy

Zickler C, Halder S, Kleih SC, Herbert C, Kübler A (2013) Brain Painting: usability testing according to the user-centered design in end users with severe motor paralysis. Artif Intell Med 59(2):99–110. https://doi.org/10.1016/j.artmed.2013.08.003

Chapter 16
Methods and Tools for Using BCI with Augmented and Virtual Reality

Felix Putze (ID)

Abstract In this chapter, we will introduce Augmented and Virtual Reality as a novel way of user interaction which holds great promises for immersive BCI art applications. We will first introduce the key terms and give an introduction to the technical challenges and possible solutions to them. Then, we will discuss a number of important examples of the combination of BCI and Augmented and Virtual Reality technology in different application domains.

Keywords Augmented reality · Virtual reality · BCI · Technical aspects

16.1 Introduction

In recent years, Augmented and Virtual Reality (AR/VR) technology has matured technically and has become widely available as a tool to create complex and immersive applications. These applications cover a wide variety of areas, for example entertainment, education, art, and therapy, among others. AR/VR technology and BCIs can mutually benefit from each other: On the one hand, the multisensory experience or augmentation through AR/VR technology allows to create scenarios which are much more stimulating and expressive than standard desktop applications. On the other hand, BCI technology can provide additional explicit or implicit input channels to manipulate or influence the virtual scenario when standard input controllers fall short or are unavailable.

As AR/VR technology provides an immersive multimodal experience, they can be used as a canvas for artistic expression. The use of this technology to create unique, often interactive pieces of art through AR/VR technology is already discussed and experimented with for decades: The installation VIDEOPLACE by Krueger and Krueger (1983), originated in 1975, featured the manipulation of virtual objects

F. Putze (✉)
Cognitive Systems Lab, University of Bremen, Enrique-Schmidt-Straße 5,
28359 Bremen, Germany
e-mail: felix.putze@uni-bremen.de
URL: http://www.uni-bremen.de/csl

© Springer Nature Switzerland AG 2019
A. Nijholt (ed.), *Brain Art*,
https://doi.org/10.1007/978-3-030-14323-7_16

433

(a) Artist Barry Blitt drawing with Google Paint[1]

(b) Screenshot of "Virtualshamanism: Towards an alternative digital reality of consciousness" by Matias Brunacci (2018).

Fig. 16.1 Examples of VR art

and is considered to be one of the first AR artistic installations. Today, a large number of dedicated exhibits and galleries shows how artists explore and exploit the technological advances to express themselves in this art form.

Bates (1992) already discussed the role of VR in art in 1992 and since then, many technical advances have opened many additional opportunities to express creativity. Several VR painting apps, such as Google Tilt Brush, Facebook Quill, or Adobe project Dali, are available or under development and allow artistic expression in VR by using the controllers as brushes to draw lines and surfaces in 3D. See for example the expression of several artists in VR in Time magazine[1] which shows the versatility of VR art. This potential even increases if we take into account the possibility of freely programmable environments using generic game engines and 3D modeling and animation tools. Figure 16.1 shows two examples of VR art using existing drawing tools and custom-created scenes.

Given this vibrant environment of AR/VR art, it seems clear that for the field of BCI art, the use of AR/VR technology comes with great promises of a high level of immersion and the creation of interactive, explorable scenes. The likely reason why the use of AR/VR is not yet more prevalent in BCI art may lie in the fact that the development of a BCI interface for AR/VR requires the combination of two technologies from different fields which are both complex on their own and challenging to combine.

In this chapter, I will define the fundamental concepts of AR/VR, give an overview of existing hardware and software toolkits, and give a primer on the technical and methodological challenges for the integration of both technologies. Furthermore, we will describe some current showcases in which BCI and AR/VR technology are combined successfully.

---

[1] http://time.com/vr-is-for-artists/.

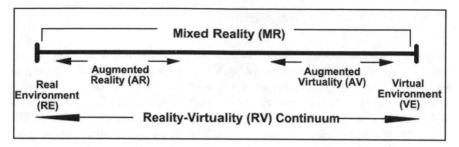

Fig. 16.2 Spectrum of mixed reality interfaces according to Milgram and Colquhoun (1999)

16.2 VR—AR—Mixed Reality

While AR and VR are often mentioned in the same breath, they are actually two relatively different concepts. Additionally, there is no clear dichotomy that exists between
these two concepts but rather a continuous spectrum. Milgram and Colquhoun (1999)
define "A Taxonomy of Real and Virtual World Display Integration" and consider different definitions of augmented reality, virtual reality, and mixed reality (Fig. 16.2).
For the concept of AR, they differentiate between see-through displays based on head-
mounted displays or glasses which add spatially situated virtual objects to the real
scene, and a broader definition which also encompasses other kinds of augmenting
displays (today, such non-HMD AR is often based on smart phones). See-through
AR can use video-based see-through or optical see-through (Si-Mohammed et al.
2017). Recent AR systems do not only provide visual augmentation but can also
address other senses, most importantly through spatial sound.

Building from that definition, the authors develop a spectrum of reality-virtuality
mixtures, ranging from augmented reality, over augmented virtuality (which are both
forms of "mixed reality"), to a full virtual environment. All types of systems involve
both real and virtual components but differ in what the primary environment is (real
or virtual) and how and to what extend this environment is mixed with the other one.
This mixing can refer to the blending of the visual scene or the physical interaction.
Interestingly, adding BCI as a control paradigm for VR systems can move these
further along the "virtuality" axis as it reduces the need for actual physical interaction
with the environment or a controller.

16.3 Combining AR/VR with BCI

In this section, we will introduce some of the technical concepts which are required to
tackle the combination of AR/VR and BCI technology. This comprises the selection
and combination of hardware components as well as the development of the software.

16.3.1 Application Development

Developing applications for AR/VR environments is most often done using a compatible game engine. While many existing VR tools for artistic expression exist or are under active development, they are designed as tools for a specific use case with a closed feature set. As they are not created with BCI input in mind, transparently replacing the designated input modalities will be very challenging to impossible (especially VR-based painting usually relies heavily on spatial operations through the standard controllers). Thus, designing applications for BCI art will in most cases require the development and programming of custom applications from scratch.

The two dominant toolkits for development of AR/VR applications are Unity3D and the Unreal game engine (Fig. 16.3). They provide functionality for scene construction, animation, and scripting. The support for modern concepts in computer graphics like shaders does not only allow to precisely create complex effects but also gives better control over refresh rates which might for example be useful for the implementation of SSVEP-based BCI. As the applications can be programmed in programming languages such as C# or C++, any desired, arbitrarily complex and interactive behavior can be implemented. Depending on the choice of AR/VR headset, one will need to import a headset-specific software development kit (SDK) or plugin to the game engine which provides methods to access the specific headset functionality. A wide array of available assets and plugins allows to rapidly bootstrap applications with advanced functionality. If available 3D models are not sufficient, they can be created through specialized modeling applications like Blender or Maya. Finally, the behavior of the engines is not limited to the 3D scene, but also involves the auditory scene and the responses to user input. Game engines and 3D modeling tools are complex to master but countless official and unofficial tutorials are available for a gentle start.

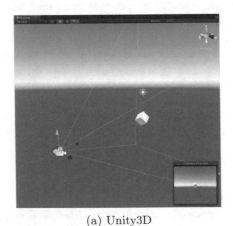

(a) Unity3D   (b) Unreal Engine

Fig. 16.3 Development environments of popular game engines

16.3.2   EEG Headsets

The choice of a suitable EEG headset already needs some consideration for a "standard" BCI, as there exists a large variety of choices from research-grade systems to consumer-ready devices. Gel-based Ag/AgCl electrodes are still considered to provide the best signal quality but come at the usability disadvantage of time-consuming and uncomfortable gel application. An alternative are dry electrodes, which come close in signal quality (Kam et al. 2019) and avoid this preparation step; however, they rely on a certain pressure to be effective (Fiedler et al. 2018) which may also lead to discomfort over longer periods of time. Many vendors now offer wireless systems which transmit data via bluetooth or another communication protocol, which greatly increases the freedom of movement of a BCI user.

When combined with an AR/VR setting, the selection process can become more challenging as a head-mounted display competes with the EEG headset for the same space. Especially electrodes at the prefrontal and occipital cortex are located in regions which are usually blocked by the display. In contrast, electrodes in central regions of the scalp are not typically problematic. While electrodes can be carefully placed below the display holder in some cases, this likely results in discomfort or signal artifacts. A solution is often to choose alternative electrode locations (exploiting the low spatial resolution of EEG). Another alternative is the use of dedicated headsets (see Fig. 16.4) which promise a convenient integration of (usually dry) electrodes at pre-determined, optimized locations for quick setup times and high user comfort. As potential disadvantages, such headsets provide only a limited number of electrodes at fixed positions, cannot be used without the display and are also often only compatible with a small selection of AR/VR devices. Independent of how headset and electrodes are mounted, it should be considered that combining EEG electrodes (with wet gel or pressure-exerting dry electrodes) with a relatively heavy AR/VR headset will strain the user's head and neck over longer periods of time, i.e. sessions should be shorter than for a standard BCI task.

(a) VR-BCI integration by Wearable   (b) EEG electrodes integrated in the
Sensing.   LooxidVR headset.

Fig. 16.4   Available commercial solutions for VR-BCI integration

16.3.3   AR/VR Headsets

Combining AR/VR with BCI technology requires the selection of employed hardware platforms. Besides the EEG headset, one also needs to consider the choice of AR/VR headset. After deciding for either AR or VR, there are different criteria which need to be considered for this selection. Most obviously, available headsets differ in their visual fidelity, i.e. the provided resolution, the refresh rate, the capability to generate realistic depth perception, or the provided field of view. But other factors should also be taken into account: For example, the weight of the headset contributes to the user's comfort during extended sessions. Additionally, most headsets are wired, i.e. they need to be connected to a computer which uses the AR/VR device as a display. This is still true if the wired connection is replaced with a wireless one. At the time of writing, only few headsets (such as the Microsoft HoloLens AR headset or the HTC Vive Focus and Oculus Quest VR headsets) are fully stand-alone, which comes at the price of other disadvantages (e.g. limited runtime).

Besides its output capabilities for 3D graphics and spatial sound, AR and VR headsets usually are also equipped with sensors which one may be able to exploit for a multimodal system or as context information. Systems which track a user's position in space (as most VR headsets do) or yield position and orientation through IMU sensors. VR headsets also usually provide a controller device which allows to manipulate virtual objects through a combination of gestures and button presses. While the popular existing headsets are not equipped with eye tracking technology out-of-the-box, there are solutions for extending them with unobtrusive mobile eye trackers, such as the Pupil platform (Kassner et al. 2014). Another popular addition to AR/VR headsets is the use of motion-based controller such as the Leap Motion device (see for example Regenbrecht et al. 2013) to perform hand tracking and gesture recognition or the use of external marker-based tracking technology for improved localization of user and important objects (even outside the Field-of-View).

16.3.4   Registration and Synchronization

For any AR headset to place virtual objects in relation to the real world, it needs to register the user's location and orientation as well as of any relevant objects in the environment. Similarly, a VR headset needs to register the user's location if it aims for a physical interaction with virtual objects or wants to have physical objects appear in the virtual scene (e.g. the game controllers). Objects can be registered to the virtual environment through tracking methods using optical infrared-based tracking markers or through methods based on computer vision approaches to recognize pre-registered visual markers. For mapping out a whole 3D space and localize the device within that space, AR can apply "Simultaneous Localization and Mapping" (SLAM) methods (Cadena et al. 2016) by using a combination of vision-based registration and alignment of key points in the scene with sensor data in a stochastic map of

the environment. Such sensor data is usually recorded through an Inertial Measurement Unit (IMU) which contains accelerometers to capture the body's specific force,
gyroscopes that measure the body's angular rate, and sometimes magnetometers that
capture the body's magnetic field.

Besides a spatial synchronization as just described, we also need synchronization in time: As for all BCIs, the EEG data stream needs to be synchronized to
the events within the application. A synchronization through the use of a light sensor is not possible due to the confined space around the display. The popular Lab
Streaming Layer middleware[2] provides functionality for streaming data across several distributed clients in a time-synchronized fashion. There exists a special version
of LSL for the Unity engine[3] and Unreal engine. For the specific case of the HoloLens,
LSL4Unity cannot be used as it cannot be combined on the Universal Windows Platform (UWP). A solution to this is the HoloLensBridge[4] which emulates the LSL
functionality for this device. LSL can then also be used for a loose coupling of the
AR/VR application with the BCI software to build close-loop systems. LSL support
is included in many existing BCI suits such as OpenViBE or can be integrated in
custom solutions through the available APIs.

16.4 Showcases

In this section, we will review a number of works which demonstrate the combination
of BCI and AR/VR for different domains. Most of these showcases make use of
established active BCI paradigms, such as the P300, SSVEP, or Motor Imagery.
While passive BCI are less prevalent, neurofeedback is a paradigm which is used
regularly in the combination of AR/VR and BCI.

16.4.1 Therapy and Rehabilitation

One area of application in which AR/VR technology has made an impact in recent
years is (psycho)therapy and rehabilitation. Here, the high degree of immersion for
such interfaces and the resulting feeling of presence allow the implementation of
innovative therapeutic paradigms which are impossible or expensive to execute in
reality. Examples of such psychotherapeutic applications include exposure therapy
for treating phobias (see for example North et al. 2015; Miloff et al. 2016; Botella
et al. 2010; Juan et al. 2005), meditation exercises for chronic pain management (see
Gromala et al. 2015; Dunn et al. 2017) or motor and cognitive exercises for stroke
rehabilitation (see Yates et al. 2016). Additionally, AR is also widely explored as

---

[2]https://github.com/sccn/labstreaminglayer.

[3]https://github.com/xfleckx/LSL4Unity.

[4]https://gitlab.csl.uni-bremen.de/fkroll/LSLHoloBridge.

a tool to treat or support patients (mostly children) with attention disorders to help them focus better, for example in learning situations (see Escobedo et al. 2014; Aziz et al. 2012).

BCI technology is also strongly rooted in these areas. Therefore, it does not come as a surprise that multiple researchers have explored the potential for combination of both techniques for applications in therapy or rehabilitation. Salisbury et al. (2016) thus explores the potential of the combination of both techniques from a medical perspective. They find promising results in pilot studies but also discuss challenges, such as high costs, limited availability, and complexity of the involved technology for medical personnel.

One central type of BCI in combination with AR/VR for therapeutic purposes is based on the neurofeedback paradigm, channeling the measured brain activity back to the user to increase awareness of ongoing cognitive processes. One of the earliest works which demonstrated the feasibility of the approach is the Mind-Mirror by Mercier-Ganady et al. (2014) which superimposed a spatial distribution of raw EEG power or an estimation of a state of relaxation on an image of the person in real-time. Users reported the mirror to be "engaging and explanatory" compared to traditional feedback technology. Besides neurofeedback training, the authors also see potential application of the mirror in education and entertainment. Kosunen et al. (2016) investigated a VR-based mindfulness training using neurofeedback on the user's level of concentration and relaxation, measured from alpha and theta activity. The estimated concentration level is then mapped to different visual effects in a virtual landscape to further support the meditative depth. Their results show that both the VR environment as well as the neurofeedback contribute to the meditative experience. Amores et al. (2018) came to similar conclusions, showing that a VR setup with an additional olfactory component increased the perception of meditation by 25% compared to a condition without and they measured similarly strong increases in the strength of physiological responses.

BCI-based rehabilitation applications to regain limp motor control (usually after stroke) often uses the motor imagery (MI) paradigm, traditionally following the Graz-MI paradigm (see Pfurtscheller et al. 2003). One critical aspect of such BCIs is the presentation of feedback presented to patients, which should give a strong sense of connection to the targeted limbs. Perez-Marcos et al. (2009) were the first to show that an MI-based interface in combination with a associated visualization of the corresponding movements in an VR environment was enough to generate an illusion of limb ownership. Škola and Liarokapis (2018) used a VR setup to create a virtual embodiment, showing the person's hands executing the detected imagined movements. In comparison to the standard Graz-MI paradigm using symbolic feedback, participants exhibit significantly higher accuracy. Spicer et al. (2017) present a similar setup, demonstrating that a system for limb motor recovery with VR feedback can be build from low-cost components. The NeuRow system by Vourvopoulos et al. (2016) uses a closed-loop BCI training system using motor imagery to control a rowing simulator presented in a VR system with a vibrotactile feedback component. Their initial user study shows state-of-the-art performance for MI control and high levels of positive affect and flow.

16.4.2 Environment Control

Compared to VR, an AR system usually involves a high level of interaction with the real environment (see Sect. 16.2). Many interesting use cases of AR arise when the virtual component of the scene is used to enhance or control this interaction. For example, an AR headset can be used to superimpose the view of a scene with context-dependent control elements. These control elements can be coupled with BCI input paradigms. Extending the communication bandwidth to give input to the device is especially relevant in AR settings where users interact both with the real and the virtual environment and traditional communication channels (e.g. hands and arms for gestures) may often be occupied and not available to give input to the AR system. In contrast to the standard input mechanisms of AR devices, BCI input is silent and can be operated hands-free, which enables the use of the device even when other modalities are not available. This can of course also be operated by people with severe disabilities who want to take control of their environment, as is one of the traditional use cases for control BCIs. A review by Si-Mohammed et al. (2017) shows that many of the established BCI paradigms are also prevalent in different AR and BCI prototypes.

One example for an BCI for environment control is shown in Fig. 16.5, developed at the Cognitive Systems Lab (Bremen, Germany) by Putze et al.[5] In this AR application, control elements for different aspects of home control (controlling window blinds, music player, television, etc.) are superimposed on small visual markers situated in the environment. Following the SSVEP-BCI paradigm, Similar systems are described by Faller et al. (2017) and Saboor et al. (2017), which shows that

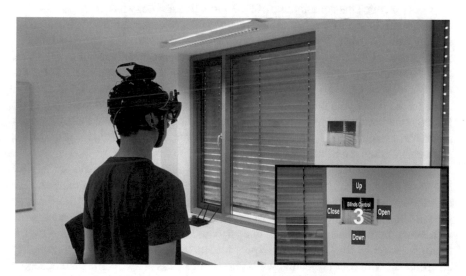

Fig. 16.5 Room control via AR using the SSVEP-BCI paradigm

[5]Thanks to Dennis Weiß for his work on the prototype and the provision of the image.

the SSVEP paradigm can routinely be used as a control paradigm in AR. Escobedo et al. (2014) overlayed a grid for a P300-based BCI over the scene for telepresence control of a robot. Coogan and He (2018) combine Unity and the BCI2000 software to provide a template for control of Internet-of-Things devices (such as smart light, television, or thermostat) via BCI interfaces in VR. The authors claim that this approach allows the rapid integration of additional tasks (due to the use of established software components) and a higher motivation for users compared to traditional BCI interfaces.

Another area of application for environment control through AR-based BCIs is the control of robotic agents. For example, Wang et al. (2018) used an SSVEP-based AR interface to control the flight of a drone. Si-Mohammed et al. (2018) used an AR interface to control the movement of a robotic platform. They systematically explored different ways of integrating SSVEP stimuli in the environment in relation to the robot and also investigated the effect of motion on the BCI performance, given that an AR scene is rarely static.

16.4.3 Games

Research in the application of BCI for game control and adaptation is relevant to people interested in BCI art as game engines offer a wide array of possibilities to create immersive and dynamic virtual scenarios, which is also often a prerequisite to implement certain artistic visions. Interestingly, BCI for game control has been discussed for many years now (see Lécuyer et al. 2008) and is thus one of the oldest domains for which the combination of BCI and VR technology has been discussed. A reason for this is likely the fact that both technologies offer much room for exploration and playful discovery which is a key factor to make games entertaining. Besides enhancing gaming experience for entertainment, there are several secondary effects of researching into game playing: On the one hand, "serious games" can provide benefits for their players, such as learning new information or training certain skills. Indeed, some of the paradigms presented in Sect. 16.4.1 can be considered serious games. On the other hand, games can serve as research platforms which are more engaging as standard experimental paradigms, leading to improved participant motivation, performance, and retention.

Kerous et al. (2018) published a recent state-of-the-art report on the use of BCI for game control and showed that several BCI paradigms have been explored for interaction. Some of these paradigms have already been explored in AR/VR based games: Guger et al. (2009) was one of the first instances to introduce VR navigation with synchronous (based on motor imagery) and asynchronous BCI (based on the P300). Leeb et al. (2013) presented a (penguin) racing game played in a virtual reality environment with four projection walls and shutter glasses for creating stereoscopic vision. A jump action of the virtual character could be triggered by issuing a MI command for the leg, mimicking the jumping movement. They could show that after short training, half of the players were able to perform the command, even in combination with joystick-based control of the penguin's lateral movement. Participants reported the BCI-based control to be more engaging than a purely controller-based one. Koo et al. (2015) used a SSVEP-BCI to control the player's movement through a virtual maze. They report that playing the game in 3D-VR increased the control performance (i.e. the information transfer rate) by 10% compared to a 2D condition and explain that improvement by the higher degree of immersion and thus the reduced susceptibility to distraction. Iidal et al. (2017) made similar observations while using a frequency-based concentration index to control an avatar through a game course with obstacles. Measured concentration was higher in the VR condition compared to a 2D control condition and also participants reported a higher level of immersion and control. Moving away from purely mapping BCI input to game character control, Mercier-Ganady et al. (2015) explored the use of BCI to give a person the "super-power" to become invisible in a virtual environment for the purpose of hiding from a ghost. Like the previous work, triggering this ability was based on concentration level. While BCI-control was less reliable and slower than pure button-control, it was reported to be more motivating and more immersive. BCI can also be used for control of other media. Brouwer et al. (2018) presented a study in which they predict head movement from EEG to reduce latency in camera adjustment for VR presentations of surround video streaming. They could show that for some users, the head rotation orientation could be predicted with an accuracy of up to 79%.

16.5   Conclusion

To summarize, this chapter showed the big potential of VR and AR as a canvas for artistic expression through BCI technology. VR is a field which is already thoroughly explored currently by digital artists and BCI technology can add another way of expression which is not only exciting to explore but also accessible to artists with disabilities and for the application of BCI art as a therapeutic tool. Several showcases illustrated how BCI was used successfully together with AR/VR interfaces already today in different application domains. We saw that a challenge of using a combination of AR/VR and BCI technology is the complexity of individual components as well as of their interplay. Future developments in this field should look into the development of standardized tools and processing pipelines (e.g. in the form of plugins to the game engine SDKs) to reduce the entry barrier for BCI art. Another existing challenge is how to deal with the spatial component of AR/VR art. Correctly connecting line strokes in virtual 3D space is already a challenge when using the standard input devices (see Yue et al. 2017). When switching to BCI input with limited input sensitivity and bandwidth, this will pose a challenge which needs to be addressed for successful BCI art in AR or VR. Friedman (2015) published a review article which concentrates on several different possibilities for navigation within VR through BCI technology: by controlling the viewpoint, by controlling a virtual avatar, or by controlling the world directly. This discussion outlines the possibilities and limitations of such approaches. For able-bodied artists, it may be an alternative to create hybrid input systems which combine BCI with manual input.

References

Amores J, Richer R, Zhao N, Maes P, Eskofier BM (2018) Promoting relaxation using virtual reality, olfactory interfaces and wearable EEG. In: 2018 IEEE 15th international conference on wearable and implantable body sensor networks (BSN), pp 98–101. https://doi.org/10.1109/BSN.2018.8329668

Aziz NAA, Aziz KA, Paul A, Yusof AM, Noor NSM (2012) Providing augmented reality based education for students with attention deficit hyperactive disorder via cloud computing: its advantages. In: 2012 14th International conference on advanced communication technology (ICACT), pp 577–581

Bates J (1992) Virtual reality, art, and entertainment. Presence: Teleoperators and Virtual Environ 1(1):133–138. https://doi.org/10.1162/pres.1992.1.1.133

Botella C, Bretón -López J, Quero S, Baños R, García-Palacios A (2010) Treating cockroach phobia with augmented reality. Behav Ther 41(3):401–413. https://doi.org/10.1016/j.beth.2009.07.002

Brouwer AM, van der Waa J, Stokking H (2018) BCI to potentially enhance streaming images to a VR headset by predicting head rotation. Front Hum Neurosci 12. https://doi.org/10.3389/fnhum.2018.00420

Cadena C, Carlone L, Carrillo H, Latif Y, Scaramuzza D, Neira J, Reid I, Leonard JJ (2016) Past, present, and future of simultaneous localization and mapping: toward the robust-perception age. IEEE Trans Robot 32(6):1309–1332

Coogan CG, He B (2018) Brain-computer interface control in a virtual reality environment and applications for the internet of things. IEEE Access 6:10840–10849. https://doi.org/10.1109/ACCESS.2018.2809453

Dunn J, Yeo E, Moghaddampour P, Chau B, Humbert S (2017) Virtual and augmented reality in the treatment of phantom limb pain: A literature review. NeuroRehabilitation 40(4):595–601. https://doi.org/10.3233/NRE-171447

Escobedo L, Tentori M, Quintana E, Favela J, Garcia-Rosas D (2014) Using augmented reality to help children with autism stay focused. IEEE Pervasive Comput 13(1):38–46

Faller J, Allison BZ, Brunner C, Scherer R, Schmalstieg D, Pfurtscheller G, Neuper C (2017) A feasibility study on SSVEP-based interaction with motivating and immersive virtual and augmented reality. arXiv:170103981 [cs] 1701.03981

Fiedler P, Möhle R, Griebel S, Pedrosa P, Fonseca C, Vaz F, Zanow F, Haueisen J (2018) Contact pressure and flexibility of multipin dry EEG electrodes. IEEE Trans Neural Syst Rehabil Eng

Friedman D (2015) Brain-computer interfacing and virtual reality. In: Nakatsu R, Rauterberg M, Ciancarini P (eds) Handbook of digital games and entertainment technologies. Springer, Singapore, pp 1–22. https://doi.org/10.1007/978-981-4560-52-8_2-1

Gromala D, Tong X, Choo A, Karamnejad M, Shaw CD (2015) The virtual meditative walk: virtual reality therapy for chronic pain management. In: Proceedings of the 33rd annual ACM conference on human factors in computing systems. ACM, CHI '15, pp 521–524. https://doi.org/10.1145/2702123.2702344

Guger C, Groenegress C, Holzner C, Edlinger G, Slater M, Sánchez-Vives MV (2009) Brain computer interface for virtual reality control. Cyberpsychology Behav 12(1):84–84

Iidal Y, Tsutsumi D, Saeki S, Ootsuka Y, Hashimoto T, Horie R (2017) The effect of immersive head mounted display on a brain computer interface game. In: Chung W, Shin CS (eds) Advances in affective and pleasurable design. Springer International Publishing, Advances in Intelligent Systems and Computing, pp 211–219

Juan MC, Alcaniz M, Monserrat C, Botella C, Banos RM, Guerrero B (2005) Using augmented reality to treat phobias. IEEE Comput Graph Appl 25(6):31–37. https://doi.org/10.1109/MCG.2005.143

Kam JWY, Griffin S, Shen A, Patel S, Hinrichs H, Heinze HJ, Deouell LY, Knight RT (2019) Systematic comparison between a wireless EEG system with dry electrodes and a wired EEG system with wet electrodes. NeuroImage 184:119–129. https://doi.org/10.1016/j.neuroimage.2018.09.012

Kassner M, Patera W, Bulling A (2014) Pupil: an open source platform for pervasive eye tracking and mobile gaze-based interaction. In: Proceedings of the 2014 ACM international joint conference on pervasive and ubiquitous computing: adjunct publication. ACM, pp 1151–1160

Kerous B, Skola F, Liarokapis F (2018) EEG-based BCI and video games: a progress report. Virtual Real 22(2):119–135. https://doi.org/10.1007/s10055-017-0328-x

Koo B, Lee H, Nam Y, Choi S (2015) Immersive BCI with SSVEP in VR head-mounted display. In: 2015 37th annual international conference of the IEEE engineering in medicine and biology society (EMBC), pp 1103–1106. https://doi.org/10.1109/EMBC.2015.7318558

Kosunen I, Salminen M, Järvelä S, Ruonala A, Ravaja N, Jacucci G (2016) RelaWorld: Neuroadaptive and immersive virtual reality meditation system. In: Proceedings of the 21st international conference on intelligent user interfaces. ACM, IUI '16, pp 208–217. https://doi.org/10.1145/2856767.2856796

Krueger MW, Krueger MW (1983) Artificial reality, vol 126. Addison-Wesley Reading, MA

Lécuyer A, Lotte F, Reilly RB, Leeb R, Hirose M, Slater M (2008) Brain-computer interfaces, virtual reality, and videogames. Computer 41(10):66–72. https://doi.org/10.1109/MC.2008.410

Leeb R, Lancelle M, Kaiser V, Fellner DW, Pfurtscheller G (2013) Thinking penguin: multimodal brain-computer interface control of a VR game. IEEE Trans Comput Intell AI Games 5(2):117–128. https://doi.org/10.1109/TCIAIG.2013.2242072

Mercier-Ganady J, Lotte F, Loup-Escande E, Marchal M, Lécuyer A (2014) The mind-mirror: See your brain in action in your head using EEG and augmented reality. In: 2014 IEEE virtual reality (VR), pp 33–38. https://doi.org/10.1109/VR.2014.6802047

Mercier-Ganady J, Marchal M, Lécuyer A (2015) B-C-invisibility power: introducing optical camouflage based on mental activity in augmented reality. In: Proceedings of the 6th augmented human international conference. ACM, AH '15, pp 97–100. https://doi.org/10.1145/2735711.2735835

Milgram P, Colquhoun H (1999) A taxonomy of real and virtual world display integration. Mixed Reality: Merging Real Virtual Worlds 1:1–26

Miloff A, Lindner P, Hamilton W, Reuterskiöld L, Andersson G, Carlbring P (2016) Single-session gamified virtual reality exposure therapy for spider phobia versus traditional exposure therapy: study protocol for a randomized controlled non-inferiority trial. Trials 17(1):60. https://doi.org/10.1186/s13063-016-1171-1

North MM, North SM, Coble JR (2015) Virtual reality therapy: an effective treatment for the feat of public speaking. Int J Virtual Real (IJVR) 03(3):1–6

Perez-Marcos D, Slater M, Sanchez-Vives MV (2009) Inducing a virtual hand ownership illusion through a brain-computer interface. NeuroReport 20(6):589. https://doi.org/10.1097/WNR.0b013e32832a0a2a

Pfurtscheller G, Neuper C, Muller GR, Obermaier B, Krausz G, Schlogl A, Scherer R, Graimann B, Keinrath C, Skliris D, Wortz M, Supp G, Schrank C (2003) Graz-BCI: state of the art and clinical applications. IEEE Trans Neural Syst Rehabil Eng 11(2):1–4. https://doi.org/10.1109/TNSRE.2003.814454

Regenbrecht H, Collins J, Hoermann S (2013) A leap-supported, hybrid AR interface approach. In: Proceedings of the 25th Australian computer-human interaction conference: augmentation, application, innovation, collaboration. ACM, pp 281–284

Saboor A, Rezeika A, Stawicki P, Gembler F, Benda M, Grunenberg T, Volosyak I (2017) SSVEP-based BCI in a smart home scenario. In: Rojas I, Joya G, Catala A (eds) Advances in Computational Intelligence. Lecture Notes in Computer Science. Springer International Publishing, pp 474–485

Salisbury DB, Dahdah M, Driver S, Parsons TD, Richter KM (2016) Virtual reality and brain computer interface in neurorehabilitation. Bayl Univ Med Cent Proc 29(2):124–127. https://doi.org/10.1080/08998280.2016.11929386

Škola F, Liarokapis F (2018) Embodied VR environment facilitates motorimagery brain-computerinterface training. Comput Graph 75:59–71. https://doi.org/10.1016/j.cag.2018.05.024

Si-Mohammed H, Argelaguet F, Casiez G, Roussel N, Lécuyer A (2017) Brain-computer interfaces and augmented reality: A state of the art. In: Proceedings of the 7th International BCI Conference, https://doi.org/10.3217/978-3-85125-533-1-82

Si-Mohammed H, Petit J, Jeunet C, Argelaguet F, Spindler F, Évain A, Roussel N, Casiez G, Lécuyer A (2018) Towards BCI-based interfaces for augmented reality: Feasibility, design and evaluation. IEEE Trans Vis Comput Graph 1–1. https://doi.org/10.1109/TVCG.2018.2873737

Spicer R, Anglin J, Krum DM, Liew S (2017) REINVENT: A low-cost, virtual reality brain-computer interface for severe stroke upper limb motor recovery. In: 2017 IEEE virtual reality (VR), pp 385–386. https://doi.org/10.1109/VR.2017.7892338

Vourvopoulos A, Ferreira A, i Badia SB (2016) NeuRow: an immersive VR environment for motor-imagery training with the use of brain-computer interfaces and vibrotactile feedback. In: Proceedings of the 3rd international conference on physiological computing systems

Wang M, Li R, Zhang R, Li G, Zhang D (2018) A wearable SSVEP-based BCI system for quad-copter control using head-mounted device. IEEE Access 6:26789–26798. https://doi.org/10.1109/ACCESS.2018.2825378

Yates M, Kelemen A, Lanyi CS (2016) Virtual reality gaming in the rehabilitation of the upper extremities post-stroke. Brain Injury 30(7):855–863. https://doi.org/10.3109/02699052.2016.1144146

Yue YT, Zhang X, Yang Y, Ren G, Choi YK, Wang W (2017) Wiredraw: 3D wire sculpturing guided with mixed reality. In: Proceedings of the 2017 CHI conference on human factors in computing systems. ACM, pp 3693–3704

Chapter 17
The BR4IN.IO Hackathons

Christoph Guger⬤, Brendan Z. Allison⬤, Martin Walchshofer
and Sarah Breinbauer

Abstract BCI hackathons are fun, collaborative activities during which teams develop and implement new BCI designs and projects. BCI hackathons have often involved artistic expression, and have led to new headwear designs and BCI systems that let users paint, make music, or play games via thought alone. In the past few years, the number of BCI hackathons worldwide has increased considerably, and we expect this trend will continue. Here, we provide an overview of BCI hackathons, then review our BR4IN.IO hackathon series. We have hosted a dozen BR4IN.IO hackathons so far, which have included programming, design, flagship, and kids' projects. At each BR4IN.IO hackathon, we provide hardware, software, and furniture for each team. Teams then have 24 intensive hours of hacking, then present their projects to a jury that decides on the winners. Like most other BCI hackathons, the BR4IN.IO hackathons have helped to teach thousands of new people about BCIs and related neurotechnologies while encouraging positive publicity for our field. In addition to increasing in number, we believe that future BCI hackathons will also entail more advanced hardware and software for hackers, a broader range of devices, new cooperation with different professional entities, more publicity, greater integration with university courses, and other improvements.

Keywords BCI · Brain-computer interface · BCI hackathon · BMI · BR41N.IO

17.1 Introduction

Brain-computer interfaces (BCIs) have been gaining attention in the academic and popular literature of late (Friedman et al. 2017; Guger et al. in press; Nam et al. 2018). This increased attention has inspired a solid rise in BCI hackathons. BCI

C. Guger (✉) · M. Walchshofer · S. Breinbauer
g.tec Medical Engineering GmbH, Schieldberg 4521, Austria
e-mail: guger@gtec.at

B. Z. Allison (✉)
Cognitive Science Department, University of California at San Diego, La Jolla 92093, USA
e-mail: ballison@ucsd.edu

© Springer Nature Switzerland AG 2019

A. Nijholt (ed.), *Brain Art*,
https://doi.org/10.1007/978-3-030-14323-7_17

hackathons are a very promising mechanism to attract new people to the field, disseminate new technologies, bolster interdisciplinary interaction, encourage hands-on experimentation, and foster positive media coverage. This chapter introduces BCI hackathons, presents projects and winners from our BR4IN.IO hackathons, presents lessons learned with suggestions for future hackathon organizers, and consider future directions. Our chapter includes numerous images and quotes to show BR4IN.IO hackathons from a personal perspective.

17.2   BCI Hackathons: What and Why

According to dictionary.com, a hackathon is "a usually competitive event in which people work in groups on software or hardware projects, with the goal of creating a functioning product by the end of the event." The term "hackathon" combines the words "hack" and "marathon," reflecting that hackathons are intensive events that often last 24 h or more. Hackathons seem to have begun about 20 years ago and have been gaining attention worldwide, with over 1000 hackathons since 2015 (Briscoe 2014; HackerEarth 2017).

A BCI hackathon is a type of hackathon focused on software, hardware, and functioning products relating to BCIs. This separates BCI hackathons from events that only include lectures, videos, demonstrations, tutorials, discussion, and other activities. These might also occur at BCI hackathons, but are neither necessary nor sufficient.

A BCI is a system that enables real-time interaction between a person and a device based on direct measures of brain activity (Wolpaw and Wolpaw 2012; Nam et al. 2018). Most BCIs rely on the EEG, but can also use other noninvasive methods (such as fNIRS, MEG, or fMRI) or invasive methods (such as depth electrodes and ECoG). BNCIs are a broader category, which differs from BCIs only in that BNCIs may also include measures of biophysical signals that are not directly generated by the brain, such as EMG, EOG, HR, or respiration. Adaptive neurotechnologies are broader still, in that they also include systems that can modify nervous system activity, rather than just read activity. Notably, all of these approaches reflect an alternative, or at least a complement, to medication, therapy, behavioral changes, and other ways to influence nervous system activity.

For decades, BCIs could only perform simple tasks like moving a cursor or spelling. BCIs were expensive, bulky systems that were not especially helpful to most patients. Today, BCIs are being explored for new applications that could help a broader range of patients, such as motor rehabilitation for stroke patients, assessment and communication for patients with a disorder of consciousness, real-time brain mapping for next-generation neurosurgery, and control of complex devices (like exoskeletons, mobile robots or advanced wheelchairs). Companies and research groups have also continued developing BCIs and related EEG-based systems that are intended for healthy users, such as gaming, neuromarketing, alertness/workload monitoring, and neurofeedback for relaxation. Electrode caps, amplifiers, laptops, and other devices keep getting smaller and cheaper, and software gets more powerful.

Therefore, BCI designers and hackers have more ideas to explore than ever before. New BCI systems or components that might have seemed pointless or impossible—especially within the 1–2 day time frame of most BCI hackathons—are instead viable options. This seems to be a major reason why BCI hackathons have been much more prevalent in the last few years. Organizers need less time and money to organize BCI hackathons, and hackers can implement a broader range of BCIs more easily.

Our BR4IN.IO hackathons, and BCI hackathons in general, do not charge the hackers for their participation. Instead, costs may be paid by the organizing institution(s), one or more sponsor(s), public funding, and/or other sources. Equipment is typically provided for free by the manufacturer and/or the organizer. Some other (non-BCI) hackathons have required fees, which can pay for organizing costs and support larger prizes for winners.

This book chapter focuses on one series of BCI hackathons called the BR4IN.IO hackathons. Other BCI hackathon series have also been successful, and have often involved a trans-disciplinary theme such as art. For example, we contributed to two BCI Hackathons through the "BrainHack" effort through Talinn University in 2016, which focused on art and science (Eigner et al. 2017; Valjamae et al. 2017). In September 2017, Prof. Azorín and colleagues organized a BCI hackathon in Spain focused on art called Your Brain On Art (YBOA). Hacking teams were each expected to develop a BCI called a "Brain To Art Interface" or B2AI.

> Some of the projects of the YBOA [Your Brain On Art] hackathon were based on BR41N.IO.
>
> —Ortiz et al., in review.

A non-profit entity called NeuroTechX reports dozens of chapters worldwide that organize and execute BCI hackathons. Large-scale networks such as these could spread BCIs and related technologies to new regions, especially less developed countries.

17.3 The BR4IN.IO Hackathons

The BR41N.IO hackathon series brings together engineers, programmers, designers, artists or enthusiasts, to collaborate intensively as an interdisciplinary team. They program or build their own fully functional EEG-based Brain-Computer Interface (BCI) to control a drone, a Sphero or e-puck robot, or an orthosis. The participation only requires basic knowledge in BCIs, machine learning, programming or designing.

We launched the BR41N.IO series in 2017. Participants from numerous countries have worked together and achieved innovative and playful BCI headsets and applications within 24 h. The BR4IN.IO hackathons have also relied on a jury. We select a jury of experts to score the projects at the end of each hackathon based on several criteria such as "originality" and "execution," then select the winners. BR41N.IO was inspired by the unique "Agent Unicorn" headpiece from Fashion-Technology-Artist Anouk Wipprecht. This accessory is shaped like a unicorn's horn, and is designed for children with ADHD or autism (see Figs. 17.1 and 17.2). Agent Unicorn is a BCI headpiece that measures brain activity to find out what might trigger shifts in attention in these children and provide a better understanding of their individual distractions. Our experiences working with Ms. Wipprecht made us appreciate how much opportunity for inspiration can emerge by collaborating with experts from different fields.

We have hosted a dozen BR4IN.IO hackathons through 2018. Our five 2017 hackathons were in Banff, Valencia, Graz, Linz, and Dublin. Our seven hackathons in 2018 occurred in Poznań, San Francisco, Prague, Honolulu, Linz, Miyazaki, and Graz. Some hackathons have been stand-alone events, while others occurred in collaboration with major conferences, festivals or other activities.

For example, our 2017 Graz hackathon and 2018 San Francisco hackathons were attached to the Seventh International BCI Conference in Graz and Seventh International BCI Meeting near San Francisco. Our 2018 hackathon in Miyazaki was part of the 2018 IEEE SMC conference. Both of our hackathons in Linz have been part of festivals hosted at the Ars Electronica Center in Linz, a major public center devoted to technology and art.

Fig. 17.1 Ms. Wipprecht presents her "Agent Unicorn" headpiece at g.tec's headquarters

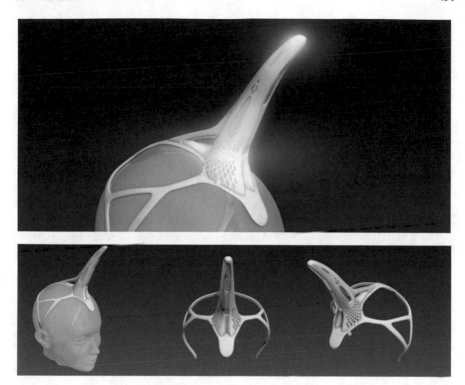

Fig. 17.2 The top panel shows the "Agent Unicorn" horn with an optional light. The bottom images show the horn from different angles

We have hosted two BR4IN.IO hackathons here at the Ars Electronica Linz, a major art and science museum. We got very positive feedback from attendees and from the media. If you plan a hackathon, I recommend contacting nearby TV stations, newspapers, magazines, universities, web-based media and more to get the word out.

—Erika Jungrithmayr, Manager, Ars Electronica Center Linz.

17.3.1 BR4IN.IO Project Types

Our BR4IN.IO hackathons have included four different types of projects.

17.3.1.1   Programming Projects

These projects challenge programmers to code an interface to control devices, robots or applications, write messages or draw paintings by using their thoughts alone. These are the most common projects in BR4IN.IO and other BCI hackathons.

17.3.1.2   Artistic Projects

BR41N.IO challenges creative minds to design and build a unique, playful and wearable headpiece that can measure useful EEG signals in real-time to create any sort of interaction. We provide 3D printers, handcraft materials and sewing machines at BR4IN.IO hackathons, so the teams can design and prototype their own BCI headpieces. Teams have also used BCIs to produce artful paintings or post a status update in their Social Media accounts.

17.3.1.3   Flagship Projects

Two BR4IN.IO hackathons have featured "Flagship projects" with special devices that are not available in most hackathons. Hackers have used BCIs to control heavy equipment for excavation and massive robot arms for drawing.

17.3.1.4   Kids' Projects

We have also had some kids who participated in our BR4IN.IO hackathons. In these projects, kids have created their own caps that are inspired by animals or mythical creatures. These projects are most similar to artistic projects.

17.3.2   BR4IN.IO Project Examples

This section presents some examples of projects that teams have executed at prior hackathons. Some of these general projects have been implemented differently in different hackathons, or even by different teams at the same hackathon. The BR4IN.IO website includes further examples of projects that teams could consider. This section is divided according to the four BR4IN.IO project types (programming, artistic, flagship, and kids').

17.3.2.1 Sample Programming Projects

This section presents twelve types of different programming projects that have been implemented. Table 17.1 summarizes the different features of these twelve programming projects. Some of them will require special software or hardware, in addition to the systems normally provided at a hackathon. These may be provided by the hackathon organizer and/or hackers. We also note skills that teams in each project had (beyond basic programming and Matlab/Simulink).

These are general examples, and different hacking teams have innovated their own implementations of them. Many projects have involved BCIs for device control. Smart Home teams have developed BCIs to control lamps, radios, and televisions with P300 BCIs. One hacking team developed a BCI using left versus right hand motor imagery to control a 3D printed orthosis or a drone as shown in Fig. 17.3.

Another team (which was popular with the other teams) brought gin and tonic and developed a BCI-controlled drink mixer. A microcontroller controlled a pump and a cooler were that communicated with the BCI system. Often teams use Unity to

Table 17.1 Each project may require software or hardware or special skills. These are just examples from prior BR4IN.IO hackathons, and future projects can (and should) explore different software, hardware, skills, and control approaches. The BCI control was implemented with the P300 evoked potential, motor imagery, mental arithmetic or steady-state visual evoked potentials (SSVEP)

Project	Software/Hardware	Skills	Control approach(es) used
Smart home control	Bring everyday devices to control	Arduino	P300
Sphero SPRK control	Robotic ball called Sphero	C#	P300
E-puck control	Mobile robot called E-puck	C#	P300
Orthosis control	Orthosis	MATLAB	Motor imagery
Flight control	Drone(s)	Java	P300, motor imagery
Camera control	Camera(s)	C#	P300, motor imagery, SSVEP
Drink mixer	Drink mixer components, drinks	MATLAB	Alpha/beta ratio
Dream painting	Painting program	MATLAB	Different frequency bands
Social media interface	Facebook, Twitter, others	C#	P300
fNIRS interface	fNIRS system	MATLAB	Motor imagery, counting, math
Music interface	Music software	Ruby	P300, motor imagery
Game interface	VR/Game software	Unity	P300, motor imagery, SSVEP

Fig. 17.3 Left: a hacking team uses a BCI to control an orthosis. Right: a BCI-controlled drone

Fig. 17.4 This developer is working within a unity environment while wearing an EEG cap

create Virtual Reality applications (see Fig. 17.4) or work with Augmented Reality gear as shown in Fig. 17.5.

Many other projects did not involve device control. Social media applications have been popular across different hackathons, often allowing people to post on Facebook or Twitter using P300s. Some groups have used BCIs to create or play music, using Sonic Pi or other applications. One team developed a tank warfare game, and three teams developed games that moved avatars through virtual environments.

Fig. 17.5 This project combines a BCI with a camera and augmented reality (AR) approach

17.3.2.2   Sample Artistic Projects

Since BCI hackathons tend to be organized by and for technical experts, they usually focus on programming projects. However, many projects have focused less on technical development and more on art, such as using BCIs to paint and creating new headsets. Table 17.2 groups these projects into three examples.

Some groups have used P300 BCIs to select and modify different art elements to create paintings on a monitor, projector or on paper as shown in Fig. 17.6.

Other groups have used 3D printers and sewing machines to make new headwear, sometimes with additional devices like cameras or lights (see Figs. 17.7 and 17.8). Some projects have developed different types of caps, while other designs used a visor, headband, or audio headphones. These projects can be challenging, because groups need to consider the intended application. For example, motor imagery BCIs require electrodes over sensorimotor areas.

Table 17.2 Three types of artistic projects

Project	Software/Hardware	Skills	Control approach(es) used
Brain painting	Unicorn Speller	Painting	P300
3D printed headsets	3D printer, 3D freeware, materials	CAD	Depends on electrode location
Sewed headsets	Sewing machine, materials	Sewing	Depends on electrode location

Fig. 17.6 The top image shows a team using the Unicorn Speller system to paint. The bottom image presents several examples of abstract art created with intendiX at a hackathon

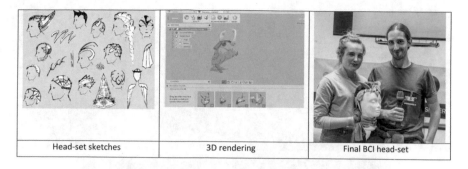

Head-set sketches	3D rendering	Final BCI head-set

Fig. 17.7 The top three panels show different phases of headset design, from sketches to 3D rendering to the final head-set. The bottom panel shows a hacker painting part of a headset that her team made

17.3.2.3 Flagship Projects

Flagship projects are special opportunities that are centered around equipment that we don't have at most hackathons. In one hackathon, we had a backhoe (see Fig. 17.9) available that teams could control with a P300 BCI. At another hackathon at the Ars Electronica Festival in Linz, we had an exposition in which the Serbian artist Dragan Ilic equipped a KUKA robot with hundreds of pencils, and used it to create numerous works of art with BCI software and the Unicorn Brain Interface (see Fig. 17.10). He selects drawing commands just by thinking, and the robot painted on a vertical and a horizontal wall guided by Dragan's mind. After the exposition, hacking teams could also use EEG activity to paint with the robot arm.

Fig. 17.8 The top left panel shows a designer working on a new cap with a sewing machine, and the top right panel shows two caps that she made. The bottom picture shows several different caps made with sewing machines

17.3.2.4 Sample Kids' Projects

The kids' projects are simpler than other projects. We often provide a "Unicorn Table" for kids with BCIs, caps, and crafting equipment and supplies, and then let the kids' imaginations do the rest (see Figs. 17.11 and 17.12). Teams can be of any size, and kids often move between teams. Like the landmark projects, kids' projects have no competition and require less time than programming or design projects. Kids can stay as long as they (and their parents) want. There are practical challenges in keeping kids in a room for 24+ h to work on projects, but kids have come up with some good designs within an afternoon.

Figure 17.13 shows the winning kids' project from the ARS Electronica Festival in Linz 2018. The team designed a good looking and functioning EEG cap.

Fig. 17.9 A hacker uses a P300 BCI to control heavy equipment. The different commands on the screen reflect different instructions

17.4 Project Presentations and Awards

The BR4IN.IO hackathons have concluded with project presentations from each team, followed by the awards ceremony (see Fig. 17.14). The jury and audience members are welcome to ask questions during each presentation. Teams are strongly encouraged to "show, not tell" through their project presentations. That is, the jury prefers presentations that demonstrate a successful project, with data and a working system, rather than just talking about a future idea. Our jury of experts views the project presentations and scores them on different criteria, such as the project's novelty, technical difficulty, practicality, and other factors.

After the project presentations, the jury tallies the scores and decides on the first, second, and third place awards (see Fig. 17.15). Depending on the hackathon, the jury may also have other awards such as "Most innovative artistic project". We then present the award certificates to the winning teams, and certificates of completion to all teams that completed a project. Some of our BR4IN.IO hackathons have also included cash prizes.

Fig. 17.10 An artist uses a KUKA robot arm with hundreds of pencils, controlled through a BCI

Fig. 17.11 A "Unicorn Table" for kids. The kids can sketch new and implement new cap designs and accessories for caps

Fig. 17.12 Two examples of caps that kids designed and implemented at a Unicorn Table. Both caps have electrodes over frontal and parietal areas, and users can also use the "Brain Bee" cap on the right to turn on a light over the forehead

Fig. 17.13 Another kids' team developed this design based on beads. The beads cover the electrodes and could make the cap more cosmetically appealing to kids, especially kids who like beads

Fig. 17.14 In the left panel, one of the teams presents its project while a jury member (Dr. Guger, second from right) asks questions. The right panel shows a jury and audience watching a presentation

Fig. 17.15 The first-place winner and other hackers celebrate after a BR4IN.IO awards ceremony

17.5 Winning Projects

This section presents some of the winning projects from BCI Hackathons in 2017 and 2018.

Fig. 17.16 These winning hackers can mix gin and tonic with a BCI

17.5.1 Winning Programming Projects

Team "Gin Tonic"

This team's BCI used the user's alpha/beta ratio to control the mixture of Gin and Tonic and thereby induce relaxation (see Fig. 17.16). They implemented the real-time parameter extraction in MATLAB and Simulink with the g.IIIsys toolbox (g.tec medical engineering GmbH) and sent the result to a microcontroller that controlled the gin pump and the tonic pump.

Team members: Theresia Grahammer, Elisabeth Böhmwalder, Stefan Zapf, Rosa Büchsenmeister, Christoph Kernstock, Bernd Reuttcrer, Lukas Traxler.

Dreamteam "Schloafn"

The Dreamteam mapped activation of the EEG onto a graphical representation while a person was sleeping, indicating how active the brain was during sleep (see Fig. 17.17). This was realized with MATLAB/Simulink and g.HIsys to extract the EEG band-power across different frequency bands with a forgetting window. The results were transferred to a graphical engine via UDP communication.

Team members: Juan Ignacio Pisula, Thomas Preindl, Tomasz Kostka, Melanie Balaz, Helene Casado, Anthulla Dushi, Billie Postlmayr.

Team "Neurity"

The team combined EEG and Virtual Reality to create an interactive tool for assisted living and an environment for telesurgery and training (see Fig. 17.18). In this case, the neurosurgeon was equipped with a HMD to see the Unity environment in 3D. The neurosurgeon can then explore the operating room with a patient. With a hand-

Fig. 17.17 An artistic representation of the sleeping brain

held controller, the user could simulate brain surgery while a BCI user selected the surgical instrument for the neurosurgeon. The BCI user had 8 EEG electrodes assembled and used the Unicorn Speller to make the necessary selections. The BCI system transmitted the information to the Unity system with UDP messages.
Team members: Rachel L'Orsa, Kourosh Zareinia, Liu Shi Gan, Steven Samoil, Usman Mir.

"The Art of War"
This team built a BCI mixed reality collaborative strategy game. The BCI user had 8 EEG electrodes and used the Unicorn Speller to direct a Sphero robot into certain directions (see Fig. 17.19). Another user had a smart phone with an Augmented Reality application that projected tanks onto the floor. This user gave commands to the BCI to destroy the tanks or to go around the tanks.
Team members: Guillermo Herrera, Daniel Pimentel, Julia Anna Adrian.

17.5.2 Winning Designer Projects

One of the winning designer teams is shown in Fig. 17.20. They used a combination of 3D printed parts and other materials to create an innovative EEG head-set. One of its appealing features is that the user does not look like a patient while wearing the headset, which can acquire high-quality EEG data.

Fig. 17.18 Team "Neurity" created a VR interface for medical applications

Fig. 17.19 A mixed-reality (MR) collaborative war strategy game based on a P300 BCI

Another team designed a fashionable EEG cap called "Chappy" with hand-made flowers using the traditional Hawaiian method. This team used more than 500 little segments to create the headset during the Hackathon (see Fig. 17.21).

Fig. 17.20 This winning design project included headphones as well as electrodes

Fig. 17.21 This winning
design used hand-made
flowers

Fig. 17.22 Two team members present the MultiCorn cap

This is a very nice example of how the BCI Hackathon just creates very useful tools within only 24 hours. This head set is great for young patients that might be afraid of standard EEG equipment.

—James Weiland, Ph.D., Professor, Biomedical Engineering Dept., University of Michigan and Jury Member.

Team member: Charlie Uyemura

The "MultiCorn" allows users to flip the horns to fill each of the EEG electrodes, which are positioned for a P300 BCI. It was one of the winning teams at the BCI Hackathon at the ARS Electronica Festival 2017 (see Fig. 17.22).

Kids win!

An important contribution of the BCI Hackathons is the integration of kids. This allows them to grow up with BCI technology and even contribute to new designs themselves (Fig. 17.23).

17.6   Lessons Learned for Hackers

Team: Assemble a team with a range of skills. Adapt to the expectations of the organizers and the jury. If the hackathon and the jury emphasize painting or new cap designs, then the team should emphasize artists and designers and not just pro-

Fig. 17.23 The winning kids' team and the jury at the ARS Electronica Festival 2018

grammers and engineers. Avoid "dead weight" and try to keep the entire team busy throughout the hackathon.

Preparation: Learn as much as possible from the hackathon website, and review examples of other BCIs. In addition to this chapter, different textbooks (Wolpaw and Wolpaw 2012; Nam et al. 2018), the BCI Awards Book Series (e.g. Guger et al. in press), and web searches will provide dozens of projects that could inspire your team and boost your "innovation" score.

Survival: Your team will get hungry and thirsty during the hackathon. Bring some food and drinks, plan ahead on ordering pizza, and/or check with the hackathon organizers about what's available.

Scoring: Based on our experience with hackathons, and discussions with other jury members, the two key criteria are innovation and execution. Most juries in our BCI hackathons and others explicitly include these criteria, or similar terms like originality, creativity, achievement, or completion. Projects that just replicate well-known studies or present only ideas don't score as well as novel BCIs that work—ideally, systems the jury could use. Your presentation to the jury should show what you accomplished during the hackathon, not just concepts or future possibilities.

Presentations: Give a clear and well-prepared presentation, because this is what the jury will see. Embed videos showing the process your team followed and successful BCI operation to convince the jury that your project is working.

Follow-up: If your team enjoyed the hackathon, you have a lot of follow-up options. Maybe you can develop your idea into a product or a course project. Consider approaching professors and companies, or launching a start-up. Contact the hackathon organizers or jury members to ask for advice. Post a video or description of your new system on support boards like bci2000.org, reddit, or other sites. Projects can also be uploaded to the Unicorn Brain Interface web-page to share the results with others (www.unicorn-bi.com).

17.7 Lessons Learned for Organizers

Synergy: Many potential hosts have seminar/lecture series, research rounds, conferences, workshops, Open Houses, and other activities that may draw possible attendees. Consider approaching professional societies, sponsors, institutions, or entities for endorsements or funding.

Planning: After identifying the host (or deciding on a stand-alone event), plans should be complete at least one month before the BCI Hackathon, to allow time to publicize the event and make travel plans.

Publicity: Try to get enthusiastic and clever hackers. Tell local media. Take photographs, make videos, and use websites, social media, and other publicity tools.

Infrastructure: Have adequate space. There should be enough power outlets, which may require extension cables and/or power splitters. There should be enough tables and chairs for the attendees, and tables should be large so people can spread out equipment on them.

Staff: Be sure that the Hack-a-thon includes staff with enough technical skill to answer questions. On the other hand, bugs will happen. Don't stress about it too much. They're engineers.

Equipment: Remember that attendees will usually be completely dependent on the organizers for BCI equipment, software, laptops, monitors, etc. The type of equipment provided will have a strong impact on the hackers' projects and satisfaction. Systems that allow 8 or more channels, with good coverage of the entire scalp, will allow hackers more options with different brain regions and signals. Software that is easy to learn and use, with clear instruction manuals and online support, will help hackers succeed in their projects and reduce the demand on technical staff. If hackers are provided with eye trackers, or software that supports painting applications, they will develop projects accordingly.

Overall, more expensive systems will lead to better hackathons, and are recommended when possible. On the other hand, organizers who cannot afford expensive systems can still organize effective hackathons; they just need to be conscious of the resulting limitations. Table 17.3 presents a broad overview of how expensive versus inexpensive systems differ across several characteristics.

This table involves some generalizations, but is meant as a broad guideline. To avoid the appearance of bias, we avoid mentioning company or product names here,

Table 17.3 Each row presents one axis of comparison. The two columns to the right show how expensive or inexpensive systems provide different options for hackers

Characteristic	Expensive systems ($2000+)	Inexpensive systems ($100–$600)
Signal quality	High	Usually low
Software quality	Usually high	Variable
# of electrodes	Typically 8 or more	Usually fewer
Scalp coverage	Entire scalp (via cap)	Varies; may just cover one site or area
Manuals and support	Usually higher quality	May be confusing, limited, or erroneous
Non-EEG sensors?	Often support many options	Few options, often just facial EMG
Appearance	Medical/research system	Often more mainstream

but a web search will reveal many manufacturers and products. Interestingly, there is a gap between inexpensive and expensive systems. We expect this gap to disappear as both types of systems become less expensive and companies fill the gap by developing high-quality systems that cost in the range of several hundred dollars.

17.8   Future Directions

BCI hackathons are still new. BCIs hardware and software continues to improve, and public interest in BCIs is certainly not waning. Hence, we expect that BCI hackathons will become much more prominent over the next several years. In a few decades, BCI hackathons and other hands-on BCI activities might become fairly routine opportunities at universities, technical schools, and other institutes that teach neuroscience, psychology, and some engineering disciplines. We see a myriad of future directions that could be explored, including:

Permanent hackathons at public venues. Many metropolitan areas have at least one Exploratorium, science museum, or similar location. These institutes could generate revenue with a BCI exhibit while providing their guests with an opportunity that is (currently) quite unique. Many Exploratoriums have exhibits where guests can make water waves to learn about tidal activities, navigate using VR or huge projection screens, or use 3D printers. To our knowledge, there are no such entities where guests could don an electrode cap or headband and see their brainwaves, perform simple BCI tasks, and even create new applications.

Hackathons at more public activities. Public expositions, maker fairs, trade shows, conferences, symposia, workshops, and other activities will include hackathons. These may be special "satellite sessions" that occur just before or after the main activity, or just an optional activity within the regular schedule.

Hackathons as coursework. Students at colleges, universities, technical schools, and high schools are often required to complete hands-on projects to learn about automobile maintenance, crafting with wood or metals, 3D printing, circuit design, software, psychology, and more. Professors and other course instructors at institutes with a BCI system could require students to complete one or more hackathon projects. More advanced courses might involve much larger, semester- or year- long term projects that could lead to new ideas, systems, and publications.

Increasingly multimodal hackathons. Modern hackathons focus heavily on the EEG, with few or no tools designed to monitor other biosignals. We expect that future hackathons will often provide a broader range of sensors, and corresponding hardware and software, to manage a broader range of biosignals. Future projects will incorporate signals such as heart rate, eye activity muscle activity, skin conductance, oxygen saturation, and perhaps other measures of brain activity.

Growing cooperation with professional entities. Our BR4IN.IO hackathon series is endorsed by IEEE Brain. We expect that different entities will organize and host hackathons, in addition to endorsements and joint agreements that address funding and publicity. In addition to public entities, private companies could also sponsor hackathons and donate equipment (such as eye-trackers or VR headsets) that could be available to hacking teams. Some BCI hackathons from different groups have relied on public funding, which has been very helpful in catalyzing the recent growth of BCI hackathons.

> The BR4IN.IO BCI Hackathons that Christoph and his colleagues put together have been a tremendous success and the IEEE Brain Initiative is happy to support them. These hackathons inspire the next generation of neural engineers and provide excellent educational and networking opportunities for all that attend.
> —Paul Sajda, Ph.D., Professor, Columbia University and Chair, IEEE BRAIN Initiative.

BCI hackathon fees. Our BR4IN.IO hackathons and most other BCI hackathons have been free to the hackers. However, some other hackathons have required fees. Conferences often charge attendees for optional tutorials or symposia, and might charge attendees a fee for hackathons attached to that conference. Public science museums or private businesses might charge guests or customers who want to use their BCI equipment and staff for a BCI hackathon. More broadly, for-profit businesses could charge for escape rooms, live roleplaying, gaming, and other themes in which using or hacking BCIs is part of the customer experience.

Different themes. BCI hackathons from different groups have included special themes such as art, gaming, or industrial applications. Future hackathons might focus on these directions, or others such as science fiction, usability, patient applications, consumer applications, or even systems for kids. The hardware and software, advertising, juries, selection criteria, and other facets might differ accordingly. For example, hackathons projects from adults, kids, or groups might be judged by kids.

Quality and impact assessment. Most of the BR4IN.IO hackathons did not employ surveys or other means to parametrically assess the quality and impact. This is a limitation, and we have begun surveying hackathon attendees. We expect that future BCI hackathons will employ surveys, follow-up emails, web-based tools, and other means to learn more about each hacker's background, experience at the hackathon, reasons for participating, impact (such as inspiring hackers to a new project, coursework, product, etc.), and other details. Similar surveys have been informative with other hackathons (Briscoe 2014). While it is currently too early to analyze patterns and trends based on BCI hackathons, this will soon change. The resulting information could be an excellent indicator of which BCI directions and applications are getting the most attention in real-world settings, much like our BCI Awards book series and its trend analyses (Guger et al. in press).

Broader media engagement. BCI hackathons have earned some local, regional, and other publicity, but they are still unknown to the public or even to most makers and tech enthusiasts. We believe this will change dramatically. Like us, organizers of other BCI hackathons will continue to seek publicity, from more sources with greater circulation. Reporters and other groups will find BCI hackathons more and more newsworthy, since we expect strong growth in the public interest in neurotechnology as well as the number, size, and quality of BCI hackathons. Networks devoted to science, science fiction, or gaming might host BCI hackathons. Many shows have featured competitions in which teams design robots that fight each other or accomplish other goals, and these shows have raised public interest in robotics and encouraged new designs.

References

Briscoe G (2014) Digital innovation: the hackathon phenomenon. http://www.creativeworkslondon.org.uk/wp-content/uploads/2013/11/. Accessed 12 Jan 2019

Eigner G, Kovács L, Némethy K (2017) The 2016 IEEE systems, man, and cybernetics conference [conference reports]. IEEE Syst Man Cybern Mag 3(2):43–51

Friedman D, Brouwer AM, Nijholt A (2017) BCIforReal: an application-oriented approach to BCI out of the laboratory. In: Proceedings of the 22nd international conference on intelligent user interfaces companion, pp 5–7

Guger C, Allison BZ, Mrachacz-Kersting N (in press) Trends in the BCI awards and BCI community. In: Brain-computer interface research: a state of the art summary 8. Springer, Cham

HackerEarth (2017) Global hackathon report. https://innovation.hackerearth.com/global-hackathon-report/?utm_source=blog&utm_medium=roleofhackathon. Accessed 12 Jan 2019

Nam CS, Nijholt A, Lotte F (eds) (2018) Brain–computer interfaces handbook: technological and theoretical advances. CRC Press

Ortiz M, Iáñez I, Guger C, Azorín JM (in press) The art, science and engineering of BCI hackathons. In: Mobile brain-body imaging technologies: how disruptive neurotechnologies are changing science, arts, and innovation

Valjamae A, Evers L, Allison BZ, Ongering J, Riccio A, Igardi I, Lamas D (2017) The Brain-Hack project: exploring art-BCI hackathons. In: Proceedings of the 2017 ACM workshop on an application-oriented approach to BCI out of the laboratory, pp 21–24

Wolpaw J, Wolpaw EW (eds) (2012) Brain-computer interfaces: principles and practice. Oxford University Press, USA

Printed in the United States
By Bookmasters